Salmonella Infections
Clinical, Immunological and Molecular Aspects

Salmonella enterica encompasses a diverse range of bacteria that cause a spectrum of diseases in many hosts. Advancements in prevention and treatment of *S. enterica* infections have at times been hampered by compartimentalization of research efforts and lack of multidisciplinary approaches. This book attempts to cover a diverse range of topics related to the biology of *S. enterica* infections, including epidemiological and clinical aspects, molecular pathogenesis, immunity to disease and vaccines. *Salmonella enterica* infections are important zoonoses and therefore material on infections of animals and public health issues have also been considered. Each chapter can be read independently, but the full contents of the book will provide the reader with up-to-date knowledge on all the key aspects of salmonellosis in humans and animals. It will therefore be of interest to graduate students and researchers, as well as to clinicians, whose research focuses on this important pathogen.

PIETRO MASTROENI is a senior lecturer in the Department of Veterinary Medicine at the University of Cambridge, where he leads the Bacterial Immunology Group.

DUNCAN MASKELL is Marks & Spencer Professor of Farm Animal Health, Food Science and Food Safety and Head of the Department of Veterinary Medicine at the University of Cambridge, where he leads the Bacterial Infection Group.

Over the past decade, the rapid development of an array of techniques in the fields of cellular and molecular biology has transformed whole areas of research across the biological sciences. Microbiology has perhaps been influenced most of all. Our understanding of microbial diversity and evolutionary biology and of how pathogenic bacteria and viruses interact with their animal and plant hosts at the molecular level, for example, have been revolutionized. Perhaps the most exciting recent advance in microbiology has been the development of the interface discipline of Cellular Microbiology, a fusion of classic microbiology, microbial molecular biology, and eukaryotic cellular and molecular biology. Cellular Microbiology is revealing how pathogenic bacteria interact with host cells in what is turning out to be a complex evolutionary battle of competing gene products. Molecular and cellular biology are no longer discrete subject areas but vital tools and an integrated part of current microbiological research. As part of this revolution in molecular biology, the genomes of a growing number of pathogenic and model bacteria have been fully sequenced, with immense implications for our future understanding of microorganisms at the molecular level.

Advances in Molecular and Cellular Microbiology is a series edited by researchers active in these exciting and rapidly expanding fields. Each volume will focus on a particular aspect of cellular or molecular microbiology and will provide an overview of the area; it will also examine current research. This series will enable graduate students and researchers to keep up with the rapidly diversifying literature in current microbiological research.

MCM ADVANCES IN MOLECULAR AND CELLULAR MICROBIOLOGY

Published titles

1. *Bacterial Adhesion to Host Tissues.* Edited by Michael Wilson 0-521-80107-9
2. *Bacterial Evasion of Host Immune Responses.* Edited by Brian Henderson and Petra Oyston 0-521-80173-7
3. *Dormancy and Low-Growth States in Microbial Disease.* Edited by Anthony R. M. Coates 0-521-80940-1
4. *Susceptibility to Infectious Diseases.* Edited by Richard Bellamy 0-521-81525-8
5. *Bacterial Invasion of Host Cells.* Edited by Richard Lamont 0-521-80954-1
6. *Mammalian Host Defense Peptides.* Edited by Deirdre Devine and Robert Hancock 0-521-82220-3
7. *Bacterial Protein Toxins.* Edited by Alistair Lax 0-521-82091-X
8. *The Dynamic Bacterial Genome.* Edited by Peter Mullany 0-521-82157-6

Forthcoming titles in the series

Bacterial Cell-to-Cell Communication. Edited by Don Demuth and Richard Lamont 0-521-84638-2

The Influence of Bacterial Communities on Host Biology. Edited by Margaret McFall Ngai, Brian Henderson and Edward Ruby 0-521-83465-1

Phagocytosis of Bacteria and Bacterial Pathogenicity. Edited by Joel Ernst 0-521-84569-6

Advances in Molecular and Cellular Microbiology 9

Salmonella Infections

Clinical, Immunological and Molecular Aspects

EDITED BY
PIETRO MASTROENI AND
DUNCAN MASKELL
University of Cambridge

CAMBRIDGE
UNIVERSITY PRESS

CAMBRIDGE UNIVERSITY PRESS
Cambridge, New York, Melbourne, Madrid, Cape Town, Singapore, São Paulo

Cambridge University Press
The Edinburgh Building, Cambridge CB2 2RU, UK

Published in the United States of America by Cambridge University Press, New York

www.cambridge.org
Information on this title: www.cambridge.org/9780521835046

First published 2006

Printed in the United Kingdom at the University Press, Cambridge

A catalogue record for this publication is available from the British Library

ISBN-13 978-0-521-83504-6 hardback
ISBN-10-0-521-83504-6 hardback

To our families and friends

TO CARLOS

Contents

The colour plates are situated between pages 206 and 207

Contributors

Helen Andrews-Polymenis
Department of Medical
Microbiology & Immunology
407 Reynolds Medical Building
Texas A&M University SHSC
College Station
TX 77843-1114
USA

Stephen Baker
Wellcome Trust Sanger Institute
Wellcome Trust Genome Campus
Hinxton
Cambs CB10 1SA
UK

Paul Barrow
Institute for Animal Health
Compton
Newbury
Berkshire RG20 7NN
UK

Andreas J. Bäumler
Department of Medical
Microbiology & Immunology
School of Medicine
University of California at Davis
One Shields Ave.
Davis
CA 95616-8645
USA

Anne L. Bishop
Wellcome Trust Sanger Institute
Wellcome Trust Genome Campus
Hinxton
Cambs CB10 1SA
UK

José A. Chabalgoity
Laboratory for Vaccine Research
Department of Biotechnology
Instituto de Higiene
Facultad de Medicina
Avda. A. Navarro 3051
Montevideo CP 11600
Uruguay

Fiona J. Cooke
Centre for Molecular Microbiology
and Infection
Imperial College of Science,
Technology and Medicine
University of London
Exhibition Road, South Kensington

London W7 2AZ
UK

Caleb W. Dorsey
Dept. Med. Microbiol. & Immunol.
407 Reynolds Medical Building
Texas A&M University SHSC
College Station
TX 77843-1114
USA

Gordon Dougan
Wellcome Trust Sanger Institute
Wellcome Trust Genome Campus
Hinxton
Cambs CB10 1SA
UK

Michael Hensel
Institut für Klinische Mikrobiologie, Immunologie und Hygiene
Universität Erlangen-Nuemberg
Wasserturmstr. 3-5
D-91054 Erlangen
Germany

Tom Humphrey
Department of Clinical Veterinary Science
University of Bristol
Langford House
Langford
North Somerset BS40 5DU
UK

Cecilia Johannson
Department of Clinical Immunology,
University of Goteborg,
Guldhedsgatan 10A
Goteborg SE-413 46
Sweden

Pietro Mastroeni
Centre for Veterinary Science
Department of Veterinary Medicine
University of Cambridge
Madingley Road
Cambridge CB3 OES
UK

Bruce D. McCollister
Department of Microbiology
University of Colorado Health Sciences Center
B175, Room 4615
4200 E. 9th Ave.
Denver, CO 80262
USA

Christopher M. Parry
Department of Medical Microbiology and Genitourinary Medicine
Duncan Building
University of Liverpool
Daulby Street
Liverpool L69 3GA
UK

Manuela Raffatellu
Dept. Med. Microbiol. & Immonol.
School of Medicine
University of California at Davis
One Shields Ave.
Davis, CA 95616-8645

Richard A. Strugnell
Department of Microbiology & Immunology
The University of Melbourne
Parkville
Victoria 3010
Australia

Malin Sundquist
Department of Clinical Immunology,
University of Goteborg,
Guldhedsgatan 10A
Goteborg SE-413 46
Sweden

Andres Vazquez-Torres
Department of Microbiology
University of Colorado Health Sciences Center
B175, Room 4615
4200 E. 9th Ave.
Denver, CO 80262
USA

Bernardo Villarreal
Institute for Animal Health
Compton
Newbury
Berkshire RG20 7NN
UK

John Wain
Wellcome Trust Sanger Institute
Wellcome Trust Genome Campus
Hinxton
Cambs CB10 1SA
UK

Timothy S. Wallis
Institute for Animal Health
Compton
Newbury
Berkshire RG20 7NN
UK

Mary Jo Wick
Department of Clinical Immunology,
University of Goteborg,
Guldhedsgatan 10A
Goteborg SE-413 46
Sweden

Paul Wigley
Institute for Animal Health
Compton
Newbury
Berkshire RG20 7NN
UK

Odilia L. C. Wijburg
Department of Microbiology & Immunology
The University of Melbourne
Parkville
Victoria 3010
Australia

Preface

Salmonella enterica encompasses a diverse range of bacteria that cause a spectrum of diseases in many hosts. Typhoid fever is still a major killer of people in the developing world and rears its ugly head whenever war or natural disaster strikes. The increase in antibiotic resistance that has been observed in *S. enterica* serovar Typhi makes the understanding of this pathogen ever more important. But typhoid fever is not the only *Salmonella*-related disease that causes concern, with human gastrointestinal disease a major problem in developed and developing countries, not forgetting salmonelloses in livestock that bring with them economic losses as well as the problems of zoonoses and food-borne disease.

The different salmonellae make up a versatile and fascinating group of organisms that have inspired both of the Editors of this book since we were scientific juveniles studying the pathogenesis and immunity of these bacteria for our Ph.D. degrees. As we have moved through the stages of our scientific careers, other bacteria and immunological questions may have caught our attention for a while, but always the salmonellae persisted, providing the bedrock of our interests and the centrepiece of our scientific enquiries.

So why edit a book on salmonellae now? The easy answer to this question is that the study of the salmonellae is entering a brave new world with the completion of the genome sequences of serovars Typhi, Paratyphi A and Typhimurium, with other sequences hot on their tail. Add to this impetus the remarkable advances in whole genome analysis that have been allied to genome science, and that have especially opened the door on so many of the secrets of how salmonellae cause disease, and it begins to look like a really exciting time to be working with salmonellae. Add again advances in the study of the cellular biology of infection that have been made recently, especially in the context of the marvellous imaging technologies that are now available, and we begin to move to a position where the diseases caused by salmonellae might be understood at a level of detail unimaginable only ten years ago.

We hope that we have covered most of the key aspects of the biology of *Salmonella* infections in this book and that we have brought out some of the excitement in the field currently being felt by researchers. We have also been intent on embedding the basic science aspects of this book in real disease states, and so we have enthusiastically included chapters on the clinical diseases and public health problems caused by this group of bacteria.

Finally, science-based vaccines against salmonellae are already a reality. Improvements in our understanding of the immunology and vaccinology of these diseases may not only lead to control of these problems in the future but may also lead in unexpected directions. In fact, this intracellular pathogen can be used as a Trojan horse to introduce antigens from other organisms to a host's immune system, or indeed deliver other immunotherapeutics that might lead to treatments for a range of non-infectious diseases. We have tried to cover these exciting advances in the book.

It has been a pleasure editing this book, and an enormous education. It would not have been possible without timely and high quality papers from our contributors, to whom we would like to say thank you, and we hope you like the end product. We also hope that you the reader like the book, find it useful and most importantly of all, are enthused by it and by these fascinating organisms.

CHAPTER 1

Epidemiological and clinical aspects of human typhoid fever

Christopher M. Parry

1.1 INTRODUCTION

Typhoid fever is an acute systemic infection caused by the bacterium *Salmonella enterica* serovar Typhi. *Salmonella enterica* serovars Paratyphi A, B, and C cause the clinically similar condition, paratyphoid fever. Typhoid and paratyphoid fevers are collectively referred to as enteric fevers. In most endemic areas, approximately 90% of enteric fever is typhoid. Typhoid is transmitted by the fecal-oral route via contaminated food and water and is therefore common where sanitary conditions are inadequate and access to clean water is limited. Although typhoid fever was common in the United States and Europe in the 19th century, it is now encountered mostly throughout the developing world. In the last fifteen years, the emergence of resistance to the antibiotics used for treatment has led to large epidemics, and complicated the management of this serious disease.

Before the 19th century, typhoid fever was commonly confused with other prolonged febrile syndromes, particularly typhus fever. Following the observations of Huxham, Louis, Bretonneau, Gerhard and William Jenner, by the middle of the 19th century the two conditions were clearly differentiated (Richens, 1996). In 1873, William Budd described the contagious nature of the disease and incriminated fecally contaminated water sources in transmission. The causative organism was visualized in tissue sections from Peyer's patches and spleens of infected patients by Eberth in 1880 and was grown in pure culture by Gaffky in 1884. The organism has been variously known as *Bacillus typhosus, Erbethella typhosa, Salmonella typhosa* and *Salmonella typhi.*

'Salmonella' Infections: Clinical, Immunological and Molecular Aspects, ed. Pietro Mastroeni and Duncan Maskell. Published by Cambridge University Press.

1.2 *SALMONELLA ENTERICA* SEROVAR TYPHI

Salmonella enterica serovar Typhi is a member of the genus *Salmonella* in the family Enterobacteriaceae. The *Salmonella* genus contains two species, *enterica* and *bongori* (Brenner *et al.*, 2000). *S. enterica* is further divided into six subspecies (*enterica, salamae, arizonae, diarizonae, houtenae* and *indica*) containing 2443 serovars. Most of the salmonellae that cause disease, with some important exceptions, are in the subspecies *Salmonella enterica* subspecies *enterica*. The agents that cause enteric fever are therefore *Salmonella enterica* subspecies *enterica* serovar Typhi (commonly referred to as *S. enterica* serovar Typhi) and serovars Paratyphi A, B and C.

When isolated from clinical specimens, colonies of *S. enterica* serovar Typhi are non-lactose fermenting and produce a characteristic biochemical pattern in Kligler iron agar (acid but without gas, an alkaline slant and a moderate amount of H_2S production). Identification is confirmed by serological demonstration of the lipopolysaccharide antigen O9, 12 (group D), protein flagellar antigen Hd and Vi polysaccharide capsular antigen. Unique flagella types, Hj and H_{z66} are present in some *S. enterica* serovar Typhi from Indonesia. *S. enterica* serovar Typhi exhibits a remarkable degree of biochemical and serological homogeneity. Vi phage typing and molecular typing by pulse field gel electrophoresis and ribotyping, differentiate strains from different geographical areas and have shown a relative diversity of strains circulating in endemic areas, but comparative uniformity in outbreak strains (Thong *et al.*, 1994).

1.3 EPIDEMIOLOGY OF TYPHOID FEVER

Typhoid fever is endemic throughout Africa and Asia and persists in the Middle East, a few southern and eastern European countries and central and South America. In the US and most of Europe, apart from occasional point source epidemics, typhoid is predominantly a disease of the returning traveler (Ackers *et al.*, 2000). A recent study estimated there to be approximately 22 million cases of typhoid each year with at least 200 000 deaths (Crump *et al.*, 2004). However, the true magnitude is difficult to quantify because the clinical picture is confused with many other febrile illnesses and most typhoid endemic areas lack facilities to confirm the diagnosis. Data from placebo groups in large-scale field trials of typhoid vaccines and population-based epidemiology studies show annual incidence rates ranging from 10 to 1000 cases per 100 000 people (Table 1.1).

Table 1.1. *Mean annual incidence of blood culture confirmed cases of typhoid fever recorded in the control groups of vaccine field trials and in population based epidemiology studies*

Country	Year	Study type	Age range (years)	Rate per 100 000 person years	Reference
Egypt	1982	Ty21a vaccine	6–7	46	Wahdan *et al.*, 1982
Chile	1983–1986	Ty21a vaccine	6–21	104	Levine *et al.*, 1987
Indonesia	1986–1989	Ty21a vaccine	3–44	810	Simanjuntak *et al.*, 1991
Nepal	1985–1987	Vi vaccine	5–44	655	Acharya *et al.*, 1987
South Africa	1985–1987	Vi vaccine	5–16	387	Klugman *et al.*, 1987
China	1995–1996	Vi vaccine	3–50	22	Yang *et al.*, 2001
Vietnam	1997–2000	Vi conjugate vaccine	2–5	414	Lin *et al.*, 2001
India	1995–1996	Epidemiology study	All	980	Sinha *et al.*, 1999
Vietnam	1995–1996	Epidemiology study	All	198	Lin *et al.*, 2000
Egypt	2001	Epidemiology study	All	13	Crump *et al.*, 2003

The incidence of typhoid in endemic areas is typically considered to be low in the first few years of life, peaking in school-aged children and young adults and then falling in middle age. Older adults are presumably relatively resistant due to frequent boosting of immunity, but the apparent low incidence in pre-school children contrasts with the high incidence of most other enteric infections at this age in these countries (Mahle and Levine, 1993). Some hospital and community-based studies have found a significant incidence of typhoid in pre-school children (Table 1.2). The character of the diseases in these studies has varied from a non-specific febrile illness or mild infection through to one that is severe and life-threatening (Bhutta, 1996a;

Table 1.2. *Mean annual age-specific incidence of typhoid fever per 100 000 population in three studies. Santiago represents an urban centre with reasonably good sanitation and clean water. Most infection is probably related to contaminated food (Ferreccio* et al.*, 1984). Dong Thap province in the Mekong river delta has poor sanitation and most people use river water for all their daily needs (Lin* et al.*, 2000). Kalkaji in New Dehli is an overcrowded urban slum with very poor levels of sanitation (Sinha* et al.*, 1999)*

Age group (years)	Santiago, Chile 1977–1981 (Ferreccio *et al.*, 1984)	Dong Thap Province, Vietnam 1995–1996 (Lin *et al.*, 2000)	Kalkaji, New Dehli India 1995–1996 (Sinha *et al.*, 1999)
0–4	89	358[1]	2730
5–9	272	531	1390[2]
10–14	333	429	860[3]
15–19	283	153	860[3]
20–24	247	149	110[4]
25–29	153	149	110[4]
All ages	166	198	980

[1] Figure is for the age range 2–4 years. No typhoid was observed in children < 2 years.
[2] Figure is the mean value for the age range ≥ 5 to 12 years.
[3] Figure is the mean value for the age range > 12 to 19 years
[4] Figure is the mean value for the age range > 19 to 40 years

Butler *et al.*, 1991; Duggan and Beyer, 1975; Ferreccio *et al.*, 1984; Topley, 1986).

Although this variation in part relates to the patchy distribution of health care facilities capable of diagnosing and treating typhoid, there do appear to be different epidemiological patterns of *S. enterica* serovar Typhi (Ashcroft, 1964). Ashcroft suggested that where hygiene and sanitation is non-existent, *S. enterica* serovar Typhi is prevalent but classical clinical typhoid fever is uncommon. Immunity is acquired in infancy or very early childhood when infection is either asymptomatic or unrecognized. Where hygiene and sanitation are poor, *S. enterica* serovar Typhi infection is common and typhoid fever is particularly frequent in school-aged children. Most infections occur in childhood and are recognizable although often mild. This is the situation in endemic typhoid regions such as the Mekong Delta region of Vietnam. Where sanitation and hygiene are a mixture of poor and good, as is the case in many of the rapidly expanding conurbations of Asia, outbreaks of typhoid fever may involve all age groups. Where hygiene is excellent, as is the situation in developed countries, *S. enterica* serovar Typhi and typhoid fever are rare.

Typhoid is usually contracted by ingestion of food or water contaminated by fecal or urinary carriers excreting *S. enterica* serovar Typhi. In addition, these bacteria can survive for prolonged periods in water, ice, dust and dried sewage and these may become sources of infection. In endemic areas, peaks of transmission occur in dry weather or at the onset of rains. Risk factors for disease include eating food prepared outside the home, such as ice creams or flavoured iced drinks from street vendors, drinking contaminated water and eating vegetables and salads that have been grown with human waste as fertilizer (Black *et al.*, 1985; Morris *et al.*, 1984; Velema *et al.*, 1997). A close contact or relative with recent typhoid fever, poor housing with inadequate food and personal hygiene and recent consumption of antimicrobials are further risk factors (Gasem *et al.*, 2001; Luby *et al.*, 1998; Luxemburger *et al.*, 2001). Transmission of typhoid has also been attributed to flies, laboratory mishaps, unsterile instruments and anal intercourse.

Chloramphenicol was introduced for the treatment of typhoid fever in 1948 (Woodward *et al.*, 1948) but resistance did not emerge as a problem until 1972. At that time outbreaks occurred in Mexico, India, Vietnam, Thailand, Korea and Peru (Rowe *et al.*, 1997). Curiously, after a few years these antibiotic resistant isolates disappeared from Mexico and Peru but persisted at low levels in Asia. Towards the end of the 1980s and 1990s, *S. enterica* serovar Typhi developed resistance simultaneously to all the first line drugs (chloramphenicol, trimethoprim, sulphamethoxazole and ampicillin) (Rowe *et al.*, 1997). Outbreaks with these strains have occurred in India, Pakistan, Bangladesh, Vietnam, the Middle East and Africa (Kariuki *et al.*, 2000). Multidrug resistant (MDR) *S. enterica* serovar Typhi are still common in many areas, although in some regions fully sensitive strains have re-emerged and in other regions *S. enterica* serovar Typhi has been overtaken by MDR *S. enterica* serovar Paratyphi A (Rodrigues *et al.*, 2003; Threlfall *et al.*, 2001; Wasfy *et al.*, 2002).

The appearance of MDR *S. enterica* serovar Typhi in Asia led to the widespread use of fluoroquinolones and extended spectrum cephalosporins for treatment. Isolates with low-level resistance to fluoroquinolones appeared within a few years of this change and have become common in Asia (Brown *et al.*, 1996; Dutta *et al.*, 2001). Large outbreaks of typhoid with such strains have occurred in Tajikistan, Vietnam and Nepal (Mermin *et al.*, 1999; Parry,

2004) and these resistant strains are causing an increasing number of infections in returning travelers (Ackers *et al.*, 2000; Threlfall *et al.*, 2001). Although reported as susceptible by disc testing using recommended breakpoints to fluoroquinolones, these isolates have smaller zones of inhibition to fluoroquinolones by disc testing and fluoroquinolone minimum inhibitory concentrations 10-fold higher than fully susceptible strains (Crump *et al.*, 2003). They are invariably resistant to nalidixic acid and this is an important laboratory marker. Infection with these isolates leads to a poor clinical response to fluoroquinolone treatment (Rupali *et al.*, 2004; Wain *et al.*, 1997). Fully fluoroquinolone resistant and fully ceftriaxone resistant isolates in Asia appear to be uncommon, although systematic data are lacking (Mehta *et al.*, 2001; Saha *et al.*, 1999).

1.4 PATHOPHYSIOLOGY OF TYPHOID FEVER

Humans are the only natural host and reservoir of infection for *S. enterica* serovar Typhi. The infectious dose in volunteers varies between 10^3–10^9 organisms (Hornick *et al.*, 1970). Vi negative strains of *S. enterica* serovar Typhi are less infectious and less virulent than Vi positive strains. *S. enterica* serovar Typhi must survive the gastric acid barrier en route to the small intestine. Achlorhydria, due to ageing, previous gastrectomy, treatment with H_2 receptor antagonists, proton-pump inhibitors, large amounts of antacids or *Helicobacter pylori* infection increase susceptibility to typhoid fever (Bhan *et al.*, 2002). In the small intestine, the bacteria adhere to mucosal cells and then invade, translocate to the intestinal lymphoid follicles and the draining mesenteric lymph nodes and some pass on to the reticuloendothelial cells of the liver and spleen (House *et al.*, 2001a). Salmonellae are able to survive and multiply within the mononuclear phagocytic cells of the lymphoid follicles, liver and spleen. After a 7- to 14-day incubation period, the onset of a sustained secondary bacteraemia results in clinical disease.

The bacteraemia of typhoid fever persists for several weeks if antibiotic therapy is not given. In this phase, the organism disseminates widely to the liver, spleen, bone marrow, gall bladder and the Peyer's patches of the terminal ileum. The symptoms and signs of typhoid fever are not thought to be entirely due to circulating endotoxin (Butler *et al.*, 1978; Hornick *et al.*, 1970). Increased levels of circulating pro- and anti-inflammatory cytokines have been demonstrated in typhoid fever as well as a reduced capacity of whole blood to produce pro-inflammatory cytokines in severe disease. Ulceration of Peyer's patches is seen where the inflammatory process has resulted in ischaemia and necrosis (Everest *et al.*, 2001). Relapse probably occurs because of persisting organisms within the reticuloendothelial system. Gall bladder infection may become chronic, particularly in those individuals who have gall bladder pathology. Carriers may shed as many as 10^9 organisms/g feces.

Typhoid fever induces systemic and local humoral and cellular immune responses but these confer incomplete protection against relapse and reinfection (Marmion *et al.*, 1953).

1.5 CLINICAL FEATURES OF TYPHOID FEVER

S. enterica serovar Typhi infections result in a clinical syndrome that varies widely in severity (Huckstep, 1962; Osler and McCrae, 1926; Stuart and Pullen, 1946). After ingestion of the bacteria, an incubation period follows usually lasting 8 to 14 days (range 3–60 days). Fever and malaise mark the onset of bacteraemia but patients do not usually present to hospital until towards the end of the first week of symptoms. Fever, flu-like symptoms with chills (although rigors are rare) and a dull frontal headache are common. The fever, initially low grade, rises progressively, and by the second week is often high and sustained (39–40 °C). Other symptoms include malaise, anorexia, poorly localized abdominal discomfort, a dry cough and myalgia. Physical signs are few, but a coated tongue, tender abdomen, hepatomegaly and/or splenomegaly may be found. The abdominal pain is usually diffuse and poorly localized, but occasionally sufficiently intense in the right iliac fossa to suggest appendicitis. Dilated loops of bowel may be palpated indicating an ileus. Nausea and vomiting are infrequent in uncomplicated typhoid but are seen with abdominal distension in severe cases. Relative bradycardia is not a consistent feature. Rose spots are reported in 5–30% of cases but are easily missed in dark-skinned patients. These rose spots are small blanching erythematous maculopapular lesions typically on the abdomen and chest. Melanesian typhoid patients may develop purpuric macules that do not blanch. Constipation is generally more common in adults, but in young children and adults with HIV infection diarrhoea predominates. *S. enterica* serovar Paratyphi causes a similar, although less severe, syndrome.

Untreated, the fever persists for two weeks or more and defervescence occurs slowly over the following 2–3 weeks. Convalescence may last for 3–4 months. If an appropriate antibiotic is given the fever gradually falls over 3–4 days. The duration of untreated illness prior to the initiation of therapy influences the severity of the disease. Those individuals infected with multi-drug-resistant (MDR) isolates of *S. enterica* serovar Typhi may also suffer more severe disease. Patients with typhoid fever in the second to fourth

Table 1.3. *Principle complications of typhoid fever*

Abdominal Intestinal perforation Intestinal haemorrhage Hepatitis Cholecystitis (usually subclinical)	*Neuropsychiatric* Obtundation Delirium Coma Psychotic states Depression Deafness Meningitis Seizures (children) Cerebellar ataxia Encephalomyelitis
Genitourinary Retention of urine Glomerulonephritis	*Haematological* Disseminated intravascular coagulation (usually subclinical) Anaemia Haemolysis
Cardiovascular Asymptomatic ECG changes Myocarditis Shock Sudden death	*Other* Focal abscesses of brain, liver, spleen, breast, thyroid, muscles, lymph nodes Metabolic acidosis Relapse
Respiratory Bronchitis Pneumonia (Rarely due to *S.enterica* serotype Typhi, more commonly a secondary infection due to *Streptococcus pneumoniae*)	

week present with accelerating weight loss, weakness, an alteration of mental state and the development of complications that occur in 10–15% of hospitalized patients and occasionally dominate the clinical picture deflecting attention from the underlying diagnosis of typhoid. Although many complications have been described (Table 1.3), gastrointestinal bleeding, intestinal perforation and typhoid encephalopathy are the most important.

Gastrointestinal bleeding occurs in up to 10% of patients and results from erosion of a necrotic Peyer's patch through the wall of an enteric vessel. Usually the bleeding is slight and resolves without the need for blood transfusion. In 1–2% of cases, bleeding is significant, and can be rapidly fatal if a large vessel is involved. Intestinal (usually ileal) perforation is the most serious complication occurring in 1–3% of hospitalized patients. Perforation may present with an acute abdomen or more covertly with simple worsening of abdominal pain, rising pulse and falling blood pressure in an already sick patient. Typhoid encephalopathy, often accompanied by shock, is associated with a high mortality. Patients may display the "typhoid" facies, a thin, flushed face with a staring, apathetic expression. Mental apathy may progress to an agitated delirium, frequently accompanied by tremor of the hands, tremulous speech and gait ataxia, and then muttering delirium, twitchings of the fingers and wrists (subsultus tendinum), agitated plucking at the bedclothes (carphology), and a staring, unrousable stupor (coma vigil) (Osler and McCrae, 1926).

Typhoid fever in pregnancy may be complicated by miscarriage, although antimicrobial treatment has made this less common (Seoud *et al.*, 1988). Vertical intra-uterine transmission from a typhoid-infected mother may lead to neonatal typhoid, a rare but severe and life-threatening complication (Reed and Klugman, 1994). Relapse occurs in 5–10% of patients, usually 2 to 3 weeks after defervescence. The illness is usually, but not invariably, milder than the original attack and the relapse *S. enterica* serovar Typhi isolate has the same susceptibility pattern as in the original episode. Reinfection may also occur (Marmion *et al.*, 1953). Up to 10% of untreated convalescent typhoid cases will excrete *S. enterica* serovar Typhi in feces for 1–3 months and between 1 and 4% become chronic carriers excreting the organism for more than one year. Chronic carriers give no prior history of typhoid fever in up to 25% of cases. Fecal carriage is more frequent in individuals with gallbladder disease and is most common in women over 40; in the Far East there is an association with opisthorchiasis. Chronic carriage carries an increased risk of carcinoma of the gallbladder, pancreas and large bowel (Caygill *et al.*, 1994). Urinary carriage is associated with schistosomiasis and nephrolithiasis.

1.6 DIAGNOSIS OF TYPHOID FEVER

The lack of specific clinical signs complicates the diagnosis of typhoid fever, which must be distinguished from other endemic acute and subacute febrile illnesses. These can include malaria, deep abscesses, tuberculosis, amoebic liver abscesses, encephalitis, influenza, dengue fever, infectious mononucleosis, infectious hepatitis, leptospirosis, endocarditis, brucellosis, typhus, visceral leishmaniasis, toxoplasmosis, lymphoproliferative disease and connective tissue diseases. A fever lasting more than one week without evident cause should be considered typhoid until proven otherwise and typhoid should always be considered when suspected malaria has not been confirmed or has not responded to antimalarial therapy. It is unusual for a patient hospitalized with typhoid fever to have no abdominal symptoms and normal bowel habit. In non-endemic countries, a travel history is crucial.

The haemoglobin, white cell and platelet count are usually within the normal range or reduced. Leucocytosis suggests either intestinal perforation or an incorrect diagnosis. Laboratory evidence of disseminated intravascular coagulation may be present but is very rarely of clinical significance. Liver transaminases are characteristically two to three times above normal. The urine may contain protein and leukocytes.

Blood cultures are the standard diagnostic method and can be positive in 60–80% of cases. In mild typhoid the number of bacteria may be as low as one colony-forming unit per ml of blood (Butler *et al.*, 1978; Wain *et al.*, 1998). The number of bacteria in the blood of children is higher than in adults and declines with increasing duration of illness. Recovery of the organism from blood cultures depends on several factors including the volume of blood cultured, the ratio of the volume of blood to the volume of culture broth in which it is inoculated (the ratio should be at least 1:8) and inclusion of anticomplementary substances in the medium (such as sodium polyethol sulfonate or bile). Culture of bone marrow is more sensitive regardless of the illness duration and is positive in 80–95% of patients despite prior antibiotic therapy (Gilman *et al.*, 1975, Hoffman *et al.*, 1986). The higher sensitivity of bone marrow cultures compared to blood in part relates to the higher concentration of organisms in bone marrow (Wain *et al.*, 2001).

Other diagnostic approaches have included culturing the organisms from the buffy coats of blood, from streptokinase-treated blood clots, from intestinal secretions using a duodenal string capsule and from skin snips of rose spots (Gilman *et al.*, 1975; Hoffman *et al.*, 1984b; Hoffman *et al.*, 1986). On average stool cultures are positive in 30% of patients with acute typhoid fever although the results should be interpreted with caution in areas with many carriers. For the detection of carriers, several samples should be examined because of irregular shedding. Isolation from urine is more common in areas with schistosomiasis.

Serological tests for typhoid fever have been used since the late 19th century. Widal and Sicard in 1896 showed that the serum of patients with typhoid fever agglutinated typhoid bacilli. The Widal test, in which O and H agglutinins are demonstrated in serum, may be performed with appropriate antigens in tubes or on slides. Typically antibodies to the O and H antigens appear during the end of the first week of disease and peak at the end of the third week but there is much variability (Levine *et al.*, 1978). The use of a single measurement of antibody titres is useful if the background levels of antibodies in the population are known (Clegg *et al.*, 1994; Parry *et al.*, 1999).

O and H antibodies may be present in the general population as a result of prior infection with *S. enterica* serovar Typhi or other *Salmonella* serotypes, or as a result of typhoid vaccination, or because of cross-reacting epitopes with other *Enterobacteriacae*. The demonstration of a four-fold rise in titre of antibodies to *S. enterica* serovar Typhi suggests typhoid but is invariably too late to help clinical decision-making. Furthermore, up to one third of typhoid patients mount no detectable antibody response or no demonstrable rise in titre. The Widal test result must therefore be interpreted with caution in endemic areas.

Serological tests that measure Vi antibody are unhelpful for the diagnosis of acute infections but are useful for screening for chronic *S. enterica* serovar Typhi carriers who usually have an elevated Vi titre (Lanata *et al.*, 1983). The reported sensitivity is 70–80% with a specificity of 80–95%. Many alternative tests for the detection of antibodies specific for *S. enterica* serovar Typhi, *Salmonella* antigens, and *Salmonella* DNA in body fluids have been described but not adopted in practice (Chaudhry *et al.*, 1997). Recently, three rapid antibody tests have shown promise and their value in routine practice is undergoing evaluation (Bhutta and Mansurali, 1999; Hatta *et al.*, 2002; House *et al.*, 2001b).

1.7 MANAGEMENT OF TYPHOID FEVER

The aims of management are to eliminate the infection with antibiotics, to restore fluid and nutritional deficits, and to monitor the patient for dangerous complications. In endemic areas, many cases of typhoid fever are managed at home with antibiotics and bed rest. For hospitalized patients, effective antibiotics, good nursing care, adequate nutrition, careful attention to fluid and electrolyte balance and the prompt recognition and treatment of complications are necessary to avoid a high fatality rate. Treatment for relapse cases should be the same as that employed for the initial infection.

Effective antibiotic therapy in typhoid reduces mortality and complications and shortens the duration of the illness. Successfully treated patients typically respond to antibiotics within 3 to 5 days with resolution of fever and other symptoms. Antimicrobial regimens should also prevent relapse, eradicate fecal carriage and be safe to use in children. Chloramphenicol was the first antibiotic to be effective in the treatment of typhoid fever and provided the standard against which subsequent antibiotics have been measured (Woodward *et al.*, 1948). Ampicillin, amoxycillin and co-trimoxazole have been shown to have comparable efficacy to chloramphenicol with less toxicity. Unfortunately, in many areas these drugs are no longer effective because of the spread of MDR *S. enterica* serovar Typhi. Antibiotics active against MDR *S. enterica* serovar Typhi include the fluoroquinolones, extended spectrum cephalosporins and azithromycin. Most active are the fluoroquinolones but resistance is again emerging. Some current treatment regimens are in Table 1.4.

Chloramphenicol, amoxycillin or co-trimoxazole are appropriate for the treatment of typhoid fever in areas of the world where the bacteria are still susceptible to these drugs. These antibiotics are inexpensive, widely available and rarely associated with side effects. They produce relief of symptoms with defervescence usually within 5 to 7 days. These antibiotics produce cure rates of 95%, relapse rates of 1–7% and convalescent carriage rates of 2–10% (Parry *et al.*, 2002). Compliance with the 2 to 3 weeks of treatment required may be variable.

Fluoroquinolones have proved safe and rapidly effective in fluoroquinolone-susceptible uncomplicated typhoid fever (Parry *et al.*, 2002). In randomized controlled trials, average fever clearance times are less than 4 days and cure rates exceed 96%. Persistent fecal carriage and relapse occurs in less than 2% of treated patients. Short course regimens (3–5 days) have proved particularly useful in epidemic containment (Hien *et al.*, 1995). For the extended spectrum cephalosporins, principally ceftriaxone and cefixime, fever clearance times average one week and treatment failure rates of 5–10% have been reported. Relapse rates are 3–6% and fecal carriage rates less than 3% (Parry *et al.*, 2002). Azithromycin has given 95% cure rates following 5 to 7 day treatment courses. Defervescence occurs in 4 to 6 days with relapse and convalescent fecal carriage rates of less than 3%(Parry *et al.*, 2002). Aztreonam and imipenem are potential third line drugs.

In areas where MDR isolates are common many physicians start with a fluoroquinolone, usually ciprofloxacin or ofloxacin. Treatment can be completed in a week or less with minimal toxicity. Some authors also suggest fluoroquinolones should be used in areas where strains sensitive to chloramphenicol are still common, because of greater efficacy and likely compliance, but others disagree with this approach. In preclinical testing, fluoroquinolones induced articular cartilage damage in young beagle dogs. As a result there have been concerns over the potential toxicity of fluoroquinolone use in children. However, when used for long periods in children with cystic fibrosis, or in short courses in children with typhoid fever or bacillary dysentery, they have not shown evidence of bone or joint toxicity, tendon rupture, or, in long term follow-up, impairment of growth (Bethel *et al.*, 1996; Doherty *et al.*, 2000).

The biggest limitation on fluoroquinolone use is the emergence of low-level resistance. For isolates with low-level fluoroquinolone resistance, short courses (<7 days) of fluoroquinolone treatment result in a high failure rate (Wain *et al.*, 1997; Parry, 2004). Treatment with maximum recommended doses for 7 days may only succeed in 70% of patients, with long fever clearance times averaging 7 days and short-term convalescent fecal carriage of up to 20% (Rupali *et al.*, 2004; Parry, 2004). These strains are frequently MDR so the choice of alternative drugs is limited to extended spectrum cephalosporins or azithromycin (Chinh *et al.*, 2000; Dutta *et al.*, 2001). Fortunately, full fluoroquinolone and extended spectrum cephalosporin resistance is still rare. Chloramphenicol, amoxycillin and ceftriaxone have been used safely in pregnancy (Seoud *et al.*, 1988). Fluoroquinolones have generally been avoided because of safety concerns, although case reports suggest they may be used if there is no alternative (Koul *et al.*, 1995).

Fluoroquinolones or injectable extended spectrum cephalosporins are suitable for the treatment of severe typhoid but should be given for a minimum of 10 days (Bhutta, 1996b; Dutta *et al.*, 1993). There have been no randomized antibiotic trials in severe typhoid and, as yet, no data on the use of azithromycin in severe disease. Adults and children with severe typhoid, characterized by delirium, obtundation, stupor, coma or shock benefit from prompt administration of dexamethasone (Hoffman *et al.*, 1984a; Punjabi *et al.*, 1988). Dexamethasone given in an initial dose of 3mg/kg by slow intravenous infusion over 30 minutes, followed by 1mg/kg dexamethasone given at the same rate every 6 hours for eight additional doses reduced the mortality from 56% to 10% in Indonesian adults and children. Hydrocortisone, given in a lower dose, was not effective (Rogerson *et al.*, 1991).

Patients with typhoid perforation require resuscitation with fluids, nasogastric suction, blood transfusions and administration of oxygen followed by surgery. Perforations are found usually in the ileum but may also occur in the caecum and proximal large bowel. Operative procedures include debridement of the ulcer with primary closure of the perforation, wedge resection and intestinal resection with primary anastomosis. A temporary ileostomy or ileocolostomy may be required. Vigorous peritoneal cavity lavage is followed by closure with or without drainage. Patients require additional parenteral antibiotics, such as metronidazole and an aminoglycoside, to cover the peritoneal fecal contamination. Early intervention is crucial, as mortality rates increase if there is a long delay between perforation and surgery. Mortality rates vary between 10 and 32% (Butler *et al.*, 1985).

In the pre-antibiotic era the typhoid case fatality rate was 10% or more (Osler and McCrae, 1926; Stuart and Pullen, 1946). With current treatment

Table 1.4. *Guidelines for drug dosages in typhoid fever*

Antibiotic	Daily dose	Route[1]	Doses/day	Duration in non-severe disease (days)	Duration in severe disease (days)[2]
Chloramphenicol[3]	50–75 mg/kg Reduce dose to 30mg/kg when fever ceases	o/(im/iv)[4]	4	14	14–21
Amoxycillin	75–100 mg/kg	o/im/iv	3	14	14–21
Co-trimoxazole[5]	8 mg/kg trimethoprim; 40 mg/kg sulphamethoxazole	o/im/iv	2–3	14	14–21
Ceftriaxone[6]	50–80 mg/kg	im/iv	2	7–10	10–14
Cefixime	20–30 mg/kg	o	2	7–10	-[7]
Ciprofloxacin[8,9]	15–25 mg/kg	o/iv	2	5–7	10–14
Ofloxacin[8,9]	10–20 mg/kg	o/iv	2	5–7	10–14
Pefloxacin[8,9]	800 mg	o/iv	2	5–7	10–14
Fleroxacin[8,9]	400 mg	o/iv	1	5–7	10–14
Azithromycin	8–10 mg/kg	o	1	5–7	-[7]

Treatment of carriers				
Amoxycillin with probenecid	100 mg/kg	o	3–4	3 months
Co-trimoxazole	8 mg/kg trimethoprim 40 mg/kg sulphamethoxazole	o	2	3 months
Ciprofloxacin	15–25 mg/kg	o	2	28 days

[1] Oral therapy is satisfactory for most patients. Parenteral therapy is generally reserved for severely ill patients.
[2] In intestinal perforation antibiotic treatment should also cover other aerobic and anaerobic bacteria contaminating the peritoneum.
[3] May cause bone marrow depression.
[4] The oral route is preferred; there are reports of lower blood levels of chloramphenicol in patients given parenteral therapy.
[5] May cause allergic reactions and nephrotoxicity. Not suitable for children <2 years or pregnant women.
[6] Short courses of ceftriaxone (3–5 days) have proved effective in non-severe typhoid in some studies but given a high failure rate in others.
[7] Not recommended in severe typhoid fever.
[8] Short courses of fluoroquinolones (3 days) have proved effective in non-severe typhoid in some studies provided the isolates did not have low-level fluoroquinolone resistance (ie they were nalidixic acid susceptible).
[9] Isolates with low-level fluoroquinolone resistance (nalidixic acid resistance) may not respond to short courses of fluoroquinolones (3–5 days). The optimum fluoroquinolone regimen for isolates with low-level fluoroquinolone resistance has not been determined, but treatment should be for at least 7 days in the maximum allowable dose.
O, oral; im, intramuscular; iv, intravenous.
(References: Parry et al., 2002; Parry, 2004; Stephens and Levine, 2002)

the rate is probably less than 1% although accurate data is lacking in many regions. In hospitalized cases, rates have been less than 2% in Pakistan (Bhutta, 1996a), 4.3% in Bangladesh (Butler *et al.*, 1991) and up to 15% in some areas of Papua New Guinea (Rogerson *et al.*, 1991). Case fatality rates are highest in the very young and the elderly, although delay in instituting effective antibiotic treatment is probably the most important contributor to a poor outcome.

The majority of intestinal carriers can be cured by a prolonged course of antibiotics, provided they do not have gallstones (Ferreccio *et al.*, 1988; Gotuzzo *et al.*, 1988). Depending on the susceptibility of the organisms, the regimens in Table 1.4 have given cure rates of approximately 80%. Patients with chronic urinary carriage related to *Schistosoma haematobium* should have the schistosomiasis treated before the treatment against *S. enterica* serovar Typhi is commenced. Cholecystectomy and nephrectomy should only be considered if antibiotics fail and there are additional indications for operation.

1.8 CONTROL AND PREVENTION OF TYPHOID FEVER

The elimination of typhoid from industrialized countries can be attributed to the provision of safe drinking water, safe disposal of sewage, legal enforcement of high standards of food hygiene, programs to detect, monitor and treat chronic carriers, and prompt investigation and intervention when these safeguards are breached. Travelers to endemic areas should drink boiled or bottled water, eat thoroughly cooked food, and avoid ice cream and fresh vegetables or fruit that have been peeled and washed in local water. In hospital, patients should be nursed using enteric precautions, and patients and convalescents advised to wash their hands after using the toilet and before preparing food.

Pfeiffer and Kolle in Germany, and Wright in England, first used inactivated (heat-killed, phenol-preserved) *S. enterica* serovar Typhi as a parenteral vaccine in 1896. By the First World War typhoid vaccination was routine in the British Army. In the 1960s randomized, placebo-controlled, field trials established the efficacy of the heat-phenolized and acetone-inactivated parenteral whole cell vaccines (Engels *et al.*, 1998). They conferred moderate levels of protection for up to 7 years. However, the local discomfort and swelling and the systemic side effects that occur in 25–50% of recipients has meant that the vaccine never became a well-accepted public health tool.

Two vaccines are currently licensed. The *S. enterica* serovar Typhi strain Ty21a vaccine is a live attenuated oral vaccine. The vaccine is given as three to four doses, in an enteric-coated capsule or liquid formulation, on alternate days and is safe for adults and children over 6 years of age. The vaccine confers protection within 10–14 days after the last dose and a booster dose is recommended every five years. Field studies have shown a protective efficacy after 3 years of 96% in Egypt (Wahdan *et al.*, 1982), 67–77% in Chile (Levine *et al.*, 1987; Levine *et al.*, 1990) and 42–53% in Indonesia (Simanjuntak *et al.*, 1991). The liquid formulation is more effective than the enteric-coated capsules. The vaccine is well tolerated but as it is a live attenuated vaccine should not be given to immunocompromised patients or patients taking antimicrobials. The parenteral Vi vaccine is suitable for adults and children over the age of 2 years and is without serious side effects. A single intramuscular dose of 0.5 mL (25 μg) confers protection within 7–10 days of inoculation and requires a booster every 3 years. A single injection of the Vi vaccine provided a protective efficacy of 72% after 17 months in Nepal (Acharya *et al.*, 1987), 55% after 3 years in South Africa (Klugman *et al.*, 1987; Klugman et al 1996) and 69% after 19 months using a locally produced Vi vaccine in Southern China (Yang *et al.*, 2001). There is no currently licensed vaccine for *S. enterica* serovar Paratyphi A. Promising new vaccine candidates are in development. The Vi polysaccharide-protein conjugate vaccine, a conjugate of the Vi capsular polysaccharide with non-toxic recombinant *Pseudomonas aeruginosa* exotoxin (rEPA), has been evaluated in Vietnam (Lin *et al.*, 2001). Two doses of the Vi-rEPA vaccine were safe and had an efficacy of 89% (95% confidence interval, 76–97%) in children aged 2–5 years after 46 months of follow-up (Lanh *et al.*, 2003). Several attenuated strains of Ty2 with defined mutations such as CVD 906-*htrA* (*aroC aroD htrA* mutants), Ty800 (*phoP phoQ* mutants) and χ4073 (*cya crp cdt* mutants) are in either phase I or phase II clinical trials (Garmory *et al.*, 2002). In each case, the intention is that they will be effective when given as a single oral dose.

The Ty21a and Vi vaccines are currently recommended for use in travelers to typhoid endemic areas, for household contacts of typhoid carriers and laboratory workers likely to handle *S. enterica* serovar Typhi. In endemic areas, mass immunization should be considered in disaster or refugee camps, in combination with adequate provision of safe water and food (Tarr *et al.*, 1999; WHO, 2003).

1.9 CONCLUSIONS

Controlling typhoid in developing countries by the provision of safe drinking water, effective sewage disposal and hygienic food preparation is a difficult and expensive proposition. The time scale for achieving these aims will not help to reduce the severe burden of disease currently seen. By default, the method of control in use is case detection and treatment. The difficulties of accurate diagnosis and emergence of resistance to multiple antimicrobials make this an unsatisfactory method for control. Mass vaccination programmes in endemic areas have rarely been used (Bodhidatta *et al.*, 1987). However, in view of the increasing morbidity, mortality and costs associated with drug resistant typhoid fever, the cost-effectiveness of mass vaccination as a public health measure to control this growing problem is now being reconsidered (Taylor *et al.*, 1999). The World Health Organization has recently extended its vaccination guidelines by recommending that a school-based immunization program should be considered in areas where typhoid is a recognized public health problem and multidrug resistant strains are particularly prevalent (WHO, 2003). It remains to be seen whether countries where the disease is endemic will be able to adopt these recommendations.

REFERENCES

Acharya, I. L., Lowe, C. U., Thapa, R. *et al.* (1987). Prevention of typhoid fever in Nepal with the Vi capsular polysaccharide of *Salmonella typhi*. A preliminary report. *N Engl J Med*, **317**, 1101–4.

Ackers, M. L., Puhr, N. D., Tauxe, R. V. and Mintz, E. D. (2000). Laboratory-based surveillance of *Salmonella* serotype Typhi infections in the United States: antimicrobial resistance on the rise. *JAMA*, **283**, 2668–73.

Ashcroft, M. T. (1964). Typhoid and Paratyphoid fever in the tropics. *J Trop Med Hyg*, **67**, 185–9.

Bethell, D. B., Hien, T. T., Phi, L. T. *et al.* (1996). Effects on growth of single short courses of fluoroquinolones. *Arch Dis Child*, **74**, 44–6.

Bhan, M. K., Bahl, R., Sazawal, S. *et al.* (2002). Association between *Helicobacter pylori* infection and increased risk of typhoid fever. *J Infect Dis*, **186**, 1857–60.

Bhutta, Z. A. (1996a). Impact of age and drug resistance on mortality in typhoid fever. *Arch Dis Child*, **75**, 214–17.

(1996b). Therapeutic aspects of typhoidal salmonellosis in childhood: the Karachi experience. *Ann Trop Paed*, **16**, 299–306.

Bhutta, Z. A. and Mansurali, N. (1999). Rapid serological diagnosis of pediatric typhoid fever in an endemic area: a prospective comparative evaluation of two dot-enzyme immunoassays and the Widal test. *Am J Trop Med Hyg*, **61**, 654–7.

Black, R. E., Cisneros, L., Levine, M. M. *et al.* (1985). Case-control study to identify risk factors for paediatric endemic typhoid fever in Santiago, Chile. *Bull WHO*, **43**, 899–904.

Bodhidatta, L., Taylor, D. N., Thisyakorn, U. and Echeverria, P. (1987). Control of typhoid fever in Bangkok, Thailand, by annual immunisation of schoolchildren with parenteral typhoid vaccine. *Rev Infect Dis*, **9**, 841–5.

Brenner, F. W., Villar, R. G., Angulo, F. J., Tauxe, R. and Swaminathan, B. (2000). *Salmonella* nomenclature. *J Clin Microbiol*, **38**, 2465–7.

Brown, J. C., Shanahan, P. M., Jesudason, M. V., Thomson, C. J. and Aymes, S. G. (1996). Mutations responsible for reduced susceptibility to 4-quinolones in clinical isolates of multi-resistant *Salmonella typhi* in India. *J Antimicrob Chemother*, **37**, 891–900.

Butler, T., Bell, W. R., Levin, J., Linh, N. N. and Arnold, K. (1978). Typhoid Fever. Studies of blood coagulation, bacteremia and endotoxaemia. *Arch Intern Med*, **138**, 407–10.

Butler, T., Islam, A., Kabir, I. and Jones, P. K. (1991). Patterns of morbidity and mortality in typhoid fever dependent on age and gender: a review of 552 hospitalised patients with diarrhoea. *Rev Infect Dis*, **13**, 85–90.

Butler, T., Knight, J., Nath, S. K. *et al.* (1985). Typhoid fever complicated by intestinal perforation: a persisting fatal disease requiring surgical management. *Rev Infect Dis*, **7**, 244–56.

Caygill, C. P. J., Hill, M. J., Braddick, M. and Sharp, J. C. M. (1994). Cancer mortality in chronic typhoid and paratyphoid carriers. *Lancet*, **343**, 83–4.

Chaudhry, R., Laxmi, B. V., Nisar, N., Ray, K. and Kumar, D. (1997). Standardisation of polymerase chain reaction for the detection of *Salmonella typhi* in typhoid fever. *J Clin Pathol*, **50**, 437–9.

Chinh, N. T., Parry, C. M., Ly, N. T. *et al.* (2000). A randomised controlled comparison of azithromycin and ofloxacin for multidrug-resistant and nalidixic acid resistant enteric fever. *Antimicrob Agents Chemother*, **44**, 1855–9.

Clegg, A., Passey, M., Omena, M., Karigifa, K. K. and Suve, N. (1994). Re-evaluation of the Widal agglutination test in response to the changing pattern of typhoid fever in the highlands of Papua New Guinea. *Acta Tropica*, **57**, 255–63.

Crump, J. A., Barrett, T. J., Nelson, J. T. and Angulo, F. J. (2003a). Reevaluating fluoroquinolone breakpoints for *Salmonella enterica* serotype Typhi and for non-Typhi salmonellae. *Clin Infect Dis*, **37**, 75–81.

Crump, J. A., Youssef, F. G., Luby, S. P. *et al.* (2003b). Estimating the incidence of Typhoid Fever and other febrile illnesses in developing countries. *Emerg Infect Dis*, **9**, 539–44.

Crump, J. A., Luby, S. P. and Mintz, E. D. (2004). The global burden of typhoid fever. *Bull World Health Org*, **82**, 346–53.

Doherty, C. P., Saha, S. K. and Cutting, W. M. (2000). Typhoid fever, ciprofloxacin and growth in young children. *Ann Trop Paed*, **20**, 297–303.

Duggan, M. B. and Beyer, L. (1975). Enteric fever in young Yoruba children. *Arch Dis Child*, **50**, 67–71.

Dutta, P., Mitra, U., Dutta, S., De, A., Chatterjee, M. K. and Bhattacharya, S. K. (2001). Ceftriaxone therapy in ciprofloxacin treatment failure in children. *Indian J Med Res*, **113**, 210–13.

Dutta, P., Rasaily, R., Saha, M. R. *et al.* (1993). Ciprofloxacin for treatment of severe typhoid fever in children. *Antimicrob Agents Chemother*, **37**, 1197–9.

Engels, E. A., Falagas, M. E., Lau, J. and Bennish, M. L. (1998). Typhoid fever vaccines: a meta-analysis of studies on efficacy and toxicity. *Br Med J*, 316, 110–16.

Everest, P., Wain, J., Roberts, M., Rook, G. and Dougan, G. (2001). The molecular mechanisms of severe typhoid fever. *Trends Microbiol*, **9**, 316–20.

Ferreccio, C., Levine, M. M., Manterola, A. *et al.* (1984). Benign bacteraemia caused by *Salmonella typhi* and *paratyphi* in children younger than two years. *J Pediatr*, **104**, 899–901.

Ferreccio, C., Morris, J. G., Valdiviseo, C. *et al.* (1988). Efficacy of ciprofloxacin in the treatment of chronic typhoid carriers. *J Infect Dis*, **157**, 1235–9.

Garmory, H. S., Brown, K. A. and Titball, R. W. (2002). *Salmonella* vaccines for use in humans: present and future perspectives. *FEMS Microbiol Rev*, **26**, 339–53.

Gasem, M. H., Dolmans, W. M. V., Keuter, M. and Djokomoeljanto, R. (2001). Poor food hygiene and housing as risk factors for typhoid fever in Semarang, Indonesia. *Trop Med Int Hlth*, **6**, 484–90.

Gilman, R. H., Terminel, M., Levine, M. M., Hernandez-Mendoze, P. and Hornick, R. B. (1975). Relative efficacy of blood, urine, rectal swab, bone marrow and rose spot cultures for recovery of *Salmonella typhi* in typhoid fever. *Lancet*, **1**, 1211–13.

Gotuzzo, E., Guerra, J. G., Benavente, L. *et al.* (1988). Use of norfloxacin to treat chronic typhoid carriers. *J Infect Dis*, **157**, 1221–5.

Hatta, M., Mubin, H., Abdoel, T. and Smits, H. L. (2002). Antibody response in typhoid fever in endemic Indonesia and the relevance of serology and culture to diagnosis. *Southeast Asian J Trop Med Hyg*, **33**, 742–51.

Hien, T. T., Bethell, D. B., Hoa, N. T. T. *et al.* (1995). Short course of ofloxacin for treatment of multidrug-resistant typhoid. *Clin Infect Dis*, **20**, 917–23.

Hoffman, S. L., Edman, D. C., Punjabi, N. H. *et al.* (1986). Bone marrow aspirate culture superior to streptokinase clot culture and 8 ml 1:10 blood-to-broth ratio blood culture for diagnosis of typhoid fever. *Am J Trop Med Hyg*, **35**, 836–9.

Hoffman, S. L., Punjabi, N. H., Kumala, S. *et al.* (1984a). Reduction of mortality in chloramphenicol-treated severe typhoid fever by high dose dexamethasone. *N Engl J Med*, **310**, 82–8.

Hoffman, S. L., Punjabi, N. H., Rockhill, R. C. *et al.* (1984b). Duodenal string-capsule culture compared with bone-marrow, blood, and rectal-swab cultures for diagnosing typhoid and paratyphoid fever. *J Infect Dis*, **149**, 157–61.

Hornick, R. B., Greisman, S. E., Woodward, T. E. *et al.* (1970). Typhoid fever: pathogenesis and immunological control. *N Engl J Med*, **283**, 686–91 and 739–46.

House, D., Bishop, A., Parry, C. M., Dougan, G. and Wain, J. (2001a). Typhoid fever: pathogenesis and disease. *Curr Opinions in Infect Dis*, **14**, 573–8.

House, D., Wain, J., Ho, V. A. *et al.* (2001b). The serology of typhoid fever in an endemic area and its relevance to diagnosis. *J Clin Microbiol*, **39**, 1002–7

Huckstep, R. L. (1962). *Typhoid fever and other "Salmonella" infections.* Edinburgh and London: Livingstone.

Kariuki, S., Gilks, C., Revathi, G., Hart, C. A. (2000). Genotypic analysis of multidrug-resistant *Salmonella enterica* serovar Typhi, Kenya. *Emerg Infect Dis*, **6**, 649–51.

Klugman, K. P., Gilbertson, I. T., Koornhof, H. J. *et al.* (1987). Protective activity of Vi capsular polysaccharide vaccine against typhoid fever. *Lancet*, **2**, 1165–9.

Klugman, K. P., Koornhof, H. J., Robbins, J. B. and Cam, N. N. L. (1996). Immunogenicity, efficacy and serological correlate of protection of *Salmonella typhi* Vi capsular polysaccharide vaccine three years after immunization. *Vaccine*, **14**, 435–8.

Koul, P. A., Wani, J. I. and Wahid, A. (1995). Ciprofloxacin for multiresistant enteric fever in pregnancy. *Lancet*, **346**, 307–8.

Lanata, C. F., Levine, M. M., Ristori, C., Black, R. E., Jiminez, L., Salcedo, M., Garcia, J. and Sotomayor, V. (1983) Vi serology in detection of chronic *Salmonella typhi* carriers in an endemic area. *Lancet*, **2**, 441–3.

Lanh, M. N., Bay, P. V., Ho, V. A. *et al.* (2003). Persistent efficacy of Vi conjugate vaccine against typhoid fever in young children. *N Engl J Med*, **349**, 1390–1.

Levine, M. M., Ferreccio, C., Black, R. E. and Germanier, R. (1987). Large-scale field trial of Ty21a live oral typhoid vaccine in enteric-coated capsule formulation. *Lancet*, **1**, 1049–52.

Levine, M. M., Ferrecio, C., Cryz, S. and Ortiz, E. (1990). Comparison of enteric-coated capsules and liquid formulation of Ty21a typhoid vaccine in randomized controlled field trial. *Lancet*, **336**, 891–4.

Levine, M. M., Grados, O., Gilman, R. H. *et al.* (1978). Diagnostic value of the Widal test in areas endemic for typhoid fever. *Am J Trop Med Hyg*, **27**, 795–800.

Lin, F. Y. C., Ho, V. A., Bay, P. V. *et al.* (2000). The epidemiology of typhoid fever in the Dong Thap province, Mekong Delta region of Vietnam. *Am J Trop Med Hyg*, **62**, 644–8.

Lin, F. Y. C., Ho, V. A., Khiem, H. B. *et al.* (2001). The efficacy of a *Salmonella typhi* Vi conjugate vaccine in two-to-five-year-old children. *N Engl J Med*, **344**, 1263–9.

Luby, S. P., Faizan, M. K., Fisher-Hoch, S. P. *et al.* (1998). Risk factors for typhoid fever in an endemic setting, Karachi, Pakistan. *Epidemiol Infect*, **120**, 129–38.

Luxemburger, C., Duc, C. N., Lanh, M. N. *et al.* (2001). Risk factors for Typhoid fever in the Mekong Delta, southern Vietnam: a case-control study. *Trans R Soc Trop Med Hyg*, **95**, 19–23.

Mahle, W. T. and Levine, M. M. (1993). *Salmonella typhi* infection in children younger than five years of age. *Pediatr Infect Dis J*, **12**, 627–31.

Marmion, D. E., Naylor, G. R. E. and Stewart, I. O. (1953). Second attacks of typhoid fever. *J Hyg (Camb)*, **51**, 260–7.

Mehta, G., Randhawa, V. S. and Mohapatra, N. P. (2001). Intermediate susceptibility to ciprofloxacin in *Salmonella typhi* strains in India. *Eur J Clin Microbiol Infect Dis*, **20**, 760–1.

Mermin, J. H., Villar, R., Carpenter, J. *et al.* (1999). A massive epidemic of multidrug-resistant typhoid fever in Tajikistan associated with consumption of municipal water. *J Infect Dis*, **179**, 1416–22.

Morris, J. G., Ferreccio, C., Garcia, J. *et al.* (1984). Typhoid fever in Santiago, Chile: a study of household contacts of pediatric patients. *Am J Trop Med Hyg*, **33**, 1198–202.

Parry, C. M. (2004). The treatment of multidrug resistant and nalidixic acid resistant typhoid fever in Vietnam. *Trans Roy Soc Trop Med Hyg*, **98**, 413–22.

Parry, C. M, Hien, T. T., Dougan, G., White, N. J. and Farrar, J. J. (2002). Typhoid Fever. *N Engl J Med*, **347**, 1770–82.

Parry, C. M., Hoa, N. T., Diep, T. S. *et al.* (1999). Value of a single-tube Widal test in diagnosis of typhoid fever in Vietnam. *J Clin Microbiol*, **37**, 2882–6.

Punjabi, N. H., Hoffman, S. L., Edman, D. C. *et al.* (1988). Treatment of severe typhoid fever in children with high dose dexamethasone. *Pediatr Infect Dis J*, **7**, 598–600.

Reed, R. P. and Klugman, K. P. (1994). Neonatal typhoid fever. *Pediatr Infect Dis J*, **13**, 774–7.

Richens, J. (1996). Typhoid. In *The Wellcome Trust Illustrated History of Tropical Diseases*, ed. F. E. G. Cox. London: Wellcome Trust.

Rodrigues, C., Shenai, S. and Mehta, A. (2003). Enteric fever in Mumbai, India: the good news and the bad news. *Clin Infect Dis*, **36**, 535.

Rogerson, S. J., Spooner, V. J., Smith, T. A. and Richens, J. (1991). Hydrocortisone in chloramphenicol-treated severe typhoid fever in Papua New Guinea. *Trans Roy Soc Trop Med Hyg*, **85**, 113–16.

Rowe, B., Ward, L. R. and Threlfall, E. J. (1997). Multidrug-resistant *Salmonella typhi*: a worldwide epidemic. *Clin Infect Dis*, **24**(suppl. 1), S106–9.

Rupali, P., Abraham, O. C., Jesudason, M. V. *et al.* (2004). Treatment failure in typhoid fever with ciprofloxacin susceptible *Salmonella enterica* serotype Typhi. *Diagn Microbiol Infect Dis*, **49**, 1–3.

Saha, S. K., Talukder, S. Y., Islam, M. and Saha, S. (1999). A highly ceftriaxone-resistant *Salmonella typhi* in Bangladesh. *Pediatr Infect Dis J*, **18**, 387.

Seoud, M., Saade, G., Uwaydah, M. and Azoury, R. (1988). Typhoid fever in pregnancy. *Obstet Gynecol*, **71**, 711–14.

Simanjuntak, C. H., Paleologo, F. P., Punjabi, N. H. *et al.* (1991). Oral immunisation against typhoid fever in Indonesia with Ty21a vaccine. *Lancet*, **338**, 1055–9.

Sinha, A., Sazawal, S., Kumar, R. *et al.* (1999). Typhoid fever in children aged less than five years. *Lancet*, **354**, 734–7.

Stephens, I. and Levine, M. M. (2002). Management of typhoid fever in children. *Pediatr Infect Dis J*, **21**, 157–9.

Stuart, B. M. and Pullen, R. L. (1946). Typhoid. Clinical analysis of three hundred and sixty cases. *Arch Intern Med*, **78**, 629–61.

Tarr, P. E., Kuppens, L., Jones, T. C. *et al.* (1999). Considerations regarding mass vaccination against typhoid fever as an adjunct to sanitation and public health measures: potential use in an epidemic in Tajikistan. *Am J Trop Med Hyg*, **61**, 163–70.

Taylor, D. N., Levine, M. M., Kuppens, L. and Ivanoff, B. (1999). Why are typhoid vaccines not recommended for epidemic typhoid fever? *J Infect Dis*, **180**, 2089–90.

Thong, K. L., Cheong, Y. M., Puthucheary, S., Koh, C. L. and Pang, T. (1994). Epidemiologic analysis of sporadic *Salmonella typhi* isolates and those from outbreaks by pulsed-field electrophoresis. *J Clin Microbiol*, **32**, 1135–41.

Threlfall, E. J., Skinner, J. A., Ward, L. R. (2001). Detection of decreased *in vitro* susceptibility to ciprofloxacin in *Salmonella enterica* serotypes Typhi and Paratyphi A. *J Antimicrob Chemother*, **48**, 740–1.

Topley, J. M. (1986). Mild typhoid fever. *Arch Dis Child*, **61**, 164–7.

Velema, J. P., van Wijnen, G., Bult, P., van Naerssen, T. and Jota, S. (1997). Typhoid fever in Ujung Pandang, Indonesia – high-risk groups and high-risk behaviours. *Trop Med Int Health*, **2**, 1088–94.

Wahdan, M. H., Sérié, C., Cerisier, Y., Sallam, S. and Germanier, R. (1982). A controlled field trial of live *Salmonella typhi* strain Ty21a oral vaccine against typhoid: three-year results. *J Infect Dis*, **145**, 292–5.

Wain, J., Bay, P. V. B., Vinh, H. *et al.* (2001). Quantitation of bacteria in bone marrow from patients with Typhoid Fever: relationship between counts and clinical features. *J Clin Microbiol*, **39**, 1571–6.

Wain, J., Diep, T. S., Ho, V. A. *et al.* (1998). Quantitation of bacteria in blood of typhoid fever patients and relationship between counts and clinical features, transmissibility, and antibiotic resistance. *J Clin Microbiol,* **36**, 1683–7.

Wain, J., Hoa, N. T., Chinh, N. T. T. *et al.* (1997). Quinolone-resistant *Salmonella typhi* in Viet Nam: molecular basis of resistance and clinical response to treatment. *Clin Infect Dis,* **25**, 1404–10.

Wasfy, M. O., Frenck, R., Ismail, T. F. *et al.* (2002). Trends of multiple-drug resistance among *Salmonella* serotype Typhi isolates during a 14-year period in Egypt. *Clin Infect Dis,* **35**, 1265–8.

Woodward, T. E., Smadel, J. E., Ley, H. L., Green, R. and Mankikar, D. S. (1948). Preliminary report on the beneficial effect of chloromycetin in the treatment of typhoid fever. *Ann Intern Med,* **29**, 131–4.

World Health Organization background document. The diagnosis, treatment and prevention of typhoid fever. *WHO*/V&B/03.07 (2003). Accessed at www.who.int/vaccines-documents/DocsPDF03/ www740.pdf on 15.03.04.

Yang, H. H., Wu, C. G., Xie, G. Z. *et al.* (2001). Efficacy trial of Vi polysaccharide vaccine against typhoid fever in South-Western China. *Bull WHO,* **79**, 625–31.

CHAPTER 2

Antibiotic resistance in *Salmonella* infections

Fiona J. Cooke and John Wain

2.1 INTRODUCTION

Salmonella infections in humans can range from a self-limiting gastroenteritis, usually associated with non-typhoidal *Salmonella* (NTS), to typhoid fever with complications such as a fatal intestinal perforation. The World Health Organization (WHO) estimates that the annual global incidence of typhoid fever is about 21 million cases with a mortality of 1% (Crump *et al.*, 2004). This may be an underestimate because typhoid is predominantly a disease of developing countries, where not all cases present to the healthcare services and data collection may be difficult. In addition, financial constraints limit outbreak investigation and antibiotics are often widely available without prescription. Not only does this compound problems with data gathering, but it is likely to add to the burden of resistant disease circulating in the community. The situation is even less clear for NTS because most patients do not need to consult the health services. Despite this, as reported in 1999 in the USA alone there were an estimated 1.4 million cases of NTS infection annually, resulting in approximately 600 deaths (Mead *et al.*, 1999).

There is no doubt that antibiotic resistance in *Salmonella* infections poses a major threat to human health, especially in cases of invasive NTS in immunocompromised patients and in typhoid fever. The cost of resistance in human terms is shown in Table 2.1. There is also a potential increase in the cost of food production. If an untreatable disease arises in food animals because of the development of antibiotic resistance, then slaughter on a scale seen during the foot and mouth epidemic in the UK is a possibility.

'Salmonella' Infections: Clinical, Immunological and Molecular Aspects, ed. Pietro Mastroeni and Duncan Maskell. Published by Cambridge University Press.

Table 2.1. *Impact on human health of resistance in Salmonella infections*

Impact	Contributing factor(s)	Examples	Reference
Increased morbidity and mortality in humans	Ineffective first choice antibiotic delays appropriate treatment	In South Africa in 1991 3 of 6 cases of MDR typhoid died because of delay in appropriate therapy.	(Coovadia *et al.*, 1992)
	Second line antibiotics may be more toxic, not available orally, less effective and more expensive	Third generation cephalosporins for the treatment of typhoid fever.	(White, 1999)
	Increased virulence of resistant infections	Danish matched cohort study showed an excess mortality associated with resistant *S. enterica* serovar Typhimurium.	(Helms *et al.*, 2002)
		Analysis of 175 reports of NTS infections (USA 1971–80) showed higher mortality rate in resistant NTS compared to susceptible NTS.	(Holmberg *et al.*, 1987)
	Untreatable infections	Emergence of full quinolone or extended spectrum cephalosporin resistance may result in some strains becoming untreatable.	(Murray, 1989) (Parry, 1998)
Increased indirect costs of treatment	Humans: increased length of hospital stay	CDC Atlanta prospective study of NTS (1989–90) found patients with a resistant organism were more likely to be hospitalized, for longer.	(Lee and Falkow, 1994)
	Humans: loss of earnings	A mathematical model estimates 5180 excess days of hospitalization per year in the USA attributable to resistant strains of *Salmonella*.	(Travers and Barza, 2002)

Table 2.2. *Definitions of resistance terms*

Minimum inhibitory concentration (MIC)	Minimum concentration of antibiotic that inhibits in vitro the growth of the organism after overnight incubation.
Biological resistance	Defined statistically, as those organisms with an MIC above the 95th percentile.
Genotypic *vs.* Phenotypic resistance	The presence of the genes that confer resistance to an antibiotic *vs* whether these genes are expressed or not.
Acquired resistance	The movement of genes or gene systems, often on mobile genetic elements, into a pathogen population from an external source.
De novo resistance	This usually develops in a stepwise manner from accumulation of mutations that individually lower susceptibility by modest increments.
Mutant selection window	The range of antibiotic concentrations from the minimum concentration needed to inhibit the growth of wild-type bacteria, up to that needed to inhibit the growth of the least susceptible, single-step mutant. (Drlica, 2003)

This chapter outlines the development of antibiotic resistance in *Salmonella enterica* serovar Typhi and NTS. Although the main focus will be on human infection, data on the resistance of isolates from animals and food products is also presented. The cost of resistance to both human and bacterial populations is discussed. Definitions relevant to the discussion of resistance are shown in Table 2.2.

2.2 ANTIBIOTIC RESISTANCE IN *S. ENTERICA* SEROVAR TYPHI

2.2.1 Chloramphenicol

Chloramphenicol was introduced in 1948 for the treatment of typhoid fever (Woodward and Smadel, 1948/Fall 2004), but only two years later chloramphenicol resistant *S. enterica* serovar Typhi was reported in England

(Colquhoun and Weetch, 1950). Nevertheless, chloramphenicol continued as the treatment of choice for over two decades (Nath and Singh, 1966). There were occasional reports of chloramphenicol resistance, which was largely sporadic, but resistance did not become a significant clinical problem until the early 1970s. In 1972, three large outbreaks were documented almost simultaneously in India (Paniker and Vimala, 1972), Mexico (Gonzalez Cortes *et al.*, 1973) and Vietnam (Brown *et al.*, 1975). Chloramphenicol resistance was also becoming a problem in Korea (Chun *et al.*, 1977), Thailand and Chile (Murray *et al.*, 1985) and Peru (Mirza *et al.*, 1996). Analysis of seven resistant strains from India demonstrated that the resistance was transferable to *E. coli* (Paniker and Vimala, 1972), and the plasmids involved were later grouped as IncHI1 (Grindley *et al.*, 1973; Taylor and Brose, 1985). The typhoid outbreak in Mexico City included over 10,000 reported cases, and in addition to chloramphenicol resistance, over 90% of the *S. enterica* serovar Typhi isolates were resistant to tetracycline, streptomycin and sulphonamides. Again, IncH plasmids were involved (Olarte and Galindo, 1973) and in all cases, they carried a gene for chloramphenicol acetyl transferase (CAT) type 1 (Taylor, 1983). This is still present on more recent *S. enterica* serovar Typhi plasmids (Shanahan *et al.*, 1998; Shanahan *et al.*, 2000).

The antibiotic resistance pattern was shared in Mexico City by *Shigella dysenteriae* and *Shigella flexneri* that caused a concurrent outbreak of bacillary dysentery (Gangarosa *et al.*, 1972). Towards the end of the Mexican outbreak of typhoid fever, seven strains were found to be resistant to ampicillin and kanamycin (Olarte and Galindo, 1973). However, this new resistance phenotype did not spread and it was clear that two different R factors were responsible. It was later shown that the *Shigella* plasmids could not be maintained in *S. enterica* serovar Typhi and that they belonged to the genetically different group IncO (Datta and Olarte, 1974).

Chloramphenicol resistance of *S. enterica* serovar Typhi in Indonesia is somewhat atypical and emerged relatively late (1978–9), along with resistance to tetracyclines and sulphonamides (Komalarini *et al.*, 1980). A more recent study found that all *S. enterica* serovar Typhi isolates were chloramphenicol sensitive, even from regions where *Shigella* isolates were resistant to all first line antibiotics (Oyofo *et al.*, 2002). Thus, in some regions of Indonesia, chloramphenicol is still useful for the first line treatment of enteric fever. Chloramphenicol sensitive isolates have also re-emerged in other countries, probably in response to a change in first-line treatment from chloramphenicol to ampicillin or co-trimoxazole (Sood *et al.*, 1999). In Southern India, chloramphenicol sensitive and resistant strains co-exist (Jesudason *et al.*, 1996).

In other areas however, chloramphenicol resistance has remained stable, despite decreased use of the antibiotic (data from the Hospital for Tropical Diseases, Ho Chi Minh City, Vietnam).

2.2.2 Multidrug resistance

Multidrug resistance (MDR) in *S. enterica* serovar Typhi is defined as resistance to the three drugs used as first line treatment, namely chloramphenicol, ampicillin and co-trimoxazole (trimethoprim-sulphamethoxazole). Driven by a change in antibiotic usage, the MDR phenotype emerged gradually in *S. enterica* serovar Typhi, as each resistance type was acquired independently. While chloramphenicol and ampicillin have been available for decades, co-trimoxazole was first described for the treatment of typhoid fever in 1981 (Mukhtar and Mekki, 1981). Until the 1980s, no single isolate harbored resistance to all three drugs even though resistance was reported to each drug independently. Early documented outbreaks of MDR *S. enterica* serovar Typhi occurred in Malaysia in 1984 (Ling and Chau, 1984) and in Kashmir, India in 1988 (Kamili *et al.*, 1993). By 1990, MDR *S.* Typhi had spread throughout the globe, including to Pakistan where a single transferable plasmid of 98Mda, encoding the MDR phenotype, was reported in 90% of isolates (Mirza and Hart, 1993). In Vietnam, 86% of all *S. enterica* serovar Typhi isolated from the south of the country by 1996 were MDR (Vinh *et al.*, 1996).

MDR *S. enterica* serovar Typhi are less prevalent in Africa compared to Asia, but have been present in South Africa for some time and recent publications have reported their presence in Ghana (Mills-Robertson *et al.*, 2002) and Kenya (Kariuki *et al.*, 2000). Like Asia, there are regional differences in Africa. In Dakar, Senegal, during 1987–1990 and 1997–2002 less than 1% of isolates were resistant to chloramphenicol. Yet, between these dates the teaching hospital in Dakar reported that 76% of strains were resistant to chloramphenicol (Dromigny and Perrier-Gros-Claude, 2003). The resistance patterns of *S. enterica* serovar Typhi therefore vary chronologically as well as geographically. The *S. enterica* serovar Typhi strains reported from South and Central America have a lower level of antibiotic resistance. In South America, 25 blood culture isolates of *S. enterica* serovar Typhi from seven different countries were tested and all were susceptible to chloramphenicol, co-trimoxazole, nalidixic acid and ciprofloxacin. Only one isolate was resistant to ampicillin (Gales *et al.*, 2002).

In some places, MDR never really became established, such as in Bangladesh where MDR typhoid fever only reached 4% at its height (Rahman *et al.*, 2002). In recent years several regions have seen a decline in MDR *S. enterica* serovar Typhi, attributed to a change in the use of antibiotics (Zahurul Haque Asna, and Ashraful Haq, 2000). It appears that in many areas the removal of the selective pressures due to antibiotic treatment, results in loss of the resistance phenotype. It has been known for many years that plasmids from *E. coli* are not stable in *S. enterica* serovar Typhi because of defects in the partition of new plasmids into new bacterial cells and this could in part explain why plasmid-mediated antibiotic resistance is lost (Mendoza-Medellin *et al.*, 2004; Mendoza-Medellin *et al.*, 1993). Following changes in the use of antibiotics in Mumbai, India, plasmid-mediated resistance initially decreased from 74% in 1990 to 40% in 1994 then stabilized at 46% until 2000 and finally fell again to 17% in 2002 (Rodrigues *et al.*, 2002). Even more striking is the situation in Cairo, Egypt where the proportion of isolates that were MDR decreased from 100% in 1993 to 5% in 2000 (Wasfy *et al.*, 2002).

Despite the percentage of MDR *S. enterica* serovar Typhi isolates decreasing in some areas, MDR typhoid fever may still be increasing in countries such as Japan (Hirose *et al.*, 2001).

2.2.3 Resistance beyond MDR

With the emergence of resistance to the three first line drugs, the fluoroquinolones have become established as the treatment of choice for typhoid fever. Since 1993 however, there has been a global epidemic of nalidixic acid resistant (Na^R) *S. enterica* serovar Typhi. These strains exhibit decreased susceptibility to ciprofloxacin, and infected patients have a poorer clinical response to fluoroquinolones compared to those infected with nalidixic acid susceptible strains (Parry *et al.*, 1998; Wain *et al.*, 1997). A large water-associated outbreak of typhoid fever occurred in Tajikistan in 1997, causing almost 9,000 cases of typhoid fever and 95 deaths in Dushanbe (Mermin *et al.*, 1999). The epidemic strain was Na^R and the outbreak subsided only when water purification treatment was improved. In India and Pakistan, resistant strains are now endemic. For example in Mumbai, 82% of 240 blood culture isolates of *S. enterica* serovar Typhi in the year 2000 were Na^R (Rodrigues *et al.*, 2002). In the UK, there has been an increase in the proportion of *S. enterica* serovar Typhi with reduced sensitivity to ciprofloxacin referred to the Health Protection Agency (HPA), Colindale. The resistance patterns of cases of *S. enterica* serovar Typhi imported into England, Scotland and Wales from 2000–2003 is summarized in Table 2.3 and most patients infected with Na^R strains had traveled to the Indian Subcontinent.

Table 2.3. *Resistance pattern of cases of* S. enterica *serovar Typhi imported into England, Scotland and Wales (2000–2003)*

Country	Resistance pattern	No. cases
Afghanistan	ACSSuTTm	1
	sensitive	2
Africa (unspecified)	ACSuTTm	1
	sensitive	2
Bangladesh	ACSSuTmNaCp$_{L}$	2
	ACSSuTTmNaCp$_{L}$	5
	Cp$_{L}$	4
	NaCp$_{L}$	5
	sensitive	21
Cambodia	ASSuTTm	1
	sensitive	1
Canary Islands	NaCp$_{L}$	1
China	sensitive	1
Ethiopia	sensitive	1
Far East	sensitive	2
France	sensitive	1
Ghana	ACSuTTm	2
	ASuTTm	1
	sensitive	4
Greece	sensitive	2
Guatemala	sensitive	1
Hong Kong	NaCp$_{L}$	1
	sensitive	1
India	ACSSuSpTTmNaCp$_{L}$	1
	ACSSuTm	1
	ACSSuTmNap$_{L}$	2
	ACSSuTTm	3
	ACSSuTTmCp$_{L}$	6
	ACSSuTTmFuNaCp$_{L}$	1
	ACSSuTTmNaCp$_{L}$	13
	Cp$_{L}$	12
	CTTmNaCp$_{L}$	1
	NaCp$_{L}$	60
	SNaCp$_{L}$	1
	sensitive	30
Indonesia	sensitive	6
Kenya	ACSSuTTmCp$_{L}$	1
	sensitive	1

(*cont.*)

Table 2.3. (*cont.*)

Country	Resistance pattern	No. cases
Malawi	$NaCp_L$	1
Malaysia	$NaCp_L$	1
Middle East	sensitive	1
Morocco	sensitive	4
Nepal	$ACSSuTTmNaCp_L$	4
	$NaCp_L$	1
	sensitive	3
Nigeria	ACSSuSpTTm	1
	ACSSuTTm	1
	ACSuTTm	1
	ACTTm	1
	ASuTTm	1
	C	1
	CSuTm	1
	sensitive	22
Pakistan	ACGSSuTTm	1
	$ACSSuSpTNaCp_L$	1
	ACSSuTm	1
	ACSSuTTm	19
	$ACSSuTTmNaCp_L$	17
	ACSSuTTmNa	1
	ACSuTTm	1
	ATm	1
	CpL	8
	CSSuTTm	2
	$CTTmNaCp_L$	1
	NaCpL	17
	sensitive	92
Philippines	sensitive	2
Spain	sensitive	1
Sri Lanka	$ACSSuTTmNaCp_L$	1
	$NaCp_L$	2
	sensitive	1
Thailand	sensitive	1
Unknown	Various	276

Key to antibiotics: A = ampicillin, S = streptomycin, T = tetracycline, C = chloramphenicol, Su = sulphathiozole, Tm = trimethoprim, Na = nalidixic acid, G = Gentamicin, Sp = spectinomycin, Fu = Furazolidone, Cp_L = low level Ciprofloxacin resistance Data from Laboratory of Enteric Pathogens, Health Protection Agency, Colindale, UK.

Reduced susceptibility to fluoroquinolones is not always reported by diagnostic laboratories because sensitivity testing with fluoroquinolone discs usually produces visible zones of inhibition of bacterial growth on solid media (interpreted as fully susceptible). To assist in the detection of isolates with reduced sensitivity to fluoroquinolones, all isolates should to be screened with nalidixic acid discs but to define the problem the Minimum Inhibitory Concentration (MIC) of the fluoroquinolone used for treatment needs to be determined. There is much discussion in the literature about either introducing a new 'intermediate' class in the interpretation of the sensitivity result, or lowering the MIC for which sensitivity to the fluoroquinolone is assumed (break point) (Crump *et al.*, 2003). High-level fluoroquinolone resistance has been described in other *S. enterica* serovars (Ling *et al.*, 2003) and *E. coli* (Webber and Piddock, 2001), but as yet there are no reports of this occurring in *S. enterica* serovar Typhi. Resistance to fluoroquinolones is usually mediated by mutations in topoisomerase genes. However, plasmid borne resistance associated with a different quinolone resistance gene (*qnr*) (Martínez *et al.*, 1998; Martínez-Martínez *et al.*, 1998) is now widespread in *Klebsiella pneumoniae* (Wang *et al.*, 2004). The resistance is associated with a class 1 integron like structure (Tran and Jacoby, 2002), which contains a common region (Partridge and Hall, 2003) when compared to integrons found on resistance plasmids of other *S. enterica* serovars (Verdet *et al.*, 2000). It therefore seems likely that the *qnr* gene is capable of spreading to the plasmids of *S. enterica* serovar Typhi.

Resistance to the third generation cephalosporins is usually mediated by extended spectrum beta-lactamases (ESBLs) on integrons that are plasmid borne (Jacoby and Sutton, 1991; Poirel *et al.*, 1999). Resistance to third generation cephalosporins has been reported in *S. enterica* serovars (Shannon and French, 1998), but to date only one report of high-level resistance to ceftriaxone (MIC 64 mg/L) in *S. enterica* serovar Typhi has been published (Saha *et al.*, 1999). However, as integrons have been detected in *S. enterica* serovar Typhi (Pai *et al.*, 2003) there is enormous potential for additional antibiotic resistance genes to be acquired.

2.2.4 Molecular analysis of resistance plasmids

Resistance to several antibiotics in *S. enterica* serovar Typhi is associated with the presence of IncHI plasmids. It appears that the IncHI plasmids from several geographical locations are highly similar, and the resistance determinants are located together in the variable regions of a conserved DNA backbone (Taylor and Brose, 1985). The IncHI1 plasmid pHCM1 present in

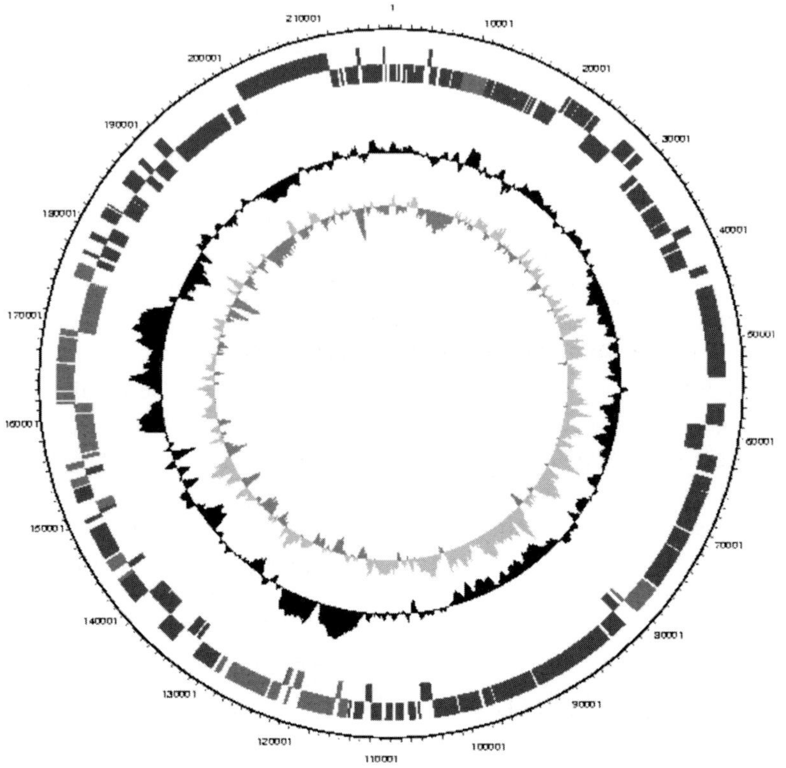

Figure 2.1. A map of plasmid pHCM1 from the sequenced strain of *S. enterica* serovar Typhi showing regions of similarity with plasmid R27. The outer ring gives the location of predicted genes, shown in the next two rings. The inner rings show G + C content and G + C bias. (See colour plate section.)

S. enterica serovar Typhi CT18 shows considerable similarity to plasmid R27 (Parkhill *et al.*, 2001; Sherburne *et al.*, 2000; Wain *et al.*, 2003). R27 was isolated from *S. enterica* serovar Typhimurium in 1961 in Winchester and confers only tetracycline resistance (Taylor, Brose, 1985). Thus, pHCM1 and R27 were isolated from different serotypes, on different continents, 33 years apart and yet show over 99% similarity at the DNA level (Figure 2.1). There are two regions of insertion in pHCM1 relative to R27. The larger insertion region (Figure 2.2) has several resistance genes flanked by insertion sequences, and contains integron-like structures including a truncated Tn*10* element (Parkhill *et al.*, 2001). The smaller insertion region contains a truncated class one integron, with a trimethoprim resistance gene, but no recognizable gene cassette and no 3'conserved sequence. Adjacent to this region is an IS*6100* element and a mercury resistance operon. The truncation in the

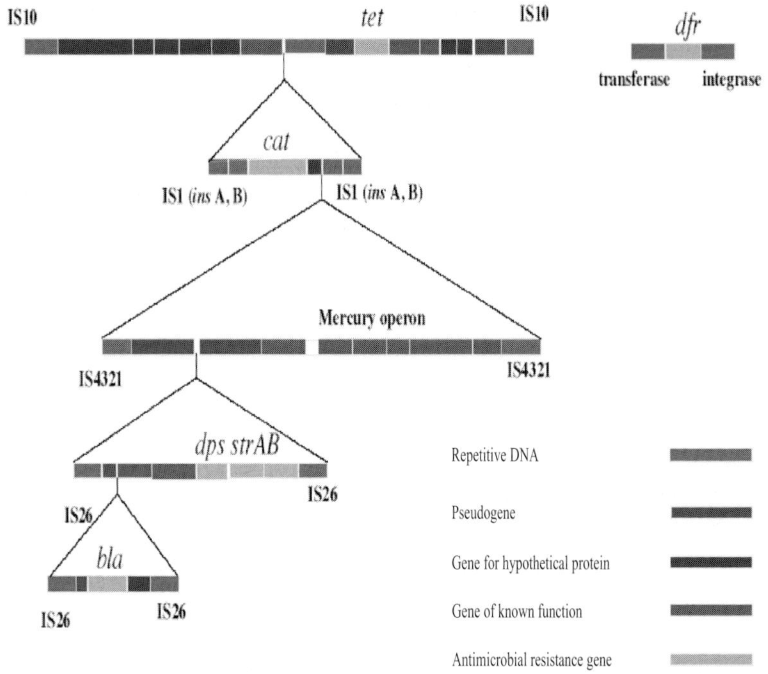

Figure 2.2. A diagram of the large insertion region of plasmid pHCM1. (See colour plate section.)

class one integron probably represents a deletion event following insertion of the IS6100 during the acquisition of mercury resistance.

Although the group of IncH plasmids may show variation (Tassios *et al.*, 1997), and there are occasional reports of IncB, IncN, IncP IncB/O (Ling and Chau, 1984) and Inc F plasmids (Mirza *et al.*, 2000), it is clear that IncHI1 plasmids conferring resistance in *S. enterica* serovar Typhi are relatively conserved, and have disseminated rapidly within the bacterial population (Hampton *et al.*, 1998; Harnett *et al.*, 1998). What is less well understood is whether the global picture is due to identical resistance plasmids spreading within the clonal expansion of a single *S. enterica* serovar Typhi host, or whether similar plasmids have spread in different bacterial hosts. In India (Shanahan *et al.*, 1998), Canada (Harnett *et al.*, 1998) and Bangladesh (Hermans *et al.*, 1996) the bacterial host seems to be of a single genotype, yet in South East Asia multiple genetic variants of MDR *S. enterica* serovar Typhi co-exist (Mirza *et al.*, 2000). Elegant studies in Kenya have shown that isolates of MDR *S. enterica* serovar Typhi (which first appeared in this country around 1997) were different from Asian strains, but similar to earlier drug susceptible Kenyan isolates (Kariuki *et al.*, 2000). This suggests that a single plasmid has spread through different bacterial hosts. It seems likely that outbreaks of MDR typhoid fever caused by clonal types occur amongst a variable background of sporadic cases (Connerton *et al.*, 2000; Thong *et al.*, 1994). If so, the spread of similar plasmids through different *S. enterica* serovar Typhi background strains represents the expansion of a successful plasmid rather than the expansion of a single serovar Typhi strain. However, variation in molecular fingerprint patterns of *S. enterica* serovar Typhi resistance plasmids is well recognized (Taylor *et al.*, 1985), and a reduction in the variation of these plasmids has been documented over time (Wain *et al.*, 2003). To explain this we could envisage an adaptation process occurring in the plasmid population, allowing one successful plasmid genotype to sweep through the whole population thus reducing any variability.

2.3 ANTIBIOTIC RESISTANCE IN ENTERIC FEVERS OTHER THAN TYPHOID

Other serovars of *S. enterica* can cause a typhoid like illness in humans, referred to under the umbrella term 'enteric fever', for example *Salmonella enterica* serovar Paratyphi A, B and C, and *Salmonella enterica* serovar Sendai. The epidemiology of these organisms is less well studied, although the pattern of disease due to *S. enterica* serovar Paratyphi A in developing countries is similar to that of *S. enterica* serovar Typhi. Most are human host restricted, so selection for resistant strains is a direct consequence of antibiotic use in human medicine, as with *S. enterica* serovar Typhi. Globally there are fewer reported cases of paratyphoid fever compared to typhoid fever (Herikstad *et al.*, 2002). The ratio has been estimated as approximately one case of paratyphoid for every four typhoid cases (Crump *et al.*, 2004) although this may be higher in India (Sood *et al.*, 1999b) where increased resistance has been reported (Chandel *et al.*, 2000). Of concern, an increase in paratyphoid fever associated with resistance has also been reported from China (Ahmad, 2002) and resistance to fluoroquinolones has been described (Harish *et al.*, 2004).

2.4 ANTIBIOTIC RESISTANCE IN NON-TYPHOID *SALMONELLA ENTERICA* SEROVARS

2.4.1 General

There are numerous serovars of non-typhoid *Salmonella* (NTS), which generally cause a self-limiting gastroenteritis in humans but may cause severe invasive disease usually in immunocompromised individuals. The most common serovars isolated from humans in the developed world are serovars Typhimurium, Enteritidis, Virchow and Hadar. These are all considered to be of zoonotic origin and all exhibit resistance to commonly used antibiotics. Whether the resistance genes found in these bacterial populations have arisen in animals and spread to humans or vice versa is far from clear (Phillips *et al.*, 2004).

In the developing world the situation is far more complex. Little is known about the bacterial populations in animal hosts. In humans the above serotypes, together with *S. enterica* serovars Dublin and Choleraesuis, tend to cause a more invasive disease, particularly in immunosuppressed individuals or those at extremes of age (Bolton *et al.*, 2000). Antibiotic therapy is essential to save life in resource-poor settings, particularly where HIV infection is common, and mortality rates can be extremely high. Both plasmid and chromosomally mediated multi-antibiotic resistance have developed in NTS and continue to spread throughout the world.

2.4.2 Antibiotic resistance in *S. enterica* serovar Typhimurium infections

MDR in the NTS is usually defined as resistance to four or more drugs. In *S. enterica* serovar Typhimurium MDR emerged in the 1960s. Over the next decade an MDR strain of *S. enterica* serovar Typhimurium definitive phage type 29 (DT29), resistant to ampicillin (A), streptomycin (S), sulfonamides (Su), tetracyclines (T) and furazolidone (Fu) (ASSuTFu), caused numerous infections in cattle and humans (Anderson, 1968). This epidemic, which resulted in several deaths, led to the formation of the Swann Committee (The Joint Committee on the Use of Antibiotics in Animal Husbandry and Veterinary Medicine (Anonymous, 1969)). Members addressed the growing concern about the use of antibiotics as growth promoters in livestock and recommended that feed antibiotics should only be used on the same terms as scheduled antibiotics. Legislation ensued and by 1971 multi-resistant *S. enterica* serovar Typhimurium DT29 had virtually disappeared from humans and bovines in the UK (Threlfall *et al.*, 2001).

Resistance in *S. enterica* serovar Typhimurium in the UK remained relatively low until the emergence of strains DT204, 193 and 240c. These strains became epidemic in calves and humans in the UK and caused infections in France, Germany, Belgium and Italy following the export of infected livestock (Rowe *et al.*, 1979). Despite the demise of phage types 240, 240c and 193, MDR in *S. enterica* serovar Typhimurium increased between 1986 and

1990. By 1994, 62% of isolates were MDR (data from the Division of Enteric Pathogens, Health Protection Agency (HPA), Colindale, UK), mainly due to the epidemic spread between 1990 and 1995 of a strain of *S. enterica* serovar Typhimurium DT104, R-type ACSSuT (Wall *et al.*, 1994). Notably, resistance to ampicillin, chloramphenicol, streptomycin, sulphonamides and tetracycline was chromosomally encoded (Briggs, Fratamico, 1999). This MDR strain was first identified in the UK in the early 1980s from exotic birds. Apart from a small human outbreak in Scotland in the mid-1980s, it was not a major human problem in the UK until the late 1980s (Threlfall *et al.*, 2001). By this time it was also common in UK cattle, and became epidemic in bovines over the following five years. However, in contrast to the epidemic of multi-resistant phage types DT29, DT204, DT193 and DT204c, the DT104 strain was also common in poultry (particularly turkeys) and in pigs and sheep. In 1996, infections with DT104 were recognized in cattle and humans in the USA, and over the next four years appeared in humans and food animals elsewhere in Europe, Canada, Japan, Turkey and Israel. DT104 is now widespread in cattle, poultry, pigs and sheep in the UK and elsewhere (Poppe *et al.*, 2002).

Fluoroquinolone resistance began to emerge in NTS in the 1990s. It was chromosomally mediated and associated with one of several single amino acid changes in the topoisomerase subunits GyrA or ParC. By 1994, low-level ciprofloxacin resistance (MIC ranging from 0.25–1.0 mg/L) in human NTS isolates from England and Wales had risen to 2.1%, from 0.3% in 1991 (HPA data). Among the most prevalent serotypes, the highest incidence was seen in *S. enterica* serovar Hadar where ciprofloxacin resistance had increased from 2% in 1991 to 40% in 1994. In 1994, 5% of *S. enterica* serovar Virchow and serovar Newport were ciprofloxacin resistant compared with 1.4% of serovar Typhimurium and 0.4% of serovar Enteritidis. Isolates with reduced sensitivity to ciprofloxacin continued to increase in *S. enterica* serovar Enteritidis and serovar Virchow, and by 1996 70% of *S. enterica* serovar Hadar strains showed low level resistance to ciprofloxacin (HPA data). Around the same time, there was a slight reduction in the proportion of isolates of *S. enterica* serovar Virchow and serovar Hadar that were MDR; 49% in 1999 compared to 56% in 1996. In 2000, 51% of isolates of *S. enterica* serovar Typhimurium in Europe were MDR (Eurosurveillance data). There was a 75% decline in MDR *S. enterica* serovar Typhimurium DT104 between 1996 and 1999 in England and Wales.

More recent concerns regarding the MDR *S. enterica* serovar Typhimurium DT104 strain center on the increase in resistance to trimethoprim and decreased susceptibility to ciprofloxacin. In 2000, 11% of 1168 human isolates of multiresistant DT104 had additional resistance to trimethoprim, 10% had low level resistance to ciprofloxacin (MIC 0.5–1.0 mg/L) and 1% had both (Eurosurveillance data). Similar patterns of ciprofloxacin and trimethoprim resistance are seen among animal isolates.

Resistance has emerged in other *S. enterica* serovar Typhimurium phage types, such as DT204b and DT204c and 4,5,12: i-,-. An international outbreak of MDR *S. enterica* serovar Typhimurium DT204b involved over 350 confirmed cases in 5 European countries in 2000 (Lindsay *et al.*, 2002). In Spain, MDR serovar Typhimurium strains with the antigenic structure 4,5,12: i:-,-, also called U302, have been associated with an increasing number of human infections since the mid-1990s (Echeita *et al.*, 2001). They have also caused infections in the UK and Denmark (Walker *et al.*, 2001). Multi-resistant U302 strains usually possess the same antibiotic resistance genes as multi-resistant DT104. However, a small number of U302 isolates with distinctive pulsed field gel electrophoresis (PFGE) patterns can have different mechanisms of resistance to ampicillin, chloramphenicol, streptomycin, sulphonamides and tetracyclines. The relationship of these isolates to the predominant strain type of U302 is not clear.

2.4.3 Antibiotic resistance in infections with other *S. enterica* serovars

The European Enter-net surveillance program (see Table 2.4 for website address) investigated over 27000 human isolates of *Salmonella* from ten European countries and found that in 2000, 43% of the isolates tested were resistant to antibiotics. Over 20% were resistant to at least one of the following drugs: ampicillin, streptomycin, sulphonamides and tetracycline. Eighteen percent of the isolates were MDR and this multi-resistance was concentrated in four serotypes (Table 2.5). Strains of *S. enterica* serovar Hadar are particularly resistant, and the Enter-net program reported that 37% of strains were MDR in 2000. England and Wales have even higher rates of MDR in *S. enterica* serovar Hadar. In fact, 70% of all strains tested exhibited reduced susceptibility to ciprofloxacin and 3% percent had high-level ciprofloxacin resistance (MIC >1.0 mg/L). At the other extreme is *S. enterica* serovar Enteritidis, which rarely exhibits antibiotic resistance. *S. enterica* serovar Enteritidis PT4 is one of the most common *Salmonella* strains isolated from humans in England and Wales, and recent outbreaks have been associated with imported eggs (HPA data). In Norway, resistance

Table 2.4. *List of useful websites*

UK information
www.defra.gov.uk/animalh/diseases/zoonoses/index.htm
www.defra.gov.uk/animalh/diseases/zoonoses/zoonoses_reports/zoonoses_2000.pdf
www.hpa.org.uk/infections/topics_az/typhoid/menu.htm
CDC
www.cdc.gov/ncidod/diseases/submenus/sub_typhoid.htm
www.cdc.gov/narms/
Enter-net European surveillance scheme
www.hpa.org.uk/hpa/inter/enter-net_menu.htm
www.eurosurveillance.org/
WHO sites
www.who.int/foodsafety/links/general/en/
www.who.int/vaccines/en/typhoid.shtml#summary
www.who.int/topics/typhoid_fever/en/
www.who.int/vaccine_research/documents/en/typhoid_diagnosis.pdf
Recent updates on drug resistance in bacteria:
news.surfwax.com/health/files/Antibiotic_Resistance.html
The history of drug resistance:
www.fda.gov/fdac/features/795_antibio.html

plasmids have been reported from *S. enterica* serovar Enteritidis but these only confer a limited resistance phenotype (Lindstedt *et al.*, 2003). These authors isolated several pentaresistant *S. enterica* serovar Typhimurium DT104 strains at the same time, but *S. enterica* serovar Enteritidis was poor at acquiring multi-drug resistance. This absence of plasmid borne resistance in *S. enterica* serovar Enteritidis is difficult to explain and it may be due to restriction systems encoded on other plasmids present in this serotype (Gregorova *et al.*, 2002).

Since 1991, multi-resistant *S. enterica* serovar Newport has emerged in North America (Zansky *et al.*, 2002). Some strains have additional resistance to third-generation cephalosporins and have caused outbreaks with reported treatment failures. The extended-spectrum cephalosporin resistance is due to a plasmid borne AmpC beta-lactamase (CMY-2). There is evidence that the

Table 2.5. *Incidence of drug resistance by serovars of NTS from Europe 2000*

Serovar	No. tested	MDR (4+ drugs)	Nalidixic acid resistance (and decreased susceptibility to Ciprofloxacin, MIC 0.25–1.0 mg/L)	Ciprofloxacin resistance (MIC ≥ 1.0 mg/L)	Cefotaxime resistance
Enteritidis	14,636	2%	13%	0.4%	0.3%
Typhimurium	6,777	51%	8%	0.6%	0.5%
Hadar	622	37%	57%	3%	0.6%
Virchow	449	36%	53%	0.9%	0.4%
Infantis	439	5%	6%	0	0.2%
Blockley	229	25%	17%	0	0
Newport	243	5%	8%	0	0.8%
Agona	206	0.5%	2%	0.5%	0
Heidelberg	175	14%	6%	0	2%
Brandenburg	160	1%	0	0	1%
Others (245 serovars)	3,123	18%	7%	0.4%	0.5%
TOTAL	27,059	18%	14%	0.5%	0.4%

(Data from Eurosurveillance website, Table 2.4.)

use of antibiotics in livestock, including cattle, has been associated with the emergence of this resistance and with the dissemination and transmission of resistant strains to humans (Angulo *et al.*, 2000). These epidemic strains from the USA have not yet been seen in human cases in Europe.

The latest serotype of *S. enterica* subspecies I to cause outbreaks is the Java biotype of *S. enterica* serovar Paratyphi B (*S.* Java) which causes gastroenteritis in humans. There has been a large increase in reported cases of *S.* Java infection in poultry in the Netherlands (Eurosurveillance data), with the number of isolates rising from less than 2% of all isolates before 1996 to 60% in 2002. In addition, *S.* Java is becoming more resistant to flumequin (quinolone) and to ciprofloxacin. Human infections with biotype Java have been documented in Scotland, but overall levels are relatively low considering the high exposure to contaminated meats (Eurosurveillance data).

Perhaps the most alarming report of resistance from a clinical viewpoint is the combined resistance to fluoroquinolones, associated with double mutations in topoisomerases, and broad spectrum cephalosporins caused by the beta-lactamase CMY-2 (Chiu *et al.*, 2004) in *Salmonella enterica* serovar Choleraesuis.

2.4.4 Transmission routes for antibiotic resistant NTS

The incidence of antibiotic resistance is higher among animal isolates of *Salmonella* than those from humans or food. In the year 2000, the most common serovars recovered from food-producing animals in the UK were Dublin (697 incidents), Typhimurium (602), Senftenberg (269) and Give (148) (data from the Veterinary Laboratories Agency, UK). In the same year a study of serovars Enteritidis, Typhimurium, Hadar and Virchow from food-producing animals in England and Wales found that resistance was most common in strains from pigs (88% resistant) and poultry (43% resistant) (Threlfall *et al.*, 2003). Only 21% of cattle isolates and 19% of sheep isolates were resistant. Certain serovars are particularly prevalent in some food-producing animals, so the overall resistance levels for a species may depend on common patterns of resistance.

The same study found that resistance was most common in *S. enterica* serovar Typhimurium from both humans and food-producing animals, particularly in strain DT104 but also in strains DT193, DT208 and U302 (Threlfall *et al.*, 2003). The antibiogram of DT104 was similar for isolates from humans, cattle, pigs and poultry, suggesting that these food-animal species may have provided the reservoir of drug-resistant strains of this phage type. However, because the spread of DT104 has been clonal, strains detected in a range of environments and species are likely to be similar. MDR *S. enterica* serovar Typhimurium DT208 was predominantly associated with pigs, and other phage types have more complex relationships between humans and food-producing animals. Isolates of *S. enterica* serovar Enteritidis, Hadar and Virchow have different resistance patterns in humans compared to food-producing animals, so food-producing animals bred in England and Wales are unlikely to be the primary source of these drug-resistant strains causing human infections.

2.5 THE CAUSES OF RESISTANCE

2.5.1 General

Antibiotics, and resistance to them, are part of the complex ecosystem inhabited by pathogenic bacteria. Antibiotics are present in many environments, including the gastrointestinal tracts of patients and animals. Resistance genes are present in bacteria that normally colonize the gastrointestinal tract. Drug resistant bacteria can be transferred between humans and animals in both directions and therefore the flow of resistance is probably dynamic with antibiotic resistance constantly being spread between humans and other animals (Phillips *et al.*, 2004).

The use of antibiotics in human medicine, veterinary medicine and crop protection is very likely to be providing the selection pressure that results in the preferential spread of resistant bacteria. It seems plausible that the use of antibiotics in a particular host or environment can provide strong selection for the appearance of antibiotic resistant bacteria that would infect different hosts. For example, the molecular determinants that confer resistance to some antibiotics in *Salmonella* might have been selected for and propagated in organisms such as *E.coli* in the gut of humans and animals, and then transferred to *Salmonella* at a later stage (O'Brien, 2002).

Conflicting opinions have been expressed on the large-scale use of antibiotics as growth promoters in domestic animals. Banning the extensive use of antibiotics in animals is believed to reduce the emergence and spread of drug resistance in bacteria. However, this could result in the increased incidence of infections in animals leading in turn to an increased need for antibiotic treatment, poorer quality food and more food-borne human infections. Evidence from Denmark supports the view that a ban on antibiotics as growth promoters is beneficial with no increase in *Salmonella* shedding in pigs and poultry and no increase in production costs (Wegener *et al.*, 2003). Mathematical models using the best available evidence propose that preventing the emergence of resistance is often easier than reducing the prevalence of resistance once it has appeared (Smith *et al.*, 2002). The use of antibiotics selects for spontaneously occurring resistant bacterial clones, so anything that reduces antibiotic use should limit the emergence of resistance.

2.5.2 Does withdrawing antibiotics limit the maintenance of resistant organisms in the population?

The acquisition and spread of antibiotic resistance genes may be associated with changes in the evolutionary pattern of a bacterial population. Fitness can be defined as a composite measure of an organism's ability to survive, reproduce and be transmitted, and may be related to growth characteristics within a host, the ability to withstand within-host and between-host environmental stresses, and the capacity to disseminate and set up residence within a new host. 'Epidemic potential' may be a useful marker for fitness, whereby the number of secondary infections or cases of disease caused by a specific genotype after its introduction into an entirely susceptible population can be estimated.

It is possible that some antibiotic resistance genes, mobile genetic elements or point mutations confer some fitness cost to the bacterial host, but that this cost can be ameliorated by subsequent adaptation of the bacteria to a new, more fit resistant genotype. In the presence of antibiotic usage, resistant isolates of *S. enterica* serovar Typhi did not appear as quickly as did resistant isolates of other enteric pathogens such as *Shigella* and *E. coli*. This observation prompted several investigators to study the effects of transferring the *E. coli* resistance plasmids into *S. enterica* serovar Typhi. This work has shown that the acquisition of plasmids encoding antibiotic resistance genes has a profound effect on the growth rate of *S. enterica* serovar Typhi with poor stability of the plasmid in the serovar Typhi population (Mendoza-Medellin *et al.*, 2004; Murray *et al.*, 1985). These data may explain why resistance plasmids from other enteric pathogens do not usually transfer readily to, or persist in, *S. enterica* serovar Typhi.

Once the antibiotic selective pressure is removed, in some cases plasmid borne resistance declines in the *S. enterica* serovar Typhi population, but in other cases the resistance persists. The decline in resistance is probably due to the loss of plasmid bearing strains in the population that would be out-competed by plasmid free strains. It is more difficult to explain why resistance sometimes persists, in the absence of direct antibiotic use, in the *S. enterica* serovar Typhi population. This could be due to an unrecognized selective pressure that would make the presence of the plasmid advantageous for the host organism, or to the fact that the acquisition and persistence of some antibiotic resistance phenotypes is not detrimental to the bacterium. Suggestions for what the unseen selective pressure may be include the unrecorded use of antibiotics and the linkage of resistance alleles to other genes that are advantageous for the bacteria. These may include the mercury resistance genes, with co-selection under the pressure of mercury from dental amalgam (Summers *et al.*, 1993). Alternatively the advantageous genes could represent an adhesion factor encoded for by the plasmid conjugation machinery itself and be connected with enhanced transmission of the host bacterium. Evidence from the Mekong Delta area in Vietnam suggests that patients infected with bacteria containing IncH plasmids have higher bacterial loads in the tissues in the absence of selective pressure (Wain *et al.*, 1998). Patients with high bacterial loads also shed organisms more frequently and therefore plasmid-carrying *S. enterica* serovar Typhi may have a higher transmission potential than sensitive isolates. The Mekong Delta is one area where plasmid borne MDR has persisted despite a change to fluoroquinolones for treatment.

Virulence plasmids are present in some *Salmonella* serotypes (Haneda *et al.*, 2001). The virulence region contains 5 genes: *spvABCD* and the positive regulator *spvR*. These plasmid encoded genes have been found in association with antibiotic resistance genes in *S. enterica* serovars Choleraesuis, Typhimurium, Dublin, and 4,5,12:I: – (Chu *et al.*, 2001) (Table 2.6). Plasmids have also acquired systems which mean that the bacterial host cannot lose the plasmid. In some cases loss of the plasmid causes the maximum cost to the host bacterial cell (*i.e.* cell death). One example is the possession of toxin/antitoxin systems that ensure survival of the plasmid. Both toxin and antitoxin are encoded by the plasmid and the two proteins bind to form a harmless dimer in the bacterial cell. The antitoxin is labile but is continually replenished from the plasmid-encoded gene. At bacterial cell division, if one cell happens to be plasmid free it will still possess the toxin-antitoxin protein. Therefore as the antitoxin deteriorates (and cannot be replenished) the toxin will be released and the plasmid free bacterial cell is killed. Thus, only plasmid containing cells can survive and therefore possession of this system is essential for continued existence of the strain (Hayes, 2003).

Table 2.6. *Resistance–virulence plasmids: association of* spv *genes with resistance genes*

Serovar of *Salmonella enterica*	Plasmid	Virulence genes	Resistance genes	Size of plasmid	Reference	Comments
Choleraesuis	OU plasmids	*spv*	*sulI, bla* $_{TEM\text{-}1}$	Variable (50 kb plus)	(Chu *et al.*, 2001)	Amalgamation of virulence and resistance plasmids
Typhimurium	pUO-*St*VR2	*spvABC*	*oxa /aadA1a catA1,tet*B, *qacEΔ1/sul1*	140 kb	(Guerra *et al.*, 2002)	IncFII plasmid
4,5,12:i: -	pUO-SVR3	*spvABC*	*tem1-like, cmlA1, aac(3)-IV, aadA2, dfrA12, tetA, qacEΔ1/sul1*	140-kb	(Guerra *et al.*, 2001)	Possible amalgamation of virulence and resistance plasmid

Table 2.7. *Examples of compensatory mutations in* S. enterica *serovar Typhimurium*

Chromosomal mutation	Resistance	Compensatory mutation	Resistance in compensated mutant	Compensation demonstrated in . . .
rpsL	Streptomycin	Intragenic, *rpsL*	M	Mice
rpsL	Streptomycin	Extragenic, *rpsD/E*	M	Lab. medium
gyrA	Nalidixic acid	Intragenic, *gyrA*	M	Mice
rpoB	Rifampicin	Intragenic, *rpoB*	M	Mice
fusA	Fusidic acid	True reversion, *fusA*	L	Mice
fusA	Fusidic acid	Intragenic, *fusA*	OM	Lab. medium

Key: M = maintained, L = lost, OM = often maintained.
Data from Andersson and Levin (1999).

Table 2.8. *Surveillance of resistance*

Surveillance program	Location	Notes
Typhinet	Global	Typhinet global surveillance system set up in 2002 at 5th International symposium on typhoid fever.
WHO Salm-Serv	Global	WHO Global *Salmonella* Surveillance Program. See www.who.int/salmsurv for more information.
ENTER-NET	15 EU countries, Japan, Australia, New Zealand, South Africa, Norway and Switzerland	International surveillance of salmonella and verocytoxin producing *E coli*. Salmnet preceded this (1994–7).
Pulsenet	USA	*Salmonella* subtyping network based at CDC, Atlanta. See http://www.cdc.gov/pulsenet/ for details.

Acquisition of resistance by point mutations may also incur a cost, which can be ameliorated by second site compensatory mutations (Table 2.7). These are known as intragenic or extragenic suppression, depending on whether the additional mutation occurs at the same locus as the resistance gene or at another locus. The physiological mechanism by which compensatory mutations restore fitness have been determined in only a few cases, for example in *rpsL* mutants the second site mutations restore the efficacy and rate of translation to wild-type or nearly wild-type levels (see Table 2.7).

2.6 CONCLUSIONS

The exact prevalence of antibiotic resistance in salmonella on a global basis is difficult to ascertain. In recent years a number of national and international surveillance schemes have been set up to monitor the changing resistance patterns in isolates from humans, animals, the environment and foodstuffs (Table 2.8). Although there are documented examples of the reduction of resistance in some areas, the overall trend is of increasing resistance problems, particularly with the fluoroquinolones and third generation cephalosporins. Alternative treatment options, including synergistic antibiotic combination regimens warrant further research. By furthering our understanding of the nature of the resistance genes involved and their mechanism of spread, both within and between different serotypes, we can suggest practical ways of trying to combat the emergence and spread of antibiotic resistance. This is likely to involve a combination of infection control and antibiotic control programs.

REFERENCES

Ahmad, K. (2002). Experts call for surveillance of drug-resistant typhoid at a global level. *Lancet*, **359**, 592.

Anderson, E. S. (1968). Drug resistance in *Salmonella typhimurium* and its implications. *Br Med J*, **3**, 333–9.

Andersson, D. I. and Levin, B. R. (1999). The biological cost of antibiotic resistance. *Curr Opin Microbiol*, **2**, 489–93.

Angulo, F. J., Johnson, K. R., Tauxe, R. V. and Cohen, M. L. (2000). Origins and consequences of antimicrobial-resistant nontyphoidal *Salmonella*: implications for the use of fluoroquinolones in food animals. *Microb Drug Resist*, **6**, 77–83.

Anonymous (1969). *Report of the Joint Committee on the Use of Antibiotics in Animal Husbandry and Veterinary Medicine*. London: HMSO.

Bolton, A. J., Osborne, M. P. and Stephen, J. (2000). Comparative study of the invasiveness of *Salmonella* serotypes Typhimurium, Choleraesuis and Dublin for Caco-2 cells, HEp-2 cells and rabbit ileal epithelia. *J Med Microbiol*, **49**, 503–11.

Briggs, C. E. and Fratamico, P. M. (1999). Molecular characterization of an antibiotic resistance gene cluster of *Salmonella typhimurium* DT104. *Antimicrob Agents Chemother*, **43**, 846–9.

Brown, J. D., Mo, D. H. and Rhoades, E. R. (1975). Chloramphenicol-resistant *Salmonella typhi* in Saigon. *JAMA*, **231**, 162–6.

Chandel, D. S., Chaudhry, R., Dhawan, B., Pandey, A. and Dey, A. B. (2000). Drug-resistant *Salmonella enterica* serotype paratyphi A in India. *Emerg Infect Dis*, **6**, 420–1.

Chiu, C. H., Su, L. H., Chu, C. *et al.* (2004). Isolation of *Salmonella enterica* serotype Choleraesuis resistant to ceftriaxone and ciprofloxacin. *Lancet*, **363**, 1285–6.

Chu, C., Chiu, C. H., Wu, W. Y. *et al.* (2001). Large drug resistance virulence plasmids of clinical isolates of *Salmonella enterica* serovar Choleraesuis. *Antimicrob Agents Chemother*, **45**, 2299–303.

Chun, D., Seol, S. Y., Cho, D. T. and Tak, R. (1977). Drug resistance and R plasmids in *Salmonella typhi* isolated in Korea. *Antimicrob Agents Chemother*, **11**, 209–13.

Connerton, P., Wain, J., Hien, T. T. *et al.* (2000). Epidemic typhoid in Vietnam: molecular typing of multiple-antibiotic-resistant *Salmonella* enterica serotype Typhi from four outbreaks. *J Clin Microbiol*, **38**, 895–7.

Coovadia, Y. M., Gathiram, V., Bhamjee, A. *et al.* (1992). An outbreak of multiresistant *Salmonella typhi* in South Africa. *Quarterly J of Med*, **82**, 91–100.

Crump, J. A., Barrett, T. J., Nelson, J. T. and Angulo, F. J. (2003). Reevaluating fluoroquinolone breakpoints for *Salmonella enterica* serotype Typhi and for non-Typhi salmonellae. *Clin Infect Dis*, **37**, 75–81.

Crump, J. A., Luby, S. P. and Mintz, E. D. (2004). The global burden of typhoid fever. *Bull WHO*, **82**, 346–53.

Datta, N. and Olarte, J. (1974). R factors in strains of *Salmonella typhi* and *Shigella dysenteriae* 1 isolated during epidemics in Mexico: classification by compatibility. *Antimicrob Agents Chemother*, **5**, 310–17.

Drlica, K. (2003). The mutant selection window and antimicrobial resistance. *J Antimicrob Chemother*, **52**, 11–17.

Dromigny, J. A. and Perrier-Gros-Claude, J. D. (2003). Antimicrobial resistance of *Salmonella enterica* serotype Typhi in Dakar, Senegal. *Clin Infect Dis*, **37**, 465–6.

Echeita, M. A., Herrera, S. and Usera, M. A. (2001). Atypical, *fljB*-negative *Salmonella enterica* subsp. enterica strain of serovar 4,5,12:i:- appears to be a monophasic variant of serovar Typhimurium. *J Clin Microbiol*, **39**, 2981–3.

Gales, A. C., Sader, H. S., Mendes, R. E. and Jones, R. N. (2002). *Salmonella* spp. isolates causing bloodstream infections in Latin America: report of antimicrobial activity from the SENTRY Antimicrobial Surveillance Program (1997–2000). *Diagn Microbiol Infect Dis*, **44**, 313–18.

Gangarosa, E. J., Bennett, J. V., Wyatt, C. *et al.* (1972). An epidemic-associated episome? *J Infect Dis*, **126**, 215–18.

Gonzalez Cortes, A., Bessudo, D., Sanchez Leyva, R. *et al.* (1973). Water-borne transmission of chloramphenicol-resistant *Salmonella typhi* in Mexico. *Lancet*, **2**, 605–7.

Gregorova, D., Pravcova, M., Karpiskova, R. and Rychlik, I. (2002). Plasmid pC present in *Salmonella enterica* serovar Enteritidis PT14b strains encodes a restriction modification system. *FEMS Microbiol Lett*, **214**, 195–8.

Grindley, N. D., Humphreys, G. O. and Anderson, E. S. (1973). Molecular studies of R factor compatibility groups. *J Bacteriol*, **115**, 387–98.

Guerra, B., Soto, S., Helmuth, R. and Mendoza, M. C. (2002). Characterization of a self-transferable plasmid from *Salmonella enterica* serotype Typhimurium clinical isolates carrying two integron-borne gene cassettes together with virulence and drug resistance genes. *Antimicrob Agents Chemother*, **46**, 2977–81

Guerra, B., Soto, S. M., Arguelles, J. M. and Mendoza, M. C. (2001). Multidrug resistance is mediated by large plasmids carrying a class 1 integron in the emergent *Salmonella enterica* serotype [4,5,12:i:-]. *Antimicrob Agents Chemother*, **45**, 1305–8.

Hampton, M. D., Ward, L. R., Rowe, B. and Threlfall, E. J. (1998). Molecular fingerprinting of multidrug-resistant *Salmonella enterica* serotype Typhi. *Emerg Infect Dis*, **4**, 317–20.

Haneda, T., Okada, N., Nakazawa, N., Kawakami, T. and Danbara, H. (2001). Complete DNA sequence and comparative analysis of the 50-kilobase virulence plasmid of *Salmonella enterica* serovar Choleraesuis. *Infect Immun*, **69**, 2612–20.

Harish, B. N., Madhulika, U. and Parija, S. C. (2004). Isolated high-level ciprofloxacin resistance in *Salmonella enterica* subsp. enterica serotype Paratyphi A. *J Med Microbiol*, **53**, 819.

Harnett, N., McLeod, S., AuYong, Y. *et al.* (1998). Molecular characterization of multiresistant strains of *Salmonella typhi* from South Asia isolated in Ontario, Canada. *Can J Microbiol*, **44**, 356–63.

Hayes, F. (2003). Toxins–antitoxins: plasmid maintenance, programmed cell death, and cell cycle arrest. *Science*, **301**, 1496–9.

Helms, M., Vastrup, P., Gerner-Smidt, P. and Molbak, K. (2002). Excess mortality associated with antimicrobial drug-resistant *Salmonella typhimurium*. *Emerg Infect Dis*, **8**, 490–5.

Herikstad, H., Motarjemi, Y. and Tauxe, R. V. (2002). *Salmonella* surveillance: a global survey of public health serotyping. *Epidemiol Infect,* **129**, 1–8.

Hermans, P. W., Saha, S. K., van Leeuwen, W. J. *et al.* (1996). Molecular typing of *Salmonella typhi* strains from Dhaka (Bangladesh) and development of DNA probes identifying plasmid-encoded multidrug-resistant isolates. *J Clin Microbiol,* **34**, 1373–9.

Hirose, K., Tamura, K., Sagara, H. and Watanabe, H. (2001). Antibiotic susceptibilities of *Salmonella enterica* serovar Typhi and *S. enterica* serovar Paratyphi A isolated from patients in Japan. *Antimicrob Agents Chemother,* **45**, 956–8.

Holmberg, S. D., Solomon, S. L. and Blake, P. A. (1987). Health and economic impacts of antimicrobial resistance. *Rev Infect Dis,* **9**, 1065–78.

Jacoby, G. A. and Sutton, L. (1991). Properties of plasmids responsible for production of extended-spectrum beta-lactamases. *Antimicrob Agents Chemother,* **35**, 164–9.

Jesudason, M. V., John, R. and John, T. J. (1996). The concurrent prevalence of chloramphenicol-sensitive and multi-drug resistant *Salmonella typhi* in Vellore, S. India. *Epidemiol Infect,* **116**, 225–7.

Kamili, M. A., Ali, G., Shah, M. Y., Rashid, S., Khan, S. and Allaqaband, G. Q. (1993). Multiple drug resistant typhoid fever outbreak in Kashmir Valley. *Indian J of Med Sci,* **47**, 147–51.

Kariuki, S., Gilks, C., Revathi, G. and Hart, C. A. (2000). Genotypic analysis of multidrug-resistant *Salmonella enterica* serovar Typhi, Kenya. *Emerg Infect Dis,* **6**, 649–51.

Komalarini, S., Njotosiswojo, S., Rockhill, R. C. and Lesmana, M. (1980). Chloramphenicol resistant strains in salmonellosis in Jakarta. *Southeast Asian J Trop Med Public Health,* **11**, 539–42.

Lee, C. A. and Falkow, S. (1994). Isolation of hyperinvasive mutants of *Salmonella. Methods Enzymol,* **236**, 531–45.

Lindsay, E. A., Lawson, A. J., Walker, R. A. *et al.* (2002). Role of electronic data exchange in an international outbreak caused by *Salmonella enterica* serotype Typhimurium DT204b. *Emerg Infect Dis,* **8**, 732–4.

Lindstedt, B. A., Heir, E., Nygard, I. and Kapperud, G. (2003). Characterization of class I integrons in clinical strains of *Salmonella enterica* subsp. enterica serovars Typhimurium and Enteritidis from Norwegian hospitals. *J Med Microbiol,* **52**, 141–9.

Ling, J. and Chau, P. Y. (1984). Plasmids mediating resistance to chloramphenicol, trimethoprim, and ampicillin in *Salmonella typhi* strains isolated in the Southeast Asian region. *J Infect Dis,* **149**, 652.

Ling, J. M., Chan, E. W., Lam, A. W. and Cheng, A. F. (2003). Mutations in topoisomerase genes of fluoroquinolone-resistant salmonellae in Hong Kong. *Antimicrob Agents Chemother*, **47**, 3567–73.

Martínez, J. L., Alonso, A., Gómez-Gómez, J. M. and Baquero, F. (1998). Quinolone resistance by mutations in chromosomal gyrase genes. Just the tip of the iceberg? *J Antimicrob Chemother*, **42**, 683–8.

Martínez-Martínez, L., Pascual, A. and Jacoby, G. A. (1998). Quinolone resistance from a transferable plasmid. *Lancet*, **351**, 797–9.

Mead, P. S., Slutsker, L., Dietz, V. *et al.* (1999). Food-related illness and death in the United States. *Emerg Infect Dis*, **5**, 607–25.

Mendoza-Medellin, A., Curiel-Quesada, E. and Camacho-Carranza, R. (2004). *Escherichia coli* R-factors unstable in *Salmonella typhi* are deleted before being segregated in this host. *Plasmid*, **51**, 75–86.

Mendoza-Medellin, A., Rios-Chavez, I. and Amaro-Robles, D. (1993). Behavior of Escherichia coli R factors in *Salmonella typhi*. *Rev Latinoam Microbiol*, **35**, 77–85.

Mermin, J. H., Villar, R., Carpenter, J. *et al.* (1999). A massive epidemic of multidrug-resistant typhoid fever in Tajikistan associated with consumption of municipal water. *J Infect Dis*, **179**, 1416–22.

Mills-Robertson, F., Addy, M. E., Mensah, P. and Crupper, S. S. (2002). Molecular characterization of antibiotic resistance in clinical *Salmonella typhi* isolated in Ghana. *FEMS Microbiol Lett*, **215**, 249–53.

Mirza, S., Kariuki, S., Mamun, K. Z., Beeching, N. J. and Hart, C. A. (2000). Analysis of plasmid and chromosomal DNA of multidrug-resistant *Salmonella enterica* serovar Typhi from Asia. *J Clin Microbiol*, **38**, 1449–52.

Mirza, S. H. and Hart, C. A. (1993). Plasmid encoded multi-drug resistance in *Salmonella typhi* from Pakistan. *Ann Trop Med Parasitol*, **87**, 373–7.

Mirza, S. H., Beeching, N. J. and Hart, C. A. (1996). Multi-drug resistant typhoid: a global problem. *J Med Microbiol*, **44**, 317–19.

Mukhtar, E. D. and Mekki, M. O. (1981). Trimethoprim-sulphamethoxazole in the treatment of enteric fever in the Sudan. *Trans R Soc Trop Med Hyg*, **75**, 771–3.

Murray, B. E. (1989). Problems and mechanisms of antimicrobial resistance. *Infect Dis Clin North Am*, **3**, 423–39.

Murray, B. E., Levine, M. M., Cordano, A. M. *et al.* (1985). Survey of plasmids in *Salmonella typhi* from Chile and Thailand. *J Infect Dis*, **151**, 551–5.

Nath, M. L. and Singh, J. (1966). Antibiotic sensitivity of *Salmonella typhi*. *Indian J Med Res*, **54**, 217–19.

O'Brien, T. F. (2002). Emergence, spread, and environmental effect of antimicrobial resistance: how use of an antimicrobial anywhere can increase resistance to any antimicrobial anywhere else. *Clin Infect Dis*, **34**(suppl. 3), S78–84.

Olarte, J. and Galindo, E. (1973). *Salmonella typhi* resistant to chloramphenicol, ampicillin, and other antimicrobial agents: strains isolated during an extensive typhoid fever epidemic in Mexico. *Antimicrob Agents Chemother*, **4**, 597–601.

Oyofo, B. A., Lesmana, M., Subekti, D. *et al.* (2002). Surveillance of bacterial pathogens of diarrhea disease in Indonesia. *Diagn Microbiol Infect Dis*, **44**, 227–34.

Pai, H., Byeon, J. H., Yu, S., Lee, B. K. and Kim, S. (2003). *Salmonella enterica* serovar Typhi strains isolated in Korea containing a multidrug resistance class 1 integron. *Antimicrob Agents Chemother*, **47**, 2006–8.

Paniker, C. K. and Vimala, K. N. (1972). Transferable chloramphenicol resistance in *Salmonella typhi*. *Nature*, **239**, 109–10.

Parkhill, J., Dougan, G., James, K. D. *et al.* (2001). Complete genome sequence of a multiple drug resistant *Salmonella enterica* serovar Typhi CT18. *Nature*, **413**, 848–52.

Parry, C., Wain, J., Chinh, N. T., Vinh, H. and Farrar, J. J. (1998). Quinolone-resistant *Salmonella typhi* in Vietnam. *Lancet*, **351**, 1289.

Parry, C. M. (1998). Untreatable infections? – The challenge of the 21st century. *Southeast Asian J Trop Med Public Health*, **29**, 416–24.

Partridge, S. R. and Hall, R. M. (2003). In34, a complex In5 family class 1 integron containing orf513 and dfrA10. *Antimicrob Agents Chemother*, **47**, 342–9.

Phillips, I., Casewell, M., Cox, T. *et al.* (2004). Does the use of antibiotics in food animals pose a risk to human health? A critical review of published data. *J Antimicrob Chemother*, **53**, 28–52.

Poirel, L., Naas, T., Guibert, M. *et al.* (1999). Molecular and biochemical characterization of VEB-1, a novel class A extended-spectrum beta-lactamase encoded by an *Escherichia coli* integron gene. *Antimicrob Agents Chemother*, **43**, 573–81.

Poppe, C., Ziebell, K., Martin, L. and Allen, K. (2002). Diversity in antimicrobial resistance and other characteristics among *Salmonella typhimurium* DT104 isolates. *Microb Drug Resist*, **8**, 107–22.

Rahman, M., Ahmad, A. and Shoma, S. (2002). Decline in epidemic of multidrug resistant *Salmonella typhi* is not associated with increased incidence of antibiotic-susceptible strain in Bangladesh. *Epidemiol Infect*, **129**, 29–34.

Rodrigues, C., Mehta, A. and Joshi, V. R. (2002). *Salmonella typhi* in the past decade: learning to live with resistance. *Clin Infect Dis*, **34**, 126.

Rowe, B., Threlfall, E. J., Ward, L. R. and Ashley, A. S. (1979). International spread of multiresistant strains of *Salmonella typhimurium* phage types 204 and 193 from Britain to Europe. *Vet Rec*, **105**, 468–9.

Saha, S. K., Talukder, S. Y., Islam, M. and Saha, S. (1999). A highly ceftriaxone-resistant *Salmonella typhi* in Bangladesh. *Pediatr Infects Dis J*, **18**, 387.

Shanahan, P. M., Jesudason, M. V., Thomson, C. J. and Amyes, S. G. (1998). Molecular analysis of and identification of antibiotic resistance genes in clinical isolates of *Salmonella typhi* from India. *J Clin Microbiol*, **36**, 1595–600.

Shanahan, P. M., Karamat, K. A., Thomson, C. J. and Amyes, S. G. (2000). Characterization of multi-drug resistant *Salmonella typhi* isolated from Pakistan. *Epidemiol Infect*, **124**, 9–16.

Shannon, K. and French, G. (1998). Multiple-antibiotic-resistant *Salmonella*. *Lancet*, **352**, 490.

Sherburne, C. K., Lawley, T. D., Gilmour, M. W. *et al.* (2000). The complete DNA sequence and analysis of R27, a large IncHI plasmid from *Salmonella typhi* that is temperature sensitive for transfer. *Nucleic Acids Res*, **28**, 2177–86.

Smith, D. L., Harris, A. D., Johnson, J. A., Silbergeld, E. K. and Morris, J. G., Jr (2002). Animal antibiotic use has an early but important impact on the emergence of antibiotic resistance in human commensal bacteria. *Proc Natl Acad Sci USA*, **99**, 6434–9.

Sood, S., Kapil, A., Das, B., Jain, Y. and Kabra, S. K. (1999a). Re-emergence of chloramphenicol-sensitive *Salmonella typhi*. *Lancet*, **353**, 1241–2.

Sood, S., Kapil, A., Dash, N. *et al.* (1999b). Paratyphoid fever in India: an emerging problem. *Emerg Infect Dis*, **5**, 483–4.

Summers, A., Wireman, J., Vimy, M. *et al.* (1993). Mercury released from dental 'silver' fillings provokes an increase in mercury- and antibiotic-resistant bacteria in oral and intestinal floras of primates. *Antimicrob Agents Chemother*, **37**, 825–34.

Tassios, P. T., Vatopoulos, A. C., Mainas, E. *et al.* (1997). Molecular analysis of ampicillin-resistant sporadic *Salmonella typhi* and *Salmonella paratyphi* B clinical isolates. *Clin Microbiol Infect*, **3**, 317–23.

Taylor, D. E. (1983). Transfer-defective and tetracycline-sensitive mutants of the incompatibility group HI plasmid R27 generated by insertion of transposon 7. *Plasmid*, **9**, 227–39.

Taylor, D. E. and Brose, E. C. (1985). Characterization of incompatibility group HI1 plasmids from *Salmonella typhi* by restriction endonuclease digestion and hybridization of DNA probes for Tn3, Tn9, and Tn10. *Can J Microbiol*, **31**, 721–9.

Thong, K. L., Cheong, Y. M., Puthucheary, S., Koh, C. L. and Pang, T. (1994). Epidemiologic analysis of sporadic *Salmonella typhi* isolates and those from outbreaks by pulsed-field gel electrophoresis. *J Clin Microbiol*, **32**, 1135–41.

Threlfall, E. J., Lawson, A. J., Walker, R. A. and Ward, L. R. (2001). *Salmonella typhimurium* DT104: the rise and fall of a multiresistant epoizootic clone. *SCIEH Weekly Report*, **35**, 142–4.

Threlfall, E. J., Teale, C. J., Davies, R. H. *et al.* (2003). A comparison of antimicrobial susceptibilities in nontyphoidal salmonellas from humans and food animals in England and Wales in 2000. *Microb Drug Resist*, **9**, 183–9.

Tran, J. H. and Jacoby, G. A. (2002). Mechanism of plasmid-mediated quinolone resistance. *Proc Natl Acad Sci USA*, **99**, 5638–42.

Travers, K. and Barza, M. (2002). Morbidity of infections caused by antimicrobial-resistant bacteria. *Clin Infect Dis*, **34**(suppl. 3), S131–4.

Verdet, C., Arlet, G., Barnaud, G., Lagrange, P. H. and Philippon, A. (2000). A novel integron in *Salmonella enterica* serovar Enteritidis, carrying the bla(DHA-1) gene and its regulator gene *ampR*, originated from Morganella morganii. *Antimicrob Agents Chemother*, **44**, 222–5.

Vinh, H., Wain, J., Vo, T. N. *et al.* (1996). Two or three days of ofloxacin treatment for uncomplicated multidrug-resistant typhoid fever in children. *Antimicrob Agents Chemother*, **40**, 958–61.

Wain, J., Diem Nga, L. T., Kidgell, C. *et al.* (2003). Molecular analysis of incHI1 antimicrobial resistance plasmids from *Salmonella serovar* Typhi strains associated with typhoid fever. *Antimicrob Agents Chemother*, **47**, 2732–9.

Wain, J., Diep, T. S., Ho, V. A. *et al.* (1998). Quantitation of bacteria in blood of typhoid fever patients and relationship between counts and clinical features, transmissibility, and antibiotic resistance. *J Clin Microbiol*, **36**, 1683–7.

Wain, J., Hoa, N. T., Chinh, N. T. *et al.* (1997). Quinolone-resistant *Salmonella typhi* in Viet Nam: molecular basis of resistance and clinical response to treatment. *Clin Infect Dis*, **25**, 1404–10.

Walker, R. A., Lindsay, E., Woodward, M. J., Ward, L. R. and Threlfall, E. J. (2001). Variation in clonality and antibiotic-resistance genes among multiresistant *Salmonella enterica* serotype Typhimurium phage-type U302 (MR U302) from humans, animals, and foods. *Microb Drug Resist*, **7**, 13–21.

Wall, P. G., Morgan, D., Lamden, K. *et al.* (1994). A case control study of infection with an epidemic strain of multiresistant *Salmonella typhimurium* DT104 in England and Wales. *Commun Dis Rep CDR Rev*, **4**, R130–5.

Wang, M., Sahm, D. F., Jacoby, G. A. and Hooper, D. C. (2004). Emerging plasmid-mediated quinolone resistance associated with the qnr gene in *Klebsiella pneumoniae* clinical isolates in the United States. *Antimicrob Agents Chemother*, **48**, 1295–9.

Wasfy, M. O., Frenck, R., Ismail, T. F. *et al.* (2002). Trends of multiple-drug resistance among *Salmonella* serotype Typhi isolates during a 14-year period in Egypt. *Clin Infect Dis*, **35**, 1265–8.

Webber, M. and Piddock, L. J. (2001). Quinolone resistance in *Escherichia coli*. *Vet Res*, **32**, 275–84.

Wegener, H. C., Hald, T., Lo Fo Wong, D. *et al.* (2003). *Salmonella* control programs in Denmark. *Emerg Infect Dis*, **9**, 774–80.

Woodward, T. E., Smadel, J. E., Lay, H. L., Jr., Green, R. and Marnkihar, D. S. (1948 / Fall 2004). Preliminary report on the beneficial effect of chloromycetin in the treatment of typhoid fever. *Wilderness Environ Med*, **15** (3), 218–20; discussion 216–17.

Zahurul Haque Asna, S. M. and Ashraful Haq, J. (2000). Decrease of antibiotic resistance in *Salmonella typhi* isolated from patients attending hospitals of Dhaka City over a 3 year period. *Int J Antimicrob Agents*, **16**, 249–51.

Zansky, S., Wallace, B., Schoonmaker-Bopp, D. *et al.* (2002). From the Centers for Disease Control and Prevention. Outbreak of multi-drug resistant *Salmonella* Newport – United States, January–April 2002. *Jama*, **288**, 951–3.

CHAPTER 3

Host-specificity of *Salmonella* infections in animal species

Timothy S. Wallis

3.1 INTRODUCTION

The bacterial species *Salmonella enterica* subspecies enterica can be divided into over 2400 antigenically distinct serovars and the pathogenicity of most of these serovars is undefined. The majority of incidents of salmonellosis in humans and domestic animals are caused by relatively few serovars and these can be subdivided into three groups on the basis of host prevalence. The first group consists of host-specific serovars. These typically cause systemic disease in a limited number of phylogenetically related species. For example, *S. enterica* serovar Typhi, serovar Gallinarum and serovar Abortusovis are almost exclusively associated with systemic disease in humans, fowl and sheep respectively. The second group consists of host-restricted strains. These are primarily associated with one or two closely related host species but may also infrequently cause disease in other hosts. For example, *S. enterica* serovar Dublin and serovar Choleraesuis are generally associated with severe systemic disease in ruminants and pigs respectively (Sojka *et al.*, 1977). However, these serovars are potentially capable of infecting other animal species and humans. The third group consists of the ubiquitous *S. enterica* serovars, such as Typhimurium and Enteritidis that usually induce gastroenteritis in a broad range of unrelated host species.

Clearly the nature and severity of *Salmonella* infections in different animal species varies enormously and is influenced by many factors including the infecting *Salmonella* serovar, strain virulence, infecting dose, host animal species, age and immune status of the host, and the geographical region. All these factors are likely to inter-relate. In the following sections the nature of

Salmonella infections in several domesticated animal species is described to illustrate the differences in the epidemiology and pathogenesis of salmonellosis in different hosts. Thereafter, the biology of *Salmonella* serovar host-specificity will be discussed in the context of our current understanding of the molecular basis of pathogenesis and the potential impact of different virulence determinants on *Salmonella* natural history.

3.2 *SALMONELLA* INFECTIONS OF CATTLE

3.2.1 Epidemiology

Salmonellosis occurs worldwide in cattle and has been associated primarily with serovars Dublin and Typhimurium. Other serovars are sporadically associated with bovine infections. During the past 20 years, approximately 100 serovars other than *S. enterica* serovar Dublin and serovar Typhimurium have accounted for less than 10% of incidents in the U.K. Salmonellosis reached a peak in the British cattle industry in the 1960s with over 4000 incidents in 1969 (Sojka *et al.*, 1977). More recently there has been a steep decline in the number of *Salmonella* outbreaks and over the last 5 years there have been only 400–500 incidents per annum, with similar numbers of incidents caused by *S. enterica* serovar Dublin and serovar Typhimurium in adult cattle and calves. *S. enterica* serovar Dublin and serovar Typhimurium are endemic in northern Europe, although the distributions of these serovars differ. In the UK *S. enterica* serovar Typhimurium occurs in all geographical regions whereas serovar Dublin is predominately found in North- and South-Western England and Wales. In the USA *S. enterica* serovar Typhimurium is endemic in cattle throughout the country. In contrast, *S. enterica* serovar Dublin, which before the 1980s occurred only to the west of the Rocky mountains, has recently spread eastwards to other states and north into Canada from where it had not previously been isolated.

The source of most outbreaks of salmonellosis in cattle is probably fecal to oral contact. Infected cattle may excrete up to 10^8 cfu *Salmonella*/g of feces and contamination of the environment in the proximity of other animals is a potent source of infection. Sub-clinical excretion of *Salmonella* exacerbates the problem of dissemination. Cattle that carry an active *Salmonella* infection but display no clinical symptoms (often convalescing animals) are known as "active carriers". These may excrete *Salmonella* continuously in concentrations greater than 10^5cfu/g of feces and thus can be detected by routine bacteriological examination. Active carriage is usually the sequel to clinical enteritis or systemic infection, and infected animals may excrete *Salmonella* for years or even for life. "Passive carriers" are immune animals that ingest *Salmonella* with feed and subsequently pass them in their feces with no active infection of the intestines. Consequently, when removed from an infected environment these animals will stop excreting *Salmonella*. In some animals, known as "latent carriers", *Salmonella* persists sub-clinically in the tissues but is only intermittently excreted in feces. Excretion may be activated by stress, for example, at parturition. Understanding the biology of this true "carrier state" is likely to be key to ultimately controlling this important pathogen in cattle and may also provide insight into, for example, the asymptomatic carriage of *S. enterica* serovar Typhi by humans.

3.2.2 Salmonellosis in cattle

As in other animal species, pathogenesis is dependent on factors such as the infecting serovar, the age of the animal and the route of infection. Cattle are most likely to be infected by the oral route, although respiratory and conjunctival infection may also occur. The immune status of the animal is also important: calves that are deprived of colostrum are particularly susceptible to infection.

In calves clinical disease is most common at 2–6 weeks of age. The peak incidence of disease with *S. enterica* serovar Typhimurium is at three weeks of age, whereas with *S. enterica* serovar Dublin it is at four weeks of age. Clinical signs vary, but typically the enteric form of the disease predominates. Salmonellosis is characterised by pyrexia, dullness and anorexia, followed by diarrhoea that may contain fibrin and mucous. Pneumonia may also occur. The feces may become blood-stained and "stringy" due to the presence of undigested milk, pseudo-membrane formation and shreds of necrotic intestinal mucosa. Calves rapidly become weak and dehydrated. Unless treated, the infected calf will die typically five to seven days after the onset of disease. Salmonellosis is very variable and in some animals, particularly the very young, septicaemia may occur and animals may die in the absence of diarrhoeal signs. Calves infected with *S. enterica* serovar Dublin, which tend to be a little older than those infected with other serovars, predominantly suffer from septicaemia and pneumonia. Calves that recover from infection do not typically grow into adult latent carriers.

In adult cattle, both acute and sub-acute forms of disease are recognised. The onset of the severe, acute form of the disease is sudden and typically accompanied by pyrexia, dullness, anorexia and reduced milk-yield. Severe diarrhoea follows, which may contain blood, mucus and necrotic intestinal mucosa. Pregnant animals may abort. High temperatures usually persist for several days and typically drop precipitously in the 24 hours just before death. The sub-acute form of disease is less severe though it may be more protracted. Animals may become emaciated, dehydrated and show signs of abdominal pain. Abortion may occur in the absence of other clinical signs, and this is particularly the case with *S. enterica* serovar Dublin infections (Hinton, 1973). Usually, abortion is preceded by pyrexia (Hall and Jones, 1976) when bacterial multiplication is likely to be occurring in the placenta. The spread of *S. enterica* serovar Dublin to reproductive tissues is not well understood and may originate either from a systemic infection or possibly from fecal contamination of the vagina.

Adult survivors of *S. enterica* serovar Dublin infections often become latent carriers, a state which may last for life. The outcome of infection with other serovars seldom results in the latent carrier state although active excretion may continue for years. The reasons for this remain unclear.

3.2.3 Molecular basis of pathogenesis of *Salmonella* in cattle

3.2.3.1 Intestinal invasion

Invasion of intestinal mucosa is a characteristic feature of *Salmonella* pathogenesis. Within minutes of contact with intestinal mucosa, *Salmonella* can invade both M cells and enterocytes that overlie domed villi associated with lymphoid follicles and absorptive villi respectively (Frost *et al.*, 1997). Subsequently, *Salmonella* elicit membrane ruffles in the apical membranes of both M cells and enterocytes that results in the uptake of bacteria into membrane-bound vesicles. Thereafter, *Salmonella* can be seen associated with reticulo-endothelial cells and polymorphonuclear neutrophils (PMNs), which are recruited to the foci of infection. The molecular mechanisms by which *Salmonella* invade eukaryotic cells have been studied intensively in vitro using cultured cells. A virulence locus was found at centisome 63 on the *S. enterica* serovar Typhimurium chromosome, which was identified as the first *Salmonella* pathogenicity island (SPI-1). SPI-1 encodes a type three secretion system termed TTSS-1, which has a major role in the invasion process (see Chapter 6). Molecular genetic approaches are now being applied to study intestinal invasion in vivo. Mutations in genes that encode essential structural proteins of TTSS-1 act to block the delivery of translocated effector proteins, and block *Salmonella* invasion of bovine intestines in vivo (Watson *et al.*, 1995).

The roles of individual TTSS-1-dependent translocated effector proteins in intestinal invasion are not clear. Mutation of *sopE1* reduced invasion of cultured cells (Wood *et al.*, 1996) although this reduction was relatively minor when compared with mutations that entirely disrupt the function of TTSS-1 and therefore block the delivery of all TTSS-1-dependent effectors. SopE1 was found to activate small Rho GTPases, which influence cytoskeletal rearrangements, thus promoting epithelial cell invasion (Hardt *et al.*, 1998). The *sopE1* gene is not present in many *Salmonella* strains. This observation, together with the weak invasion phenotype of the *sopE1* mutant, suggests that SopE1 may have functions other than those that influence invasion; therefore other secreted invasins await characterization. One such invasin is the SopE2 protein, which is highly homologous to SopE1. Not only is *sopE2* more prevalent than *sopE1* in *Salmonella* strains, but mutation of *sopE2* appears to have a major effect on the invasion phenotype (Bakshi *et al.*, 2000). SopE1 is carried by the SopE**Φ** bacteriophage and epidemiological evidence shows that the *sopE1* gene is associated with *S. enterica* serovar Typhimurium phage types that cause epidemics in cattle. Horizontal transfer of the *sopE1* gene by lysogenic conversion with SopE**Φ** increased the enteropathogenicity of *S. enterica* serovar Typhimurium in the bovine ligated ileal loop model (Zhang *et al.*, 2002a). These observations support the hypothesis that phage mediated horizontal transfer of the *sopE1* gene contributes to the emergence of epidemic cattle-associated *S. enterica* serovar Typhimurium clones.

It is generally assumed that *Salmonella* invasiveness is important in virulence. However, the actual requirement for intestinal invasion in the induction of enteritis has not been demonstrated. In orally infected calves, extracellular bacteria are present in the intestinal lumen in high numbers. Similarly, in ligated ileal loop studies over 90% of the inoculum remains gentamicin sensitive, suggesting that bacteria are primarily located in an extracellular niche (Watson *et al.*, 1995). Extracellular bacteria use TTSS-1 to deliver effector proteins required for enteritis into epithelial cells. This questions the requirement for invasion of *Salmonella* into intestinal mucosa for the induction of enteritis. To date it has not been possible to assess directly the role of invasion in enteropathogenesis as mutation of TTSS-1 results in pleiotropic effects. However, the observation that different, but equally invasive, serovars of *Salmonella* induce different levels of enteropathogenic response (Watson *et al.*, 1998; Watson *et al.*, 1995) suggests that there is no direct correlation between the magnitude of invasion and enteropathogenicity. This is further supported by the observation that invasion of epithelial cells can be uncoupled from *Salmonella*-induced PMN transmigration in vitro (Gewirtz *et al.*, 1999).

3.2.3.2 *Salmonella*-induced enteritis in cattle

In addition to cellular invasion, TTSS-1 also plays a major role in the induction of enteropathogenesis. Disruption of TTSS-1 abolishes enteropathogenicity in bovine ligated ileal loops (Ahmer *et al.*, 1999; Watson *et al.*, 1998) and orally infected calves (Tsolis *et al.*, 1999; Watson *et al.*, 1998).

Some of the translocated effector proteins have important roles in the induction of intestinal inflammation and fluid secretion. The SopB protein of *S. enterica* serovar Dublin (also known as SigD in serovar Typhimurium (Hong, Miller, 1998)) is translocated into eukaryotic cells via a TTSS-1-dependent mechanism (Galyov *et al.*, 1997). Inactivation of *sopB* has little or no effect on intestinal invasiveness, yet it significantly reduces enteropathogenesis (Galyov *et al.*, 1997; Reis *et al.*, 2003). SopB is a phosphatase capable of hydrolysing several inositol phosphates (Norris *et al.*, 1998). This is of particular interest since it is known that *Salmonella* infection of intestinal epithelial cells in vitro results in elevated intracellular levels of Ins(1,4,5,6)P_4, which in turn can antagonise the closure of chloride channels (Eckmann *et al.*, 1997), thus influencing electrolyte and fluid secretion in intestinal mucosa. The increase in Ins(1,4,5,6)P_4 levels in *Salmonella* infected cells could be directly or indirectly attributed to SopB activity (Norris *et al.*, 1998). Together these observations implicate SopB as a novel bacterial enterotoxin.

Unlike mutations that disrupt TTSS-1 and abolish enteropathogenesis entirely, mutation of *sopB* only reduces enteropathogenesis (Galyov *et al.*, 1997; Watson *et al.*, 1998). This implicates other TTSS-1 secreted effector proteins in the induction of enteritis and indeed several have already been demonstrated to be involved in this process (Jones *et al.*, 1998; Zhang *et al.*, 2002b).

3.2.3.3 Role of other genes in *Salmonella* pathogenesis in cattle

A second type three secretion system (TTSS-2) encoded on *Salmonella* pathogenicity island 2 (SPI-2) is pivotal in influencing systemic disease in mice (Shea *et al.*, 1996). Whereas extracellular bacteria express TTSS-1, TTSS-2 is expressed by intracellular *Salmonella*, and TTSS-2 dependent secreted effectors are thought to modulate vesicular trafficking inside *Salmonella*- infected cells (see Chapter 6). Through modulation of different cellular compartments TTSS-2 secreted effectors are thought to stabilize the membranes surrounding intracellular bacteria (Beuzon *et al.*, 2000), promote intracellular replication rates in epithelial cells (Shea *et al.*, 1999) and block oxygen-dependent and nitric oxide-dependent killing mechanisms by macrophages (Chakravortty *et al.*, 2002; Vazquez-Torres *et al.*, 2000). TTSS-2 also influences the pathogenesis of *Salmonella* in cattle. A mutation in *sseD*, which encodes a putative TTSS-2 translocon protein, thought to mediate the translocation of other effector proteins through the target cell membrane, attenuates *S. enterica* serovar Dublin in calves following infection by the intravenous or oral routes (Bispham *et al.*, 2001). The *sseD* mutant was fully invasive for bovine intestinal mucosa but was unable to proliferate to the same extent as the parental strain in vivo. The *sseD* mutant and a second SPI-2 mutant, with a transposon insertion in the *ssaT* gene, induced significantly weaker secretory and inflammatory responses in bovine ligated ileal loops than did the parental strain. These results demonstrate that TTSS-2 is required by *S. enterica* serovar Dublin for the induction of both systemic and enteric phases of salmonellosis in cattle.

Other genes have been identified that also directly affect *Salmonella* enteropathogenicity. The region around *sopB* in *S. enterica* serovar Dublin contains a *Salmonella* specific DNA fragment, termed *Salmonella* pathogenicity island-5 (SPI-5) (Wood *et al.*, 1998). Mutations in SPI-5-encoded *pip* genes (*P*athogenicity *I*sland encoded *P*roteins) produced a phenotype similar to that of a *sopB/sigD* mutant, with partial attenuation of enteropathogenicity in bovine ligated ileal loops and no effect on systemic virulence in mice. PipC *(SigE)* was identified as a SopB/SigD-specific chaperone (Hong and Miller, 1998). However, the functions and sites of action of other Pips are not well defined. Recently it was shown that SPI-5-encoded effectors are induced by distinct regulatory cues and targeted to different TTSSs. As stated above, SPI-5 encodes the TTSS-1 translocated effector SopB/SigD. In contrast, an adjacently encoded effector PipB is part of the *Salmonella* pathogenicity island 2 (SPI-2) regulon. PipB is translocated by TTSS-2 to the *Salmonella*-containing vacuole and *Salmonella*-induced actin filaments (Knodler *et al.*, 2002). SPI-5 is not conserved in all *Salmonella* species. Although *sopB/sigD* is present in all *Salmonella* species, *pipB* is not found in *Salmonella bongori*, which also lacks a functional TTSS-2. Thus there appears to be functional and regulatory cross-talk between three chromosomal pathogenicity islands, SPI-1, SPI-2 and SPI-5, which has significant implications for the evolution of bacterial pathogenesis and host-specificity.

Large molecular weight plasmids can influence the systemic virulence in mice of several *Salmonella* serovars, including Typhimurium (Jones *et al.*, 1982), Enteritidis (Nakamura *et al.*, 1985) and Dublin (Baird *et al.*, 1985; Terakado *et al.*, 1983). However, the role of plasmid genes in *Salmonella* pathogenesis in cattle is somewhat controversial. In one study, the virulence of plasmid-bearing and plasmid-cured strains of *S. enterica* serovar Dublin was compared in calves. The plasmid-bearing strains were highly virulent, causing severe enteric and systemic disease with high mortality. In contrast, the plasmid cured strain caused diarrhoea but only low mortality. The strains were equally invasive for intestinal mucosa and elicited comparable secretory and inflammatory responses in ligated ileal loops. It was concluded that virulence plasmid genes are not involved in either the enteric phase of infection or the systemic dissemination of *S. enterica* serovar Dublin, but probably mediate the persistence of *S. enterica* serovar Dublin at systemic sites (Wallis *et al.*, 1995). More recently, a study assessed the role of the virulence plasmid-encoded *spv* operon in the induction of salmonellosis in cattle by *S. enterica* serovar Dublin. SpvR is the transcriptional regulator required for expression of the *spvABCD* operon. The virulence of wild type and an isogenic *spvR* knockout mutant was compared in calves. As in the previous study, calves that were infected with the *spvR* mutant showed little or no clinical signs of systemic salmonellosis. However, in contrast to the previous study, calves infected with the *spvR* mutant developed only mild diarrhoea compared to those infected with the wild type. The intracellular survival and growth of the wild-type strain and the *spvR* mutant were determined by using blood-derived bovine monocytes. Wild-type *S. enterica* serovar Dublin survived and grew inside cells, while the *spvR* mutant did not proliferate. These results suggest that the *spv* genes of *S. enterica* serovar Dublin promote intracellular proliferation both in intestinal tissues and at extra intestinal sites in calves (Libby *et al.*, 1997).

3.3 *SALMONELLA* INFECTIONS OF PIGS

3.3.1 Epidemiology

The serovars of *Salmonella* associated with clinical disease in pigs can be divided into two groups: the host-restricted serovars typified by *S. enterica* serovar Choleraesuis and the ubiquitous serovars typified by *S. enterica* serovar Typhimurium. In the 1950s and 1960s, the predominant serovar isolated from pigs in the UK was *S. enterica* serovar Choleraesuis representing a major problem for the pig industry, affecting pig health and welfare (Sojka *et al.*, 1977). Since then the occurrence of *S. enterica* serovar Choleraesuis has fallen dramatically and it is now only isolated sporadically. In contrast, this serovar remains a major threat to the pig industry in the USA. The decline of serovar Choleraesuis in the UK was not associated with any specific intervention measure. *S. enterica* serovar Typhimurium is the most common serovar isolated from pigs both in Europe and in the USA. Likewise, *S. enterica* serovar

Derby has a strong association with pigs on both sides of the Atlantic Ocean, and for the past 15 years it has been the second most prevalent serovar in pigs in the UK.

Oral ingestion is thought to be an important route of infection as *Salmonella* are shed in high numbers in the feces of clinically infected pigs. In experimental infections, high doses of between 10^8 and 10^{11} cfu are required to reproducibly cause disease in pigs via the oral route (Gray *et al.*, 1995; Gray *et al.*, 1996). Reproducible results are only obtained using a lower dose if the gastric pH is first neutralised with antacids (Watson *et al.*, 2000). This demonstrates that the low pH of the stomach is an effective barrier to infection by *Salmonella*. Inhalation of infected material into the upper respiratory tract is another potential route of infection. Pneumonia is a common feature of *S. enterica* serovar Choleraesuis infections in pigs (Baskerville and Dow, 1973) and several studies have shown that pigs can be experimentally infected by intranasal inoculation. Pigs infected with *S. enterica* serovar Choleraesuis via the intranasal route develop more severe clinical signs than those infected via the oral route (Gray *et al.*, 1995). Intranasal inoculation with *S. enterica* serovar Typhimurium results in rapid dissemination of *Salmonella* to intestinal sites, even in pigs in which an oesophagotomy has been performed (Fedorka-Cray *et al.*, 1995). Together these observations suggest that the tonsils and lungs are likely to be important sites of invasion.

3.3.2 Salmonellosis in pigs

The ability of *Salmonella* to cause disease in pigs depends on numerous factors including the infecting serovar and the age of the pig. Clinical salmonellosis in pigs is typically of two forms; septicaemia caused by host-restricted *S. enterica* serovars such as Choleraesuis, and enterocolitis caused by broad host range serovars such as Typhimurium. Unsurprisingly, weaned pigs that are intensively reared are most frequently affected by *Salmonella* infections. Like other host-specific serovars, *S. enterica* serovar Choleraesuis has the ability to cause disease in both young and older animals, whereas *S. enterica* serovar Typhimurium typically causes disease in pigs aged between 6 and 12 weeks, but rarely in adult animals. In older animals sub-clinical infections with *S. enterica* serovar Typhimurium are frequent, leading to high transmission rates if active carrier animals are not detected.

S. enterica serovar Choleraesuis typically causes septicaemic forms of infection. Affected pigs are lethargic, pyrexic and often have respiratory symptoms, including coughing. Diarrhoea may or may not be present, cyanosis of the extremities is common and mortality is high. Gross lesions typically include colitis, swollen mesenteric lymph nodes, splenomegaly, hepatomegaly and lung congestion. Foci of necrosis are often seen on the liver. Histopathological examination of such areas of necrosis reveals clusters of histiocytes in foci of acute coagulative hepatocellular necrosis. Other lesions may include fibrinoid thrombi in venules of the gastric mucosa, in cyanotic skin and pulmonary vessels.

S. enterica serovar Typhimurium typically causes enterocolitis. Here, watery diarrhoea is the initial clinical sign of infection. Pigs become anorexic, lethargic and febrile, but mortality is typically low. Gross lesions typically include necrotic colitis and pseudo-membranous typhlitis. Mesenteric lymph nodes are typically swollen. Intestinal necrosis is often seen as distinct button ulcers. Histopathological examination of intestinal mucosa reveals necrosis of enterocytes on the mucosal surface and within crypts, resulting in enterocyte extrusion and villous blunting. During acute infection infiltrating neutrophils can be seen in the intestinal lumen, lamina propria and submucosa. However, at later stages of infection increased cellularity may be due to macrophages and lymphocytes.

3.3.3 *Salmonella* pathogenesis in pigs

Our understanding of the pathogenesis of *Salmonella* infections in pigs is limited, due to the relatively few studies carried out using pigs as an experimental infection model. Furthermore, the different forms of infection caused by the different serovars complicate insights into pathogenesis. Several studies have attempted to gain further understanding of pathogenesis by comparing and contrasting pathogenic processes of different serovars in experimental infections. Inoculation of pigs with *S. enterica* serovar Typhimurium results in acute enterocolitis, whereas inoculation with *S. enterica* serovar Choleraesuis initially produces septicaemia followed by necrosis in the colonic mucosa (Reed *et al.*, 1986; Watson *et al.*, 2000a). There is some evidence that serovars Typhimurium and Choleraesuis may invade intestinal mucosa by distinct routes. Following oral infection, *S. enterica* serovar Typhimurium had a low tendency to invade the enteric mucosa and did not reveal any tropism for a specific intestinal location. However, in the same experiment *S. enterica* serovar Choleraesuis was found predominantly in the colon and on the luminal surface of ileal M cells of Peyer's patches and had a tendency to invade epithelial cells (Pospischil *et al.*, 1990). In contrast, a study using a polarised in vitro organ culture system, demonstrated that *S. enterica* serovars Typhimurium and Choleraesuis invaded ileal mucosa, with and without Peyer's patches, in equal numbers (Bolton *et al.*, 1999).

More recently it was reported that serovars Choleraesuis and Typhimurium invaded enterocytes, goblet cells and M cells in porcine ileal mucosa, but that serovar Choleraesuis was found more frequently within M cells than *S. enterica* serovar Typhimurium. In addition, serovar Choleraesuis appeared to induce less damage to the mucosa than serovar Typhimurium (Meyerholz and Stabel, 2003).

It has long been understood that stressed animals are more susceptible to diarrhoeal diseases such as salmonellosis (Wills, 2000). There is increasing evidence that bacterial pathogens are able to respond to the host environment by detecting host-derived neurotransmitters (Lyte, 2004). Recently it was reported that *S. enterica* serovar Choleraesuis invasion of porcine Peyer's patches could be promoted by the stress hormone norepinephrine and that this phenotype was blocked by the α-adrenergic agonist phentolamine (Green *et al.*, 2003). Understanding the mechanisms by which *Salmonella* uses stress hormones to modulate virulence gene expression may well provide further insight into *Salmonella* pathogenesis.

3.3.4 Molecular basis of pathogenesis of salmonellosis in pigs

Little is known about which virulence genes contribute to the pathogenesis of *S. enterica* in pigs.

Recently it was shown that mutation of *hilA*, a regulator of TTSS-1, influenced the virulence of *S. enterica* serovar Choleraesuis in pigs following oral but not intraperitoneal challenge (Lichtensteiger and Vimr, 2003). This implicates a role for TTSS-1 in the enteric but not the systemic phase of infection.

S. enterica serovar Choleraesuis carries a large virulence plasmid that influences systemic pathogenesis in pigs (Danbara *et al.*, 1992). The complete nucleotide sequence of pKDSC50, a plasmid from *Salmonella enterica* serovar Choleraesuis strain RF-1, has been determined and forty eight open reading frames (ORFs) are predicted to be encoded by the plasmid (Haneda *et al.*, 2001). Analysis of the genetic organization of pKDSC50 suggests that the plasmid is composed of several virulence-associated genes,which include the *spvRABCD* genes, together with genes needed for plasmid replication and maintenance. A second virulence-associated region including the *pef* (plasmid-encoded fimbria) operon and *rck* (resistance to complement killing) gene, which has been identified on the virulence plasmid of *S. enterica* serovar Typhimurium, was absent. Comparative analysis of the nucleotide sequences of the 50-kb virulence plasmid of serovar Choleraesuis and the 94-kb virulence plasmid of serovar Typhimurium revealed high levels of homology suggesting a common ancestry.

3.4 *SALMONELLA* INFECTIONS OF DOMESTIC FOWL AND OTHER AVIAN SPECIES

3.4.1 Epidemiology

The prevalence of *Salmonella* serovars in domestic fowl varies in different countries and with time. Certain serovars have been known to emerge within a country or region for a period and then disappear with no obvious cause or intervention measure. Historically, *S. enterica* serovar Typhimurium has been amongst the most prevalent serovars isolated from poultry. In the UK between the years 1968 and 1973, *S. enterica* serovar Typhimurium accounted for over 40% of all *Salmonella* isolations associated with poultry, followed by *S. enterica* serovar Enteritidis (6%), serovar Pullorum (4%) and serovar Gallinarum (3%) (Sojka *et al.*, 1977). However, during the 1980s, *S. enterica* serovar Enteritidis phage type 4 (PT4) emerged as the predominant serovar exceeding the isolation rates of serovar Typhimurium. The reasons for this epidemic remain unclear, but the introduction of intensive screening and control measures including an active immunization program in the UK poultry industries, have contributed to the recent decline of PT4 over the past few years. In recent years, high numbers of *Salmonella* isolations have being reported in the UK poultry industries. After *S. enterica* serovar Typhimurium and serovar Enteritidis the most commonly isolated serovars are Livingstone, Senftenberg, Kedougou and Montevideo. However, the isolation of these serovars, together with the isolation of other rarer serovars, is probably a reflection of the improved surveillance activity rather than due to a real increase in clinical disease.

The poultry specific *S. enterica* serovars Gallinarum and Pullorum have largely been eradicated from the industries of Europe and North America. However, in regions of the world with less developed industries, and particularly in facilities with poor bio-security, these serovars still represent major threats to bird health and welfare. Although chickens are the natural hosts of *S. enterica* serovars Gallinarum and Pullorum, natural outbreaks caused by these serovars have been described in turkeys, guinea fowl and other avian species.

There are many sources of infection in poultry including vertical transmission, contaminated feed and the environment. Asymptomatic shedding of *Salmonella* from the intestine leads to the contamination of eggs resulting in vertical transmission. Immediately after hatching, oral ingestion by the chicks results in very high numbers of *Salmonella* in the gut and extensive shedding in the feces. This leads to rapid horizontal spread around the hatchery.

3.4.2 Salmonellosis in poultry

The ability of *Salmonella* to cause disease in poultry is closely related to the infecting serovar and the age and genetic background of the bird.

Fowl typhoid (FT) is a disease caused by *S. enterica* serovar Gallinarum that is usually transmitted by the oro-fecal route and mainly affects adult birds. The birds show signs of distress with ruffled feathers, drop in food consumption and increased mortality. Livers and spleens are enlarged with yellowish lesions and liver bronzing. Similar necrotic foci may be found in the pancreas, lung, gizzard and wall of the caecum indicating severe systemic dissemination of infection. The pullorum disease (PD) is caused by *S. enterica* serovar Pullorum, is egg transmitted and occurs primarily in the first few days of life. High numbers of dead-in-shell chicks are seen, and the surviving animals show anorexia, sleepiness and white viscous droppings on the feathers and around the vent (white bacillary diarrhoea).

The ability of serovars other than Gallinarum and Pullorum to cause disease is relatively poorly understood. *S. enterica* serovar Typhimurium is primarily known for producing clinical salmonellosis in very young birds. Infections are typified by acute enteropathogenic responses, characterized by expression of CXC chemokines and a PMN influx (Withanage *et al.*, 2004). Mortality rates vary enormously, from less than 10% to more than 80% in severe outbreaks. Resistance to infection develops rapidly over the first 72 hours of life and has been attributed to maturation of macrophages and the development of a commensal flora in the gut leading to competitive exclusion of *Salmonella* (Barrow, 2000). Strains of *S. enterica* serovar Enteritidis are also highly virulent for young chicks (Halavatkar, Barrow, 1993; Desmidt *et al.*, 1997). *S. enterica* serovar Enteritidis, and in particular strains of phage type 4 (PT4) can also cause asymptomatic and chronic infections in older birds including commercial layers and broiler breeders (Hinton *et al.*, 1989; Hopper and Mawer, 1988; Lister, 1988). Epidemiological data demonstrate a clear association between food poisoning caused by serovar Enteritidis PT4 and the consumption of undercooked eggs (Coyle *et al.*, 1988). The extent to which egg contamination occurs before or after egg formation is unclear.

3.4.3 Pathogenesis of *Salmonella* infections in poultry

Following oral ingestion of *S. enterica* serovar Typhimurium, bacterial multiplication occurs within the intestines and the microorganisms can invade the intestinal mucosa (Barrow *et al.*, 1987). In young chicks invasion is followed by a severe systemic infection and death is likely to result from a combination of anorexia and dehydration due to diarrhoea. Several experimental infection studies have shown that strains of *S. enterica* serovar Enteritidis PT4 appear to be more virulent in chicks than other phage types of this serovar (Dhillon *et al.*, 1999; Poppe *et al.*, 1993). The virulence of PT4 strains in chicks can approach that of *S. enterica* serovar Pullorum strains (Dhillon *et al.*, 1999; Gast and Benson, 1995).

The potential for *Salmonella* to invade the intestines of poultry has been assessed in several studies. The passage of *S. enterica* serovar Enteritidis and serovar Thompson across the cecal mucosa of freshly hatched chicks was visualized by electron microscopy (Popiel and Turnbull, 1985). The uptake of *Salmonella* by macrophages was observed in the cecal lumen; the macrophages then became abnormal and often ruptured to release organisms back into the lumen. Epithelial cell death was related to large numbers of bacteria. Bacteria were never observed in large numbers below the basement membrane, and there was no significant pathology in the lamina propria. Macrophages appeared to contain bacteria and were observed spanning the epithelial and lamina propria regions through breaks in the basement membrane. It is suggested that the passage of bacteria from the epithelium to the lamina propria is primarily the result of capture and transport within host macrophages.

Two studies in chickens directly compared the invasiveness of different serovars of *Salmonella*. In ligated jejunal loops, zoonotic serovars of *Salmonella* were more invasive than serovars normally considered to be horizontally transmitted (Aabo *et al.*, 2002). Invasiveness of different serovars was more comprehensively assessed following oral inoculation of one week old chicks, in addition to inoculation onto cecal tonsils and into ligated jejunal loops. *S. enterica* serovar Typhimurium was more invasive than *S. enterica* serovar Gallinarum at all sites tested, demonstrating that primary invasion does not correlate with systemic pathogenesis in chickens (Chadfield *et al.*, 2003).

A study of orally infected adult birds showed that *S. enterica* serovar Enteritidis and *S. enterica* serovar Typhimurium strains can both colonise the reproductive tract and eggs that are forming in the oviduct prior to ovipositon (Keller *et al.*, 1997). Unfortunately, *S. enterica* serovar Enteritidis PT4 strains were not included in the study. More recently, in a comparison of six serovars injected intravenously into adult birds, *S. enterica* serovar Enteritidis PT4 colonized the reproductive organs of mature laying hens most efficiently (Okamura *et al.*, 2001), which may well explain the association of this serovar with infected eggs. Another study suggests that the tropism of the host-restricted *S. enterica* serovar Pullorum for the reproductive tract is much greater than for ubiquitous *S. enterica* serovars. *S. enterica* serovar Pullorum readily infects the reproductive tract and developing eggs following oral infection, with particularly high numbers in the oviduct at the point of lay (Wigley *et al.*, 2001). The molecular basis for this tissue tropism remains unknown.

Many *S. enterica* serovars have been associated with food poisoning in humans, however the potential for such serovars to infect poultry has been little studied in controlled experiments. A chick isolate of *S. enterica* serovar Kedougou colonized the gut, but did not invade the mucosa of experimentally infected day old chicks (Brito *et al.*, 1995). Similarly, strains of serovars Heidelberg, Senftenberg, Infantis, Montevideo and Menston all efficiently colonized the intestines of young birds, but were less invasive than a strain of serovar Typhimurium (Barrow *et al.*, 1988). More recently, the virulence of several different serovars of *Salmonella* was assessed in day old specific pathogen free chicks. The host-specific serovar Pullorum proved to be the most virulent, followed by the ubiquitous serovars Typhimurium and Enteritidis. Three out of the four strains of serovar Heidelberg caused low levels of mortality, whereas birds infected with isolates of Montevideo, Hadar and Kentucky all survived. However, these latter serovars all colonized the intestines efficiently and caused a reduction in body weight, indicating that sub-clinical *Salmonella* infections can still be detrimental to bird health, welfare and productivity (Roy *et al.*, 2001). The reasons why such serovars are apparently much less virulent in chicks, yet retain the ability to cause human food poisoning are not understood.

3.4.4 Molecular basis of *Salmonella* pathogenesis in poultry

Our understanding of the virulence mechanisms of *Salmonella* for poultry is restricted by the limited number of studies that have been carried out comparing the behaviour of wild type strains, of defined virulence, with isogenic strains carrying defined mutations. The subject is further complicated by the differing virulence of different serovars for poultry, together with the profound effects that bird age, intestinal flora and genetic resistance has on *Salmonella* pathogenesis. However, what is clear is that different serovars rely to varying degrees on different virulence gene clusters to infect poultry. This is best illustrated by *Salmonella* pathogenicity islands 1 and 2, which encode type III secretion systems TTSS-1 and TTSS-2 respectively (see Chapter 6). TTSS-1 and TTSS-2 influence different stages of *Salmonella* pathogenesis. Different *Salmonella* serovars induce different forms of disease in poultry and, unsurprisingly, they show a different reliance on these virulence factors during infection. *S. enterica* serovar Gallinarum required a functional TTSS-2 for infection of 3 week old birds via the oral route, but surprisingly, TTSS-1 was unnecessary (Jones *et al.*, 2001). TTSS-1 mutants (insertional inactivation of *spaS*) were unable to invade avian epithelial cells in vitro, but persisted within chicken macrophages as well as wild type bacteria. In contrast, TTSS-2 mutants (insertional inactivation of *ssaU*) were fully invasive but less able to persist within chicken macrophages. These observations confirm a key role for TTSS-2 in influencing the systemic pathogenesis of *S. enterica* serovar Gallinarum and suggest that serovar Gallinarum penetrates the intestinal mucosa by a mechanism independent of TTSS-1. A similar role for these virulence factors was observed for *S. enterica* serovar Pullorum in one week old birds (Wigley *et al.*, 2002). Again, a functional TTSS-2 was key for systemic pathogenesis. However, in contrast to what is seen in infections with *S. enterica* serovar Gallinarum, disruption of TTSS-1 in serovar Pullorum was mildly attenuating.

In poultry, a role for large molecular weight plasmids in virulence has been found for some, but not all serovars. *S. enterica* serovar Gallinarum carries an 85Kb plasmid that influences bacterial virulence in day old and two week old birds following the oral and intravenous routes of infection (Barrow *et al.*, 1987). An 85Kb plasmid influenced the virulence of *S. enterica* serovar Pullorum for newly hatched chicks (Barrow and Lovell, 1988). In contrast a 54Kb plasmid of *S. enterica* serovar Enteritidis, required for virulence in mice, did not influence the pathogenesis of serovar Enteritidis PT4 for young chicks or adult laying birds by any route of infection (Halavatkar and Barrow, 1993). This result may be attributable to differences in the pathogenicity of these serovars in poultry. Serovars Gallinarum and Pullorum are true systemic pathogens, which require a full repertoire of virulence determinants to overcome the avian immune system. In contrast, serovars Enteritidis and Typhimurium appear to be opportunists exploiting the immunodeficiencies of young birds or immunocompromised older birds.

3.4.5 *Salmonella* infections of pigeons

Pigeons are susceptible to infection by *Salmonella*. Clinical signs vary enormously: at post mortem, intestinal inflammatory responses are prevalent in young birds, whereas systemic disease is evident in adult birds, as indicated by pneumonia, brain abscesses and congestion of the liver (Faddoul and Fellows, 1965). Phage types 2 and 99 of serovar Typhimurium variant Copenhagen (O5-negative) are almost exclusively associated with infections of pigeons. A certain degree of host "adaptation" of pigeon strains of serovar Typhimurium variant Copenhagen has been postulated (Rabsch *et al.*, 2002). A recent study documented the host specificity of pigeon-derived serovar Typhimurium variant Copenhagen strains by determining the bacterial characteristics which were associated with any adaptation of these strains for pigeons (Pasmans *et al.*, 2003). Pulsed-field gel electrophoresis patterns of 38 pigeon strains were compared with those obtained from 89 porcine, poultry and human strains of variant Copenhagen. The pigeon strains were very closely related, whereas the strains from the other host species were relatively un-related. Pigeon-derived strains were much more cytotoxic for pigeon macrophages than were the porcine strains. After experimental infection of pigeons with a pigeon strain, clinical symptoms, fecal shedding, and colonization of internal organs were more pronounced than after infection with a porcine strain. Together the data suggest that the PT 99 strains are adapted to pigeons. The molecular basis of the specificity of this serovar for pigeons remains unknown.

3.5 WHAT ARE THE DETERMINANTS OF *SALMONELLA*-SEROVAR HOST-SPECIFICITY?

3.5.1 Environmental and genetic differences

There are two possible explanations to account for the apparent host-specificity of certain *Salmonella* serovars. Environmental factors may increase exposure of particular animal species to certain serovars. For example, ruminants may be more exposed to *S. enterica* serovar Dublin than serovar Choleraesuis and *vice versa* for pigs. Alternatively, there may be genetic differences between these serovars, which allow them to survive and/or grow in specific niches only found within ruminants or pigs. These two explanations are not mutually exclusive.

The role of environmental factors in influencing *Salmonella* host-specificity can be investigated by comparing the virulence of different serovars in different animal species in controlled experimental infection studies. In a co-ordinated study, the virulence of serovars Typhimurium, Dublin, Choleraesuis, Gallinarum and Abortusovis were compared in orally infected cattle, pigs, sheep, poultry and mice. In calves, serovars Typhimurium, Dublin and Choleraesuis were all highly virulent causing severe diarrhoeal disease. In contrast, serovars Abortusovis and Gallinarum were avirulent (Paulin *et al.*, 2002; Wallis *et al.*, 1995; Watson *et al.*, 1998). In pigs, *S. enterica* serovar Typhimurium caused self-limiting diarrhoea, and serovar

Choleraesuis caused severe systemic disease, whilst serovars Dublin, Gallinarum and Abortusovis were avirulent (Watson *et al.*, 2000b; Wallis and Paulin unpublished observations). In sheep, only serovars Dublin, Abortusovis and Gallinarum were compared and none of these serovars caused diarrhoeal disease. Only serovar Dublin caused a prolonged pyrexial response, although both Dublin and Abortusovis were recoverable in significant numbers from systemic tissues (Uzzau *et al.*, 2001). In one week old poultry only serovar Gallinarum was virulent (John Olsen, personal communication). In mice, serovars Typhimurium and Dublin were highly virulent, and serovars Choleraesuis and Abortusovis were moderately virulent (Farrant *et al.*, 1997; Uzzau *et al.*, 2000); serovars Abortusovis and Gallinarum were of low virulence (Bernard *et al.*, 2002). These studies illustrate that representative strains of these serovars do exhibit clear differences in virulence for different host species and as such show host-specificity. Serovars Abortusovis and Gallinarum were only virulent for sheep and poultry respectively, and these represent examples of host-specific serovars. Serovars Dublin and Choleraesuis were virulent for several distinct host species, but not for all, and therefore represent host-restricted serovars. Finally, serovar Typhimurium has a broad host range typical of the ubiquitous serovars. These strains represent a unique collection and a valuable resource for studying further the factors that influence host-specificity.

3.5.2 Do early interactions of *Salmonella* with mucosal surfaces influence host-specificity?

The host and bacterial factors that determine whether *Salmonella* serovars remain restricted to the gastrointestinal tract or penetrate beyond the mucosa and cause systemic disease remain largely undefined. *Salmonella* must first survive passage through the stomach and then adhere to epithelial cells and ultimately invade the gut mucosa. Much effort has been made to assess the repertoire of fimbrial and non-fimbrial adhesins of *Salmonella*, as bacterial adhesins are frequently highly host-specific, and therefore, an obvious factor that can potentially influence host range. The repertoire of fimbrial operons encoded by the sequenced strain of *S. enterica* serovar Typhi CT18 has been compared with other clinical isolates of *S. enterica* serovar Typhi and other serovars of *Salmonella* by DNA hybridizations. The *S. enterica* serovar Typhi CT18 genome contains a type IV fimbrial operon, an orthologue of the *agf* operon, and 12 putative fimbrial operons of the chaperone-usher assembly class. Hybridization analysis showed that all fourteen putative fimbrial operons of serovar Typhi were also present in a number of non-typhoidal *Salmonella* serovars. Thus, a simple correlation between host range and the presence of a single fimbrial operon is not likely. However, the *S. enterica* serovar Typhi genome differed from that of all other serovars investigated in that it contained a unique combination of putative fimbrial operons (Townsend *et al.*, 2001). These observations do not rule out a role for adhesins in influencing host range and pathogenesis. DNA hybridization analysis provides no information about gene expression nor about point mutations. Clearly it is important to look at the repertoire of fimbrial expression by different serovars during infections of different host species. Such an analysis may provide a greater insight into how different serovars use adhesins during interactions with different hosts and cell types during pathogenesis.

3.5.3 Does the host intestinal inflammatory response restrict *Salmonella* infections to the gut?

The predominant pathology associated with broad host range serovars like Typhimurium is the characteristic intestinal inflammatory response. Enteropathogenic *Salmonella* serovars have acquired, through horizontal gene transfer, the apparatus and effector proteins necessary to elicit such responses. Epithelial cells play a key role in orchestrating the intestinal inflammatory responses to intestinal pathogens. The interaction of *Salmonella* with epithelial cells in vitro leads to the basolateral release of chemokines and apical secretion of pathogen-elicited epithelial chemoattractants (Kohler *et al.*, 2003). These substances are, at least in part, responsible for directing the recruitment and trafficking of PMNs across intestinal epithelial cells, and a functional TTSS-1 is required for the induction of PMN transmigration in vitro (McCormick *et al.*, 1995). *Salmonella*-induced PMN influxes in vivo are also TTSS-1 dependent (Galyov *et al.*, 1997; Tsolis *et al.*, 1999b; Watson *et al.*, 1998). Recently, *S. enterica* serovar Typhimurium was shown to induce the up-regulation of CXC chemokines in bovine (Santos *et al.*, 2003) and chick (Withanage *et al.*, 2004) intestinal mucosa in vivo, which is likely to be responsible for the PMN influx. Thus, through the activity of TTSS-1, *S. enterica* serovar Typhimurium actively induces an acute inflammatory cell influx that presumably confers some evolutionary advantage. However, PMN that have transmigrated across model intestinal epithelia appear to have an enhanced ability to kill *S. enterica* serovar Typhimurium (Nadeau *et al.*, 2002), thus the strategy of evoking an inflammatory response does not come without risks.

TTSS-1 has little impact on the systemic virulence of *S. enterica* serovar Typhimurium, as assessed in mice (Galan and Curtiss, 1989), thus it is likely that the primary functions of effector proteins secreted by TTSS-1 enable the pathogen to induce diarrhoeal disease in host animals. This ability greatly enhances *Salmonella* dissemination into the external environment, and significantly increases the likelihood of transmission via the fecal-oral route. Therefore, the induction of an acute intestinal inflammatory response by broad-host range serovars such as *S. enterica* serovar Typhimurium may restrict *Salmonella* infections to the gut but facilitate rapid transmission through the induction of diarrhea.

3.6 DO HOST-SPECIFIC SEROVARS USE A STRATEGY OF STEALTH TO CAUSE SYSTEMIC DISEASE?

There is increasing evidence that the host specific serovars are less capable of inducing intestinal inflammatory responses. For example, strains of *S. enterica* serovar Dublin are reproducibly less enteropathogenic than strains of serovar Typhimurium in bovine ligated ileal loops (Watson *et al.*, 1998). Similarly, strains of *S. enterica* serovar Choleraesuis are less enteropathogenic than strains of serovar Typhimurium in porcine ligated ileal loops (Wallis and Paulin, unpublished observations), and serovar Gallinarum is less enteropathogenic than serovar Typhimurium in young birds (P. Wigley, personal communication). In a study using chicken non-polarized primary kidney cells, cytokine responses induced by the serovars Enteritidis, Typhimurium and Gallinarum were measured by quantitative RT-PCR and bioassays. Invasion of cells by *S. enterica* serovar Typhimurium and serovar Enteritidis caused an eight- to ten-fold increase in production of the pro-inflammatory cytokine IL6, whilst invasion by serovar Gallinarum caused no such increase (Kaiser *et al.*, 2000). However, caution is required with this interpretation as flagellin is a potent stimulator of pro-inflammatory signals in non-polarized cell cultures (Gewirtz *et al.*, 2001), and unlike serovars Typhimurium and Enteritidis, serovar Gallinarum is non-flagellate. Thus, these potentially important observations need to be confirmed in a polarized cell culture model or preferably in vivo.

Avoidance of intestinal inflammatory responses could facilitate immune-evasion and aid systemic spread through tissues. This may be achieved either passively by loss of effector proteins involved in eliciting pro-inflammatory responses, or actively through the evolutionary acquisition of effector proteins involved in immuno-suppression. Consistent with the model of systemic salmonellosis developing as a result of a "passive" strategy of stealth, is the observation that serovar Typhi has lost the ability to express some pro-inflammatory TTSS-1-dependent effector proteins including SopA, SopE2 and SopD2, by gene decay (Parkhill *et al.*, 2001). Evidence for an "active" strategy of stealth comes from the observation that the TTSS-1 dependent effector AvrA inhibits activation of the key proinflammatory NF-κβ transcription factor (Collier-Hyams *et al.*, 2002). However, AvrA is expressed (at least in vitro) by *S. enterica* serovar Typhimurium, which is arguably one of the most-pro-inflammatory pathogens known. This questions a role for this protein in suppression of inflammatory responses. Further doubt is engendered by the observation that a *S. enterica* serovar Dublin *avrA* mutant was no more (or less) enteropathogenic than the wild-type parental strain (Schesser *et al.*, 2000). Thus the role of AvrA in pathogenesis remains unclear and the regulation of expression of this protein in vivo awaits clarification. Further genomic analysis of the repertoire and in vivo expression profile of TTSS-dependent genes, which are involved in inflammatory responses induced by *S. enterica* serovars may provide more insight into the role of inflammation in the spread of *Salmonella* to systemic sites.

3.7 DISSEMINATION OF *SALMONELLA* TO SYSTEMIC TISSUES – AN EVOLUTIONARY DEAD-END OR AN ALTERNATIVE MEANS OF INTER-ANIMAL SPREAD?

The host-specific serovars are typically associated with systemic infections and higher levels of mortality in adult animals, than are the broad host range serovars. A common feature of host-specific *Salmonella* serovars is infection of reproductive tissues, resulting in infection of eggs in birds and induction of abortion in mammalian species. This is particularly the case for *S. enterica* serovar Pullorum in poultry, serovar Dublin in cattle, serovar Abortusovis in sheep and serovar Abortusequi in horses. *Salmonella* replicates to high numbers in fetal tissues, which leads to abortion and subsequent dissemination of high numbers of bacteria into the environment, facilitating further infection of other animals in the near vicinity. Thus, systemic infections have the potential to promote animal to animal spread of *Salmonella*, by a mechanism that is quite distinct from diarrhoeal disease.

Studies assessing early host/pathogen interactions in the gut, using strains of defined virulence, have found no correlation between host-specificity and the magnitude of intestinal invasion in cattle (Watson *et al.*, 1995), sheep (Uzzau *et al.*, 2001; Watson *et al.*, 1995) pigs (Bolton *et al.*, 1999) or poultry (Chadfield *et al.*, 2003). During the invasion process, *Salmonella* can be seen within M cells, enterocytes, gut macrophages and dendritic cells (Frost *et al.*, 1997; Norimatsu *et al.*, 2003) in the intestinal mucosa. As a consequence, it is generally believed that *Salmonella* disseminates from the gut within an intracellular niche. Thus, if *Salmonella* is to infect successfully systemic tissues there is a requirement to survive this initial interaction with hostile cells.

Salmonella can kill macrophages in vitro and TTSS-1 is involved in this process (Chen *et al.*, 1996; Guilloteau *et al.*, 1996; Watson *et al.*, 1998). Several groups have reported that macrophage killing is mediated by *Salmonella*-induced apoptosis (Chen *et al.*, 1996; Monack *et al.*, 1996), and this occurs by SipB mediated activation of Caspase-1 (Hersh *et al.*, 1999). However, more recently it has been shown that *Salmonella* can kill macrophages by TTSS-1 dependent mechanisms that do not resemble apoptosis (Watson *et al.*, 2000), and are instead termed pyroptosis (Brennan and Cookson, 2000). Despite the cytotoxicity of *Salmonella* for macrophages, no correlation has been found between host-specificity and the magnitude of *Salmonella*-induced lysis of porcine and avian macrophages, at least not in vitro (Chadfield *et al.*, 2003; Watson *et al.*, 2000).

Having traversed the intestinal mucosa of mammalian species, the interactions of *Salmonella* with cells in the mesenteric lymph nodes (MLN) are also likely to influence whether or not a systemic infection results. The precise route and mechanism by which pathogenic *Salmonella* disseminate to extra-intestinal sites or tissues remains poorly defined. This is due in part to the experimental difficulty in assessing such parameters of the infectious process in vivo. Translocating *Salmonella* can be found in high numbers within MLN following infection of ligated ileal loops (Wells CL *et al.*, 1988) and after oral inoculation of different host species, including calves (Villarreal-Ramos *et al.*, 2000; Wallis *et al.*, 1995; Watson *et al.*, 1998). As such, it is likely that *Salmonella* disseminate from the intestinal mucosa via the draining lymphatics. Conversely, the venous drainage represents an alternative means of dissemination, as bacteria may gain access to blood vessels if damage to the intestinal mucosa is extensive. Recently a model has been developed with which to study the route and mechanisms of bacterial translocation from the intestinal mucosa to systemic tissues in vivo. By separately cannulating both venules and lymphatics that drain from infected mucosa and MLN, it was possible to shown that *S. enterica* serovar Dublin primarily disseminates from the bovine gut in lymph, rather than blood, at early stages of infection. Furthermore, the intera ctions between *Salmonella* and the host within the MLN appear pivotal in determining the outcome of pathogenesis. For example, *S. enterica* serovar Gallinarum, which does not cause disease in calves, passes through the MLN in significantly lower numbers compared with the virulent serovar Dublin (Paulin *et al.*, 2002).

The mechanisms of survival and persistence of *Salmonella* within the liver and spleen of an infected host have been extensively studied, particularly in the murine typhoid model of infection. At present, comparatively little is known about the niches in which bacterial replication takes place.

Results from detailed histological and microscopic studies suggest that serovar Typhimurium is likely to reside within murine splenic or hepatic PMNs, hepatocytes and/or Kupffer cells during the early stages of disease (Conlan and North, 1992; Dunlap *et al.*, 1991), and within macrophages during the later stages of the infectious process (Richter-Dahlfors *et al.*, 1997). Furthermore, it has been suggested that the ability of specific serovars to persist within host macrophages of a particular animal species may correlate with serovar-host-specificity. For example, a comparison of the uptake and persistence of *S. enterica* serovar Typhi in human and murine macrophages correlates with the virulence of this serovar in humans but not mice (Alpuche-Aranda *et al.*, 1995; Ishibashi and Arai, 1990). Furthermore, *S. enterica* serovar Typhimurium can persist in higher numbers than serovar Typhi in primary murine macrophages in vitro (Vladoianu *et al.*, 1990). However, factors such as bacterial uptake, intracellular persistence, bacterial killing or production of cytokines in bovine or porcine macrophages in vitro did not correlate with the virulence of different *Salmonella* serovars for cattle and pigs (Watson *et al.*, 2000). Thus, the role of *Salmonella*/macrophage interactions in determining host range and the severity of systemic disease remains unclear.

A recent study using multicolour fluorescence microscopy to visualize individual serovar Typhimurium bacteria within the livers and spleens of mice enabled the interactions between different bacterial populations within the same animal to be studied (Sheppard *et al.*, 2003). The results demonstrated that an increase in bacterial load within an organ could be attributed to the establishment of new foci of infection, rather than increased numbers of bacteria per phagocyte. This suggests that *Salmonella* do not replicate freely within phagocytes in systemic tissues and highlights how little we know about the very fundamental aspects of *Salmonella* pathogenesis.

3.8 CONCLUSIONS

Salmonella enterica subspecies enterica is a fascinating pathogen varying in its pathogenesis and virulence in different animal species. Some serovars have a broad host range and typically cause sub-clinical intestinal infections, and/or acute enteritis. In contrast host-restricted and host-specific serovars have narrower host ranges and associated infections tend to be of the more severe systemic form.

By targeting the intestines and/or reproductive tracts of animals, *Salmonella* are disseminated between animals in high numbers resulting in high levels of transmission and disease. High costs are met annually by farming industries and public health services in monitoring and trying to control these pathogens.

Knowledge of the pathogenesis of *Salmonella* infections in different animal species would help to find measures to hinder the spread of these pathogens between animals. The mechanisms of pathogenicity of a *S. enterica* serovar have been mainly studied in rodent models of infection. However, the behaviour of these micro-organisms in one particular animal species is not necessarily predictive of its behaviour in another host species. Therefore, the application of modern molecular genetics to strains of defined virulence, together with infection studies in natural target animal species will enable of a more comprehensive understanding of the determinants *Salmonella* serovar host-specificity and of the biology of these pathogens in individual animal species.

3.9 ACKNOWLEDGEMENTS

I would like to thank Paul Wigley and particularly Jennie Bispham for the interesting discussions, constructive criticisms and help during the preparation of this manuscript.

REFERENCES

Aabo, S., Christensen, J. P., Chadfield, M. S. *et al.* (2002). Quantitative comparison of intestinal invasion of zoonotic serotypes of *Salmonella enterica* in poultry. *Avian Pathol*, **31**, 41–7.

Ahmer, B. M., van Reeuwijk, J., Watson, P. R., Wallis, T. S. and Heffron, F. (1999). *Salmonella* SirA is a global regulator of genes mediating enteropathogenesis. *Mol Microbiol*, **31**, 971–82.

Alpuche-Aranda, C. M., Berthiaume, E. P., Mock, B., Swanson, J. A. and Miller, S. I. (1995). Spacious phagosome formation within mouse macrophages correlates with *Salmonella* serotype pathogenicity and host susceptibility. *Infect Immun*, **63**, 4456–62.

Baird, G. D., Manning, E. J. and Jones, P. W. (1985). Evidence for related virulence sequences in plasmids of *Salmonella dublin* and *Salmonella typhimurium*. *J Gen Microbiol*, **131**, 1815–23.

Bakshi, C. S., Singh, V. P., Wood, M. W. *et al.* (2000). Identification of SopE2, a *Salmonella* secreted protein which is highly homologous to SopE and involved in bacterial invasion of epithelial cells. *J Bacteriol*, **182**, 2341–4.

Barrow, P. A. (2000). The paratyphoid salmonellae. *Rev Sci Tech*, **19**, 351–75.

Barrow, P. A. and Lovell, M. A. (1988). The association between a large molecular mass plasmid and virulence in a strain of *Salmonella pullorum*. *J Gen Microbiol*, **134**, 2307–16.

Barrow, P. A., Huggins, M. B., Lovell, M. A. and Simpson, J. M. (1987). Observations on the pathogenesis of experimental *Salmonella typhimurium* infection in chickens. *Res Vet Sci*, **42**, 194–9.

Barrow, P. A., Simpson, J. M. and Lovell, M. A. (1988). Intestinal colonisation in the chicken by food-poisoning *Salmonella* serotypes; microbial characteristics associated with fecal excretion. *Avian Pathology*, **17**, 571–88.

Baskerville, A. and Dow, C. (1973). Pathology of experimental pneumonia in pigs produced by *Salmonella cholerae-suis*. *J Comp Pathol*, **83**, 207–15.

Bernard, S., Boivin, R., Menanteau, P. and Lantier, F. (2002). Cross-protection of *Salmonella abortusovis*, *S. choleraesuis*, *S. dublin* and *S. gallinarum* in mice induced by *S. abortusovis* and *S. gallinarum*: bacteriology and humoral immune response. *Vet Res*, **33**, 55–69.

Beuzon, C. R., Meresse, S., Unsworth, K. E. *et al.* (2000). *Salmonella* maintains the integrity of its intracellular vacuole through the action of SifA. *Embo J*, **19**, 3235–49.

Bispham, J., Tripathi, B. N., Watson, P. R. and Wallis, T. S. (2001). *Salmonella* pathogenicity island 2 influences both systemic salmonellosis and *Salmonella*-induced enteritis in calves. *Infect Immun*, **69**, 367–77.

Bolton, A. J., Osborne, M. P., Wallis, T. S. and Stephen, J. (1999). Interaction of *Salmonella choleraesuis*, *Salmonella dublin* and *Salmonella typhimurium* with porcine and bovine terminal ileum in vivo. *Microbiology*, **145**, 2431–41.

Brennan, M. A. and Cookson, B. T. (2000). *Salmonella* induces macrophage death by Caspase-1-dependent necrosis. *Mol Microbiol*, **38**, 31–40.

Brito, J. R., Xu, Y., Hinton, M. and Pearson, G. R. (1995). Pathological findings in the intestinal tract and liver of chicks after exposure to *Salmonella* serotypes Typhimurium or Kedougou. *Br Vet J*, **151**, 311–23.

Chadfield, M. S., Brown, D. J., Aabo, S., Christensen, J. P. and Olsen, J. E. (2003). Comparison of intestinal invasion and macrophage response of *Salmonella* Gallinarum and other host-adapted *Salmonella enterica* serovars in the avian host. *Vet Microbiol*, **92**, 49–64.

Chakravortty, D., Hansen-Wester, I. and Hensel, M. (2002). *Salmonella* pathogenicity island 2 mediates protection of intracellular *Salmonella* from reactive nitrogen intermediates. *J Exp Med*, **195**, 1155–66.

Chen, L. M., Kaniga, K. and Galan, J. E. (1996). *Salmonella* spp. are cytotoxic for cultured macrophages. *Mol Microbiol*, **21**, 1101–15.

Collier-Hyams, L. S., Zeng, H., Sun, J. *et al.* (2002). Cutting edge: *Salmonella* AvrA effector inhibits the key proinflammatory, anti-apoptotic NF-kappa B pathway. *J Immunol,* **169**, 2846–50.

Conlan, J. W. and North, R. J. (1992). Early pathogenesis of infection in the liver with the facultative intracellular bacteria *Listeria monocytogenes, Francisella tularensis,* and *Salmonella typhimurium* involves lysis of infected hepatocytes by leukocytes. *Infect Immun,* **60**, 5164–71.

Coyle, E. F., Palmer, S. R., Ribeiro, C. D. *et al.* (1988). *Salmonella enteritidis* phage type 4 infection: association with hen's eggs. *Lancet,* **2**, 1295–7.

Danbara, H., Moriguchi, R., Suzuki, S. *et al.* (1992). Effect of 50 kilobase-plasmid, pKDSC50, of *Salmonella choleraesuis* RF-1 strain on pig septicemia. *J Vet Med Sci,* **54**, 1175–8.

Desmidt, M., Ducatelle, R. and Haesebrouck, F. (1997). Pathogenesis of *Salmonella* enteritidis phage type four after experimental infection of young chickens. *Vet Microbiol,* **56**, 99–109.

Dhillon, A. S., Alisantosa, B., Shivaprasad, H. L. *et al.* (1999). Pathogenicity of *Salmonella enteritidis* phage types 4, 8, and 23 in broiler chicks. *Avian Dis,* **43**, 506–15.

Dunlap, N. E., Benjamin, W. H., Jr, McCall, R. D., Jr, Tilden, A. B. and Briles, D. E. (1991). A "safe-site" for *Salmonella typhimurium* is within splenic cells during the early phase of infection in mice. *Microb Pathog,* **10**, 297–310.

Eckmann, L., Rudolf, M. T., Ptasznik, A. *et al.* (1997). D-myo-Inositol 1,4,5,6-tetrakisphosphate produced in human intestinal epithelial cells in response to *Salmonella* invasion inhibits phosphoinositide 3-kinase signaling pathways. *Proc Natl Acad Sci USA,* **94**, 14456–60.

Faddoul, G. P. and Fellows, G. W. (1965). Clinical manifestations of paratyphoid infection in pigeons. *Avian Dis,* **22**, 377–81.

Farrant, J. L., Sansone, A., Canvin, J. R. *et al.* (1997). Bacterial copper- and zinc-cofactored superoxide dismutase contributes to the pathogenesis of systemic salmonellosis. *Mol Microbiol,* **25**, 785–96.

Fedorka-Cray, P. J., Kelley, L. C., Stabel, T. J., Gray, J. T. and Laufer, J. A. (1995). Alternate routes of invasion may affect pathogenesis of *Salmonella typhimurium* in swine. *Infect Immun,* **63**, 2658–64.

Frost, A. J., Bland, A. P. and Wallis, T. S. (1997). The early dynamic response of the calf ileal epithelium to *Salmonella typhimurium. Vet Pathol,* **34**, 369–86.

Galan, J. E. and Curtiss, R., III (1989). Cloning and molecular characterization of genes whose products allow *Salmonella typhimurium* to penetrate tissue culture cells. *Proc Natl Acad Sci USA,* **86**, 6383–7.

Galyov, E. E., Wood, M. W., Rosqvist, R. *et al.* (1997). A secreted effector protein of *Salmonella dublin* is translocated into eukaryotic cells and mediates inflammation and fluid secretion in infected ileal mucosa. *Mol Microbiol*, **25**, 903–12.

Gast, R. K. and Benson, S. T. (1995). The comparative virulence for chicks of *Salmonella* enteritidis phage type 4 isolates and isolates of phage types commonly found in poultry in the United States. *Avian Dis*, **39**, 567–74.

Gewirtz, A. T., Navas, T. A., Lyons, S., Godowski, P. J. and Madara, J. L. (2001). Cutting edge: bacterial flagellin activates basolaterally expressed TLR5 to induce epithelial proinflammatory gene expression. *J Immunol*, **167**, 1882–5.

Gewirtz, A. T., Siber, A. M., Madara, J. L. and McCormick, B. A. (1999). Orchestration of neutrophil movement by intestinal epithelial cells in response to *Salmonella typhimurium* can be uncoupled from bacterial internalization. *Infect Immun*, **67**, 608–17.

Gray, J. T., Fedorka-Cray, P. J., Stabel, T. J. and Ackermann, M. R. (1995). Influence of inoculation route on the carrier state of *Salmonella choleraesuis* in swine. *Vet Microbiol*, **47**, 43–59.

Gray, J. T., Stabel, T. J. and Fedorka-Cray, P. J. (1996). Effect of dose on the immune response and persistence of *Salmonella choleraesuis* infection in swine. *Am J Vet Res*, **57**, 313–19.

Green, B. T., Lyte, M., Kulkarni-Narla, A. and Brown, D. R. (2003). Neuromodulation of enteropathogen internalization in Peyer's patches from porcine jejunum. *J Neuroimmunol*, **141**, 74–82.

Guilloteau, L. A., Wallis, T. S., Gautier, A. V. *et al.* (1996). The *Salmonella* virulence plasmid enhances *Salmonella*-induced lysis of macrophages and influences inflammatory responses. *Infect Immun*, **64**, 3385–93.

Halavatkar, H. and Barrow, P. A. (1993). The role of a 54-kb plasmid in the virulence of strains of *Salmonella enteritidis* of phage type 4 for chickens and mice. *J Med Microbiol*, **38**, 171–6.

Hall, G. A. and Jones, P. W. (1976). An experimental study of *Salmonella dublin* abortion in cattle. *Br Vet J*, **132**, 60–5.

Haneda, T., Okada, N., Nakazawa, N., Kawakami, T. and Danbara, H. (2001). Complete DNA sequence and comparative analysis of the 50-kilobase virulence plasmid of *Salmonella enterica* serovar Choleraesuis. *Infect Immun*, **69**, 2612–20.

Hardt, W. D., Chen, L. M., Schuebel, K. E., Bustelo, X. R. and Galan, J. E. (1998). *S. typhimurium* encodes an activator of Rho GTPases that induces membrane ruffling and nuclear responses in host cells. *Cell*, **93**, 815–26.

Hersh, D., Monack, D. M., Smith, M. R., Ghori, N., Falkow, S. and Zychlinsky, A. (1999). The *Salmonella* invasin SipB induces macrophage apoptosis by binding to Caspase-1. *Proc Natl Acad Sci USA*, **96**, 2396–401.

Hinton, M., Pearson, G. R., Threlfall, E. J., Rowe, B., Woodward, M. and Wray, C. (1989). Experimental *Salmonella enteritidis* infection in chicks. *Vet Rec*, **124**, 223.

Hinton, M. H. (1973). *Salmonella dublin* abortion in cattle. *Vet Rec*, **93**, 162.

Hong, K. H. and Miller, V. L. (1998). Identification of a novel *Salmonella* invasion locus homologous to *Shigella ipgDE*. *J Bacteriol*, **180**, 1793–802.

Hopper, S. A. and Mawer, S. (1988). *Salmonella enteritidis* in a commercial layer flock. *Vet Rec*, **123**, 351.

Ishibashi, Y. and Arai, T. (1990). Roles of the complement receptor type 1 (CR1) and type 3 (CR3) on phagocytosis and subsequent phagosome–lysosome fusion in *Salmonella*-infected murine macrophages. *FEMS Microbiol Immunol*, **2**, 89–96.

Jones, G. W., Rabert, D. K., Svinarich, D. M. and Whitfield, H. J. (1982). Association of adhesive, invasive, and virulent phenotypes of *Salmonella typhimurium* with autonomous 60-megadalton plasmids. *Infect Immun*, **38**, 476–86.

Jones, M. A., Wigley, P., Page, K. L., Hulme, S. D. and Barrow, P. A. (2001). *Salmonella enterica* serovar Gallinarum requires the *Salmonella* pathogenicity island 2 type III secretion system but not the *Salmonella* pathogenicity island 1 type III secretion system for virulence in chickens. *Infect Immun*, **69**, 5471–6.

Jones, M. A., Wood, M. W., Mullan, P. B. *et al.* (1998). Secreted effector proteins of *Salmonella dublin* act in concert to induce enteritis. *Infect Immun*, **66**, 5799–804.

Kaiser, P., Rothwell, L., Galyov, E. E. *et al.* (2000). Differential cytokine expression in avian cells in response to invasion by *Salmonella typhimurium, Salmonella enteritidis* and *Salmonella gallinarum*. *Microbiology*, **146**, 3217–26.

Keller, L. H., Schifferli, D. M., Benson, C. E., Aslam, S. and Eckroade, R. J. (1997). Invasion of chicken reproductive tissues and forming eggs is not unique to *Salmonella enteritidis*. *Avian Dis*, **41**, 535–9.

Knodler, L. A., Celli, J., Hardt, W. D. *et al.* (2002). *Salmonella* effectors within a single pathogenicity island are differentially expressed and translocated by separate type III secretion systems. *Mol Microbiol*, **43**, 1089–103.

Kohler, H., McCormick, B. A. and Walker, W. A. (2003). Bacterial-enterocyte crosstalk: cellular mechanisms in health and disease. *J Pediatr Gastroenterol Nutr*, **36**, 175–85.

Libby, S. J., Adams, L. G., Ficht, T. A. *et al.* (1997). The *spv* genes on the *Salmonella dublin* virulence plasmid are required for severe enteritis and systemic infection in the natural host. *Infect Immun*, **65**, 1786–92.

Lichtensteiger, C. A. and Vimr, E. R. (2003). Systemic and enteric colonization of pigs by a *hilA* signature-tagged mutant of *Salmonella choleraesuis*. *Microb Pathog*, **34**, 149–54.

Lister, S. A. (1988). *Salmonella enteritidis* infection in broilers and broiler breeders. *Vet Rec,* **123**, 350.

Lyte, M. (2004). Microbial endocrinology and infectious disease in the 21st century. *Trends Microbiol,* **12**, 14–20.

McCormick, B. A., Miller, S. I., Carnes, D. and Madara, J. L. (1995). Transepithelial signaling to neutrophils by salmonellae: a novel virulence mechanism for gastroenteritis. *Infect Immun,* **63**, 2302–9.

Meyerholz, D. K. and Stabel, T. J. (2003). Comparison of early ileal invasion by *Salmonella enterica* serovars Choleraesuis and Typhimurium. *Vet Pathol,* **40**, 371–5.

Monack, D. M., Raupach, B., Hromockyj, A. E. and Falkow, S. (1996). *Salmonella typhimurium* invasion induces apoptosis in infected macrophages. *Proc Natl Acad Sci USA,* **93**, 9833–8.

Nadeau, W. J., Pistole, T. G. and McCormick, B. A. (2002). Polymorphonuclear leukocyte migration across model intestinal epithelia enhances *Salmonella typhimurium* killing via the epithelial derived cytokine, IL6. *Microbes Infect,* **4**, 1379–87.

Nakamura, M., Sato, S., Ohya, T., Suzuki, S. and Ikeda, S. (1985). Possible relationship of a 36-megadalton *Salmonella enteritidis* plasmid to virulence in mice. *Infect Immun,* **47**, 831–3.

Norimatsu, M., Harris, J., Chance, V. *et al.* (2003). Differential response of bovine monocyte-derived macrophages and dendritic cells to infection with *Salmonella typhimurium* in a low-dose model in vitro. *Immunology,* **108**, 55–61.

Norris, F. A., Wilson, M. P., Wallis, T. S., Galyov, E. E. and Majerus, P. W. (1998). SopB, a protein required for virulence of *Salmonella dublin,* is an inositol phosphate phosphatase. *Proc Natl Acad Sci USA,* **95**, 14057–9.

Okamura, M., Miyamoto, T., Kamijima, Y. *et al.* (2001). Differences in abilities to colonize reproductive organs and to contaminate eggs in intravaginally inoculated hens and *in vitro* adherences to vaginal explants between *Salmonella enteritidis* and other *Salmonella* serovars. *Avian Dis,* **45**, 962–71.

Parkhill, J., Dougan, G., James, K. D. *et al.* (2001). Complete genome sequence of a multiple drug resistant *Salmonella enterica* serovar Typhi CT18. *Nature,* **413**, 848–52.

Pasmans, F., Van Immerseel, F., Heyndrickx, M. *et al.* (2003). Host adaptation of pigeon isolates of *Salmonella enterica* subsp. enterica serovar Typhimurium variant Copenhagen phage type 99 is associated with enhanced macrophage cytotoxicity. *Infect Immun,* **71**, 6068–74.

Paulin, S. M., Watson, P. R., Benmore, A. R. *et al.* (2002). Analysis of *Salmonella enterica* serotype-host specificity in calves: avirulence of *S. enterica* serotype Gallinarum correlates with bacterial dissemination from mesenteric lymph nodes and persistence in vivo. *Infect Immun,* **70**, 6788–97.

Popiel, I. and Turnbull, P. C. (1985). Passage of *Salmonella enteritidis* and *Salmonella thompson* through chick ileocecal mucosa. *Infect Immun*, **47**, 786–92.

Poppe, C., Demczuk, W., McFadden, K. and Johnson, R. P. (1993). Virulence of *Salmonella enteritidis* phagetypes 4, 8 and 13 and other *Salmonella* spp. for day-old chicks, hens and mice. *Can J Vet Res*, **57**, 281–7.

Pospischil, A., Wood, R. L. and Anderson, T. D. (1990). Peroxidase-antiperoxidase and immunogold labeling of *Salmonella typhimurium* and *Salmonella choleraesuis* var kunzendorf in tissues of experimentally infected swine. *Am J Vet Res*, **51**, 619–24.

Rabsch, W., Andrews, H. L., Kingsley, R. A. *et al.* (2002). *Salmonella enterica* serotype Typhimurium and its host-adapted variants. *Infect Immun*, **70**, 2249–55.

Reed, W. M., Olander, H. J. and Thacker, H. L. (1986). Studies on the pathogenesis of *Salmonella typhimurium* and *Salmonella choleraesuis* var kunzendorf infection in weanling pigs. *Am J Vet Res*, **47**, 75–83.

Reis, B. P., Zhang, S., Tsolis, R. M. *et al.* (2003). The attenuated *sopB* mutant of *Salmonella enterica* serovar Typhimurium has the same tissue distribution and host chemokine response as the wild type in bovine Peyer's patches. *Vet Microbiol*, **97**, 269–77.

Richter-Dahlfors, A., Buchan, A. M. and Finlay, B. B. (1997). Murine salmonellosis studied by confocal microscopy: *Salmonella typhimurium* resides intracellularly inside macrophages and exerts a cytotoxic effect on phagocytes *in vivo*. *J Exp Med*, **186**, 569–80.

Roy, P., Dhillon, A. S., Shivaprasad, H. L. *et al.* (2001). Pathogenicity of different serogroups of avian salmonellae in specific-pathogen-free chickens. *Avian Dis*, **45**, 922–37.

Santos, R. L., Tsolis, R. M., Baumler, A. J. and Adams, L. G. (2003). Pathogenesis of *Salmonella*-induced enteritis. *Braz J Med Biol Res*, **36**, 3–12.

Schesser, K., Dukuzumuremyi, J. M., Cilio, C. *et al.* (2000). The *Salmonella* YopJ-homologue AvrA does not possess YopJ-like activity. *Microb Pathog*, **28**, 59–70.

Shea, J. E., Beuzon, C. R., Gleeson, C., Mundy, R. and Holden, D. W. (1999). Influence of the Salmonella typhimurium pathogenicity island 2 type III secretion system on bacterial growth in the mouse. *Infect Immun*, **67**, 213–19.

Shea, J. E., Hensel, M., Gleeson, C. and Holden, D. W. (1996). Identification of a virulence locus encoding a second type III secretion system in *Salmonella typhimurium*. *Proc Natl Acad Sci USA*, **93**, 2593–7.

Sheppard, M., Webb, C., Heath, F. *et al.* (2003). Dynamics of bacterial growth and distribution within the liver during *Salmonella* infection. *Cell Microbiol*, **5**, 593–600.

Sojka, W. J., Wray, C., Shreeve, J. and Benson, A. J. (1977). Incidence of *Salmonella* infection in animals in England and Wales 1968–1974. *J Hyg (Lond)*, **78**, 43–56.

Terakado, N., Sekizaki, T., Hashimoto, K. and Naitoh, S. (1983). Correlation between the presence of a fifty-megadalton plasmid in *Salmonella dublin* and virulence for mice. *Infect Immun*, **41**, 443–4.

Townsend, S. M., Kramer, N. E., Edwards, R. *et al.* (2001). *Salmonella enterica* serovar Typhi possesses a unique repertoire of fimbrial gene sequences. *Infect Immun*, **69**, 2894–901.

Tsolis, R. M., Adams, L. G., Ficht, T. A. and Baumler, A. J. (1999a). Contribution of *Salmonella typhimurium* virulence factors to diarrheal disease in calves. *Infect Immun*, **67**, 4879–85.

Uzzau, S., Gulig, P. A., Paglietti, B. *et al.* (2000). Role of the *Salmonella abortusovis* virulence plasmid in the infection of BALB/c mice. *FEMS Microbiol Lett*, **188**, 15–18.

Uzzau, S., Leori, G. S., Petruzzi, V. *et al.* (2001). *Salmonella enterica* serovar-host specificity does not correlate with the magnitude of intestinal invasion in sheep. *Infect Immun*, **69**, 3092–9.

Vazquez-Torres, A., Xu, Y., Jones-Carson, J. *et al.* (2000). *Salmonella* pathogenicity island 2-dependent evasion of the phagocyte NADPH oxidase. *Science*, **287**, 1655–8.

Villarreal-Ramos, B., Manser, J. M., Collins, R. A. *et al.* (2000). Susceptibility of calves to challenge with *Salmonella typhimurium* 4/74 and derivatives harbouring mutations in *htrA* or *purE*. *Microbiology*, **146**, 2775–83.

Vladoianu, I. R., Chang, H. R. and Pechere, J. C. (1990). Expression of host resistance to *Salmonella typhi* and *Salmonella typhimurium*: bacterial survival within macrophages of murine and human origin. *Microb Pathog*, **8**, 83–90.

Wallis, T. S., Paulin, S. M., Plested, J. S., Watson, P. R. and Jones, P. W. (1995). The *Salmonella dublin* virulence plasmid mediates systemic but not enteric phases of salmonellosis in cattle. *Infect Immun*, **63**, 2755–61.

Watson, P. R., Galyov, E. E., Paulin, S. M., Jones, P. W. and Wallis, T. S. (1998). Mutation of *invH*, but not *stn*, reduces *Salmonella*-induced enteritis in cattle. *Infect Immun*, **66**, 1432–8.

Watson, P. R., Gautier, A. V., Paulin, S. M. *et al.* (2000a). *Salmonella enterica* serovars Typhimurium and Dublin can lyse macrophages by a mechanism distinct from apoptosis. *Infect Immun*, **68**, 3744–7.

Watson, P. R., Paulin, S. M., Bland, A. P., Jones, P. W. and Wallis, T. S. (1995). Characterization of intestinal invasion by *Salmonella typhimurium* and *Salmonella dublin* and effect of a mutation in the *invH* gene. *Infect Immun*, **63**, 2743–54.

Watson, P. R., Paulin, S. M., Jones, P. W. and Wallis, T. S. (2000b). Interaction of *Salmonella* serotypes with porcine macrophages in vitro does not correlate with virulence. *Microbiology*, **146**, 1639–49.

Wells, C. L., Maddaus, M. A., Erlandsen, S. L. and R. L. Simmons (1988). Evidence for the phagocytic transport of intestinal particles in dogs and rats. *Infect Immun*, **56**, 278–82.

Wigley, P., Berchieri, A., Jr., Page, K. L., Smith, A. L. and Barrow, P. A. (2001). *Salmonella enterica* serovar Pullorum persists in splenic macrophages and in the reproductive tract during persistent, disease-free carriage in chickens. *Infect Immun*, **69**, 7873–9.

Wigley, P., Jones, M. A. and Barrow, P. A. (2002). *Salmonella enterica* serovar Pullorum requires the *Salmonella* pathogenicity island 2 type III secretion system for virulence and carriage in the chicken. *Avian Pathol*, **31**, 501–6.

Wills, R. W. (2000). Diarrhea in growing-finishing swine. *Vet Clin North Am Food Anim Pract*, **16**, 135–61.

Withanage, G. S. K., Kaiser, P., Wigley, P. *et al.* (2004). Rapid expression of chemokines and proinflammatory cytokines in newly hatched chickens infected with *Salmonella enterica* serovar Typhimurium. *Infect Immun*, **72**, 2152–9.

Wood, M. W., Jones, M. A., Watson, P. R. *et al.* (1998). Identification of a pathogenicity island required for *Salmonella* enteropathogenicity. *Mol Microbiol*, **29**, 883–91.

Wood, M. W., Rosqvist, R., Mullan, P. B., Edwards, M. H. and Galyov, E. E. (1996). SopE, a secreted protein of *Salmonella dublin*, is translocated into the target eukaryotic cell via a *sip*-dependent mechanism and promotes bacterial entry. *Mol Microbiol*, **22**, 327–38.

Zhang, S., Santos, R. L., Tsolis, R. M. *et al.* (2002a). Phage mediated horizontal transfer of the sopE1 gene increases enteropathogenicity of *Salmonella enterica* serotype Typhimurium for calves. *FEMS Microbiol Lett*, **217**, 243–7.

Zhang, S., Santos, R. L., Tsolis, R. M. *et al.* (2002b). The *Salmonella enterica* serotype Typhimurium effector proteins SipA, SopA, SopB, SopD, and SopE2 act in concert to induce diarrhea in calves. *Infect Immun*, **70**, 3843–55.

CHAPTER 4

Public health aspects of *Salmonella enterica* in food production

Tom Humphrey

4.1 INTRODUCTION AND HISTORICAL PERSPECTIVE

Salmonella enterica is a food-borne pathogen of global significance and no population is spared. It has been reported that there are over 1.3 billion cases of human salmonellosis annually worldwide, with three million deaths (Pang *et al.*, 1995). More recent data suggest that each year in the USA there are 1.3 million cases of human *S. enterica* infections with 600 deaths. Other recent work from Denmark claims that mortality rates of people who have had *S. enterica* are three times those of controls, in the year after infection (Helms *et al.*, 2003). Infection with *S. enterica* can lead to long-term sequelae such as irritable bowel syndrome and reactive arthritis (Rees *et al.*, 2004). There are clear public health benefits to be had from better control of these bacteria in the food chain. In addition, *S. enterica* serovars have an enormous economic impact (Roberts *et al.*, 2003; Voetsch *et al.*, 2004), and are the most important foodborne pathogens in terms of deaths caused (Adak *et al.*, 2002; Kennedy *et al.*, 2004; Mead *et al.*, 1999).

Gaffky may have been the first person to isolate *S. enterica* micro-organisms. These were *S. enterica* serovar Typhi from an infected patient, in 1884. Around this time it was also known that similar bacteria could cause non-typhoid disease in both animals and humans. In 1885, two American veterinarians, Salmon and Smith, isolated the bacterium causing 'hog cholera' from infected pigs. The name *Salmonella* was subsequently adopted in honour of Dr. Salmon. Over the decades following the pioneering work of Salmon and Smith, many other salmonellae were isolated from both animals and humans.

'Salmonella' Infections: Clinical, Immunological and Molecular Aspects, ed. Pietro Mastroeni and Duncan Maskell. Published by Cambridge University Press.

Serovars of *S. enterica* can be divided into two broad groups, depending on their colonization behaviour and propensity to produce systemic disease in healthy, adult hosts. The first group is host-adapted and its members primarily infect their hosts *via* the fecal-oral route but bacterial multiplication mainly occurs intracellularly within macrophages. Bacteria may be released from lymphoid tissue into the gut and are thus transmitted to humans *via* fecal contamination of foods. In the absence of clinical disease these bacteria colonize the gut of the host animal very poorly (Barrow *et al.*, 1987 and 1988). These serovars, such as *S. enterica* serovar Typhi in humans, serovar Dublin in cattle, and serovar Gallinarum in poultry, typically have a low infective dose, long incubation period and cause systemic infection (Rice *et al.*, 1997). An asymptomatic carrier state is common following the resolution of acute infection. Infection of other animal species by these strains is uncommon. These serovars belong to subspecies *enterica* but are of diverse lineage (Selander *et al.*, 1990). Subspecies *arizonae* is similar in its behaviour but it is primarily found in reptiles and rarely causes infections in humans (Baumler *et al.*, 1998). The second group comprises most of the remaining *S. enterica* serovars. They are generally unable to produce systemic disease in non-stressed, healthy, adult animals and efficiently colonize the alimentary tract of food animals without causing disease. Importantly, this means that these serovars frequently enter the food chain, thereby triggering either sporadic cases or outbreaks of human salmonellosis. A number may also cause gastroenteritis in calves and young pigs (Wray and Wray, 2000) and, more importantly in public health terms, can be naturally invasive in key food animals like pigs and chickens (see below). The two most important serovars that can show this behaviour and which are most frequently associated with human disease in the UK and Europe, are serovar Typhimurium and serovar Enteritidis. This chapter will largely focus on these major zoonotic pathogens.

4.2 RECENT TRENDS IN *S. ENTERICA* INFECTIONS

Most *S. enterica* serovars can cause illness in humans. The epidemiology of human disease is currently dominated, however, by only a few serovars, principally *S. enterica* serovar Enteritidis and serovar Typhimurium. In England and Wales in 2002 the reported infection rate with serovar Enteritidis was 18.8 cases per 100,000 people, whilst serovar Typhimurium had a rate of 3.7 cases per 100,000 people. The rate for the next most common serovar, serovar Virchow, was 0.5 cases per 100,000 (Anonymous, 2004a). The history of *S. enterica* infections in the UK, as in other countries, shows that different serovars have been prevalent at different times. During the late 1970s,

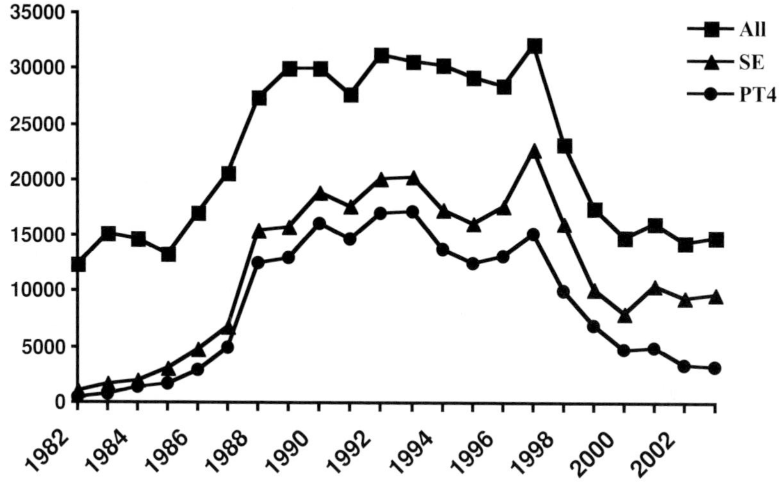

Figure 4.1. Cases of *S. enterica* infections England and Wales between 1982 and 2003. Squares: all serovars; Triangles: all *S. enterica* serovar Enteritidis. Circles: *S. enterica* serovar Enteritidis PT4.

S. enterica serovar Typhimurium was the most commonly isolated, and serovar Agona was prevalent before this time (Harbour *et al.*, 1977). Many of these peaks of infection have been associated with a particular food or animal vehicle, as *S. enterica* serovar Enteritidis is with poultry meat and eggs.

Over the last 20 years, *S. enterica* infections in humans in much of the developed world have shown a marked increase (Angulo and Swerdlow 1999; Humphrey 1994; Riemann *et al.*, 2000; Ward *et al.*, 2000). For example, in England and Wales, which have human infection patterns typical of much of the rest of Western Europe, reported cases rose from just over 10,000 in 1981 to a peak of nearly 33,000 cases in 1997 (Figure 4.1). It should be borne in mind that the reported cases represent only a small fraction of the true number. Estimates for the difference between reported and actual cases vary between approximately 3-fold (Wheeler *et al.*, 1999) and 37-fold (Patrick *et al.*, 2004). *S. enterica* serovar Enteritidis, and in particular organisms belonging to phage type (PT) 4 of this serovar, were largely responsible for the increase in cases of human salmonellosis in Western Europe mentioned above, although other PTs were also involved. In Eastern Europe PT1 was more prevalent and in North America PTs 8 and 13a were more frequently isolated than other PTs from cases of human infection (Patrick *et al.*, 2004). In England and Wales in 1981, *S. enterica* serovar Enteritidis accounted for 10% of human cases of salmonellosis; by 1997 this figure had risen to 70%. During the past seven years, to 2004, in England and Wales, as in some other countries, the number of reported cases has fallen. 14,427 cases were reported in 2001, approximately 45% of the number seen in 1997. This decline appears to be largely due to a decrease in *S. enterica* serovar Enteritidis infections, although this serovar still accounts for a large proportion of human *S. enterica* cases (Figure 4.1), and this is discussed below. The number of confirmed cases of *S. enterica* serovar Typhimurium infection is also now at its lowest level since 1981 and has declined over 3-fold from a peak in 1995.

There has, however, been a quite marked recent increase in cases of *S. enterica* serovar Enteritidis infection in England, which is believed to be associated with the consumption of imported eggs from countries, yet to apply the strict controls in place in the UK (see below; Anonymous, 2001).

4.3 HUMAN DISEASE CAUSED BY *S. ENTERICA* AND VEHICLES FOR ITS TRANSMISSION TO HUMANS

S. enterica can cause serious disease, particularly in the more vulnerable, and in the UK there are approximately 100 deaths per year (Anonymous, 1997). Those most at risk of infection and in whom the disease will be most serious are the elderly, the very young and people who have under-lying disease and/or who are immunocompromized. With most people, however, recovery is uncomplicated and can be achieved with bed rest and maintenance of fluid balance. Common clinical signs include diarrhoea (87% of cases), abdominal pain (84%), fever (75%), nausea and muscle pain (65%). Around a quarter of cases will experience vomiting and headache (Stevens *et al.*, 1989). Bloody diarrhoea is rare and, in a large egg-associated outbreak of *S. enterica* serovar Enteritidis infection investigated by Stevens *et al.*, (1989), it was only reported in 6% of cases. The incubation period for salmonellosis is between 12–72 hours but much longer times have been reported. In some outbreaks, where it is believed that large quantities of contaminated food were consumed, incubation times as short as 2.5 hours were seen (Stevens *et al.*, 1989). Extra-intestinal infection is seen in about 1% of cases. There are a few serovars, however, where more serious disease is more common. In 25% of infections with *S. enterica* serovar Dublin the bacterium could be recovered from blood. The comparable figures for two other invasive serovars, Choleraesuis and Virchow were 75 and 4% respectively (Threlfall *et al.*, 1992).

The most important vehicle by which people become infected with *S. enterica* is through the consumption of contaminated food. Much of this is related to the ubiquity of *S. enterica* in the food production environment, which means that almost any food can act as a vehicle for human infection.

Products as diverse as chocolate, potato crisps and bean sprouts have been identified as vehicles in outbreaks, some of which were on an international scale (Gill *et al.*, 1983; Killalea *et al.*, 1996; van Duynhoven *et al.*, 2002). These are largely chance events and are frequently associated with mistakes either in food production or catering. It should be remembered that *S. enterica* are not overly fastidious in their growth requirements and will grow on a wide range of foodstuffs at temperatures between 7–45 °C. The risk factors in domestic and commercial catering have been well documented (Roberts, 1986) and include incomplete cooking and storage of foods at temperatures which allow *S. enterica* to grow and cross-contaminate, particularly where raw food comes into contact with cooked food. Eating food prepared by others is a risk factor for many types of food borne disease and the generally low wages and high rates of staff turnover in the catering sector exacerbate the problems caused by a lack of proper training in food hygiene. There are also many examples of how frank stupidity has led to outbreaks of salmonellosis. Luby *et al.*, (1993) described a large outbreak where people had been infected with either *S. enterica* serovar Agona or serovar Hadar. Investigation revealed that under-cooked turkey was held at ambient temperature in a kitchen for several hours, by which time it was showing organoleptic evidence of microbial spoilage. Rather than the meat being discarded, it was rinsed with water to remove the spoilage flora and the odours they caused and served without any lethal heat treatments being applied.

It has been estimated that cross-contamination is a major contributory factor in about 30% of domestic outbreaks of salmonellosis (Roberts, 1986). This can take a number of forms but the most important is contact between raw and cooked foods, either directly or indirectly. There was a large outbreak of *S. enterica* serovar Typhimurium infection in a psycho-geriatric ward in Northern England in August 1984 (Anonymous, 1986). The vehicle of infection was cooked roast beef. It transpired that the cooked meat had been sliced and placed on trays, which had been used to hold thawing chicken carcasses and which had not been cleaned. The sliced meat was held overnight in a 'refrigerator' operating at 20–21 °C. The *S. enterica* cells, which originated from the raw chicken, grew rapidly and achieved very high numbers. The meat was served without proper heating with the result that over half the patients at the hospital became infected and 19 died (Anonymous, 1986). This was a landmark case in the UK as it removed Crown Immunity from hospital kitchens and thus made them open to inspection by local authorities.

It is the current fashion for regulatory authorities to advocate a 'farm-to-fork' approach in the control of food-borne disease. Thus consumers are seen as having an important role in protecting themselves and their families. Their ability to do this will be influenced by the nature of the food, pathogen levels and food handling practices. There can be particular difficulties in controlling cross contamination and this is the subject of some controversy. The handling of high-risk foods like raw meat and poultry can result in the spread of pathogenic bacteria such as *S. enterica*. Studies in the Netherlands (de Wit *et al.*, 1979) with chicken carcasses labelled with *E. coli* K12 as a model for *S. enterica* and used to prepare meals in 60 households, demonstrated that the marker organism was spread widely in the kitchen and could be recovered from a range of sites, even after cleaning. Similarly, the whisking of eggs artificially contaminated with high levels of *S. enterica* serovar Enteritidis led to extensive contamination of the immediate environment, especially when electric blenders were used (Humphrey *et al.*, 1994). Homogenized egg preserves the viability of bacteria, and therefore *S. enterica* cells could be recovered many weeks after contamination. Work of this type, and sampling of domestic kitchens, has revealed that *S. enterica* can spread and persist. Does this pose a quantifiable risk to human health? The many kitchen hygiene products on sale throughout the world would suggest that a contaminated kitchen environment presents a significant challenge to human health. The circle has not yet been completed and intuitively one would believe that the presence of *S. enterica* in the kitchen raises the risk of infection. The risks posed by a contaminated dishcloth or the presence of a few cells of *S. enterica* on a cupboard door handle have yet to be quantified. Similarly, consumers need to be given clear and concise messages about risks and how they can best be reduced.

4.4 ANIMAL RESERVOIRS OF *S. ENTERICA* INFECTION

The greatest risk to public health is posed by the presence of *S. enterica* in food animals. It is of interest that many of the *S. enterica* serovars common in food animals are only rarely found in humans. Thus, in the UK in 2002 (Anonymous, 2004a), the two most common serovars in cattle were *S. enterica* serovar Dublin, accounting for 80.7% of isolates and *S. enterica* serovar Typhimurium (11.4% of isolates). In sheep, *S. diarizonae* and *S. enterica* serovar Dublin accounted for 59.4 and 21.2% of isolates respectively. In chickens, the two most common were serovar Livingstone (13.4% of isolates) and serovar Senftenberg (11.8%). In pigs, the most common serovars were Typhimurium (72.2%) and Derby (7.4%). With the exception of serovar Typhimurium, none of the above serovars were found in the top 10 of human isolates (Anonymous, 2004b).

Table 4.1. *Vehicles of S. enterica infection in outbreaks in England and Wales 1998–2002*

Food type	Number of outbreaks
Eggs and dishes containing egg	69
Chicken and other poultry	39
Other foods*	41
No vehicle identified or no information available	110

*Foods included cakes and red meat dishes

The more common vehicles of infection in England and Wales are shown in Table 4.1. The pattern is dominated by foods of animal origin, particularly poultry meat and eggs, as it is in much of the rest of Europe. Most cases of human salmonellosis are sporadic and outbreaks are comparatively rare, especially in countries such as the UK where there have been concerted efforts to control infection in poultry. For example, in the UK in 2002 there were only 22 outbreaks of human salmonellosis (Anonymous, 2004b), which compares favourably with data given in Table 4.1. One of the principal reasons for the recent fall in the number of outbreaks is the marked reduction in the contamination levels of *S. enterica* in poultry meat and eggs. A survey in 1989 found that about 80% of UK-produced poultry carcasses on retail sale were contaminated with *S. enterica*. The most recent survey undertaken in 2001 demonstrated that this figure had fallen to about 5% (Anonymous, 2003). Similarly, the vaccination of laying hens under the Lion scheme in the UK has been cited as the major reason for the marked fall in cases of infection with *S. enterica* serovar Enteritidis (Anonymous, 2001). A recent survey in England and Wales gives practical evidence of the success of the vaccination programme. In the autumn of 2002, the Public Health Laboratory Service examined 691 pooled samples of six eggs (4145 eggs in total). Thirty six pools (5.2%) were *S. enterica*-positive. There was a marked contrast in contamination rates between UK-produced eggs and those from abroad. For example, 18% of pools of imported eggs from unspecified European countries were *S. enterica*-positive and 5.1% of pools of Spanish eggs yielded the bacterium. In contrast, *S. enterica* was not isolated from any eggs produced under the British Egg Industry Council (BEIC) Lion Scheme and from only 1.3% of UK non-Lion code eggs (Anonymous, 2004b). With the increasing globalization of trade it is important to ensure that control measures in one country are not compromised by the importation of contaminated products from another.

Table 4.2. S. enterica *serovars implicated in outbreaks in England and Wales 1998–2001*

S. enterica serovar	Number of outbreaks
Enteritidis	177 (75%)
Typhimurium	22 (9.3%)
Hadar	4
Heidelberg	5
Newport	4
Bredeney	4
Montevideo	4
Others*	17

Serovars Agona, Give, Weltevreden, Agama, Haifa, Hindmarsh, Ibadan, Meunster, Blockley, Kottbus and Brandenburg caused one outbreak each. Java and Thompson caused two outbreaks each.

S. enterica serovar Enteritidis remains the most common serovar involved in outbreaks (Table 4.2) and sporadic cases of infection with *S. enterica* serovar Typhimurium a distant second (Table 4.2).

It is important to differentiate between short-term problems caused by a chance infection of food animals, which leads to food contamination and the underlying background of human infection, resulting from sustained persistence of *S. enterica* in key food animals such as pigs and chickens.

4.5 MILK AND MILK PRODUCTS AS VEHICLES OF INFECTION

There are many examples of infection/intestinal carriage in dairy cows leading to outbreaks of salmonellosis. This is usually because transient intestinal carriage of *S. enterica* in a dairy cow (Gay *et al.*, 1994) may lead to contamination of milk at milking. The greatest risk factor is the consumption of unpasteurized milk or its products (Maguire *et al.*, 1992; Ryan *et al.*, 1987). There have been some outbreaks associated with 'pasteurised milk' but this is due to the process not being properly applied or post-heating cross contamination with raw milk. A particularly spectacular example of the latter was a large multi-state outbreak in the USA where treated milk was contaminated with the raw product. Many thousands of people were infected (Ryan *et al.*, 1987). In another, even larger outbreak, heat-treated ice cream mix was transported in a lorry previously used to carry liquid raw egg. The lorry had not been cleaned properly, the ice cream became contaminated and many tens of thousands of people were infected (Vought and Tatini, 1998). Other milk-based products have also been implicated in outbreaks and this is, in part, because certain *S. enterica* strains can be quite robust in their ability to survive food treatments. There was an outbreak of *S. enterica* serovar Ealing infection in babies in the UK in the mid 1980s associated with dried milk powder, which is believed to have become contaminated by *S. enterica* cells surviving in cracks in the surface of the spray drier (Rowe *et al.*, 1987). The second highest risk milk product after raw milk is cheese, particularly when it is made from raw milk. Many *S. enterica* strains are capable of surviving in cheese even ones with a relatively low pH such as Cheddar. In 1984 there was a large outbreak in Canada, infecting over 1500 people. The causative bacterium was *S. enterica* serovar Typhimurium and it is believed that the cheese was contaminated with raw milk. The outbreak is notable because of the very low infectious dose reported (see below; D'Aoust, 1985). Other notable outbreaks include one caused by *S. enterica* serovar Dublin where the vehicle was a soft raw milk cheese (Maguire *et al.*, 1992). Similarly, contaminated goats' milk cheese was responsible for a large outbreak of *S. enterica* serovar Java infection in France in 1993 (Desenclos *et al.*, 1996).

4.6 MEAT AND MEAT PRODUCTS AND *S. ENTERICA*

There are numerous examples of contaminated meat products being identified as sources/vehicles of infection in outbreaks and sporadic cases (Alban *et al.*, 2002; Berends *et al.*, 1998; Daniels *et al.*, 2002; Haeghebaert *et al.*, 2001). *S. enterica* usually contaminates meat at slaughter as a result of fecal spillage. The extent to which this occurs and levels of contamination will be governed by slaughter hygiene, and this has generally improved as a consequence of measures aimed at combating contamination with *Escherichia coli* O157: H7. The health status of the animal and the conditions it has experienced pre-slaughter will also play a role. The prevalence of *S. enterica*-positive animals will be increased by the mixing of animals at livestock markets and during transport (Grau and Smith, 1974; Williams and Spencer, 1973; Wray *et al.*, 1991;). This may be associated with the release of catecholamines by the stressed animals which in turn can stimulate the growth of *S. enterica* (Nietfeld *et al.*, 1999). Meat products such as sausages can pose a particular risk (Barrell, 1987) as contamination can be introduced into the body of the product. A case-control study in England and Wales in 1994 (Wall *et al.*, 1994) identified sausages consumed outside the home as vehicles for infection with *S. enterica* serovar Typhimurium definitive type (DT) 104 infections. This is probably associated with under-cooking but it is also possible that attachment of *S. enterica* cells to meat tissues enhanced bacterial heat tolerance sufficiently to allow survival in apparently well-cooked sausages (Humphrey *et al.*, 1997).

4.7 CONTAMINATION OF POULTRY MEAT WITH *S. ENTERICA*

Poultry meat and eggs continue to be the most important vehicles of human salmonellosis internationally. Most chickens reared for meat in the developed world are produced under the broiler system, although there is an increasing trend towards extensive production. In broiler production, birds are housed intensively, often with many thousands of birds in each house, and are given water and food *ad libitum*. Chicks reach slaughter weight in approximately 6 weeks. This concentration of potential hosts provides *S. enterica* with greater opportunities for causing infection, and spread can be rapid through infected flocks. Poultry processing is similarly intensive. Birds are transported from the farm to the processing plant in crates that can hold between 20 and 30 birds each. Crates will be stacked, which means that birds in the lower cages may become contaminated with the feces of birds in the cages above them. Crates are also recycled during the working day, washing is often unsatisfactory and crates can be contaminated with *S. enterica* and other zoonotic pathogens like *Campylobacter* spp. (Corry *et al.*, 2002; Slader *et al.*, 2002). A proportion of live birds will arrive at the processing plant already testing positive for *S. enterica*, although infection rates have shown a marked decline in recent years in many countries (see below for data from the UK). Carcass contamination rates frequently exceed those of carriage in the live bird (McBride *et al.*, 1980; Refregier-Petton *et al.*, 2003). This is a manifestation of the speed and intensity of poultry processing. Modern slaughter lines can process up to 200 birds per minute. The slaughter operations are highly efficient and the emphasis has largely been on cost-effectiveness, driven by fierce commercial competition in the retail market and the poultry industry. Cross-contamination is frequent and can occur at almost any stage of slaughter, but is more likely at evisceration, where feces can be spilled, and particularly during immersion scalding, where birds are passed through a tank of 'hot' water so that feathers will be loosened prior to mechanical plucking. Fecal material on the birds' feathers will be removed into the scald water and involuntary defecation can occur, which will introduce more feces and *S. enterica* into the water. Chickens to be sold chilled are scalded using 'soft' scalding water, where water temperatures are between 50 and 52 °C. At this temperature, particularly with the high levels of organic matter in the scalding water, death rates of S. *enterica* can be slow and the bacteria can be isolated from scald-tank water with high frequency (Humphrey and Lanning, 1987). This may lead to contamination of previously negative birds (Mulder *et al.*, 1978) and scalding can also facilitate the attachment of S. *enterica* to chicken skin, making them subsequently more difficult to remove further along the processing line (Notermans *et al.*, 1975).

Sources of infection for chickens will vary depending on the S. *enterica* serovar and poultry production system but the most common are contaminated feed (Williams, 1981), vermin including rodents, and other wild animals (Bailey *et al.*, 2001; Liebana *et al.*, 2003). A more important route of infection in public health terms is vertical transmission from infected breeding stock. This is confined to serovars that are naturally invasive in poultry and that are able to infect and persist in reproductive tissues. From there the S. *enterica* can infect forming eggs, which in poultry meat production leads to infected broiler chicks (McIlroy *et al.*, 1989). Currently the most important serovar in this respect is Enteritidis (see below).

In general, S. *enterica* are present on carcasses, either on the outer surface or in the abdominal cavity. S. *enterica* can be present in parts of the carcass where heat transfer during cooking may be slow, such as in the skin between the legs and breast. In these environments, the bacteria may survive incomplete cooking, particularly where cooking is inherently inefficient such as with barbecues. Undercooked chicken meat is an internationally important vehicle for S. *enterica* infection (Wall *et al.*, 1994).

There are potential forms of carcass contamination different from vertical transmission or gut spillage. One of these is the presence of S. *enterica* in muscle tissue. S. *enterica* serovar Enteritidis PT4 has been isolated from muscle tissues of carcasses sampled at retail sale in England, Wales and Northern Ireland (Humphrey, 1991; Wilson *et al.*, 1996). There are two possible routes for this contamination. S. *enterica* serovar Enteritidis PT4 can be highly invasive in commercial broilers, causing septicaemia, pericarditis and death (Lister, 1988; Rampling *et al.*, 1989). It is possible that, when septicaemic birds are bled, bacteria will lodge in the small blood vessels and thus be present in deep muscle tissues. The airborne route of infection with S. *enterica* can lead to high levels of muscle contamination and frequent septicaemia, even with low infectious doses (Leach *et al.*, 1999). Muscle contamination could also occur during poultry processing. Studies with *Clostridium perfringens* (Lillard, 1973) have shown that, when this bacterium is in scald water, it can subsequently be isolated from the edible parts of chicken carcasses. Given the frequency with which S. *enterica* can be isolated from scald-tank water (Humphrey and Lanning, 1987), the above route of contamination may also be important with this bacterium. The public health impact of non-surface contamination has not been investigated but it is, perhaps, relevant that eating cooked take-away chicken was an important risk factor in sporadic cases of *S.* Enteritidis PT4 infection early in the pandemic in England and Wales, when carcass contamination rates with this bacterium were high (Cowden *et al.*, 1989). The ability of *S. enterica* to attach to meat tissues (Firstenberg-Eden, 1981; Humphrey *et al.*, 1997) together with the higher heat tolerance of some isolates of PT4 (Humphrey *et al.*, 1995), mean that it is possible that some cells of the bacterium survive in the muscle of chicken which, on cursory examination, appears well cooked. The bacteria within the muscle tissue would be protected from gastric acidity and perhaps would be better able to cause infection. The proper cooking of chicken is important in the control of poultry-associated *S. enterica* infection.

With the exception of the international *S. enterica* serovar Enteritidis pandemic, outbreaks and incidents of food-borne human salmonellosis usually show a typical pattern in that there is often a marked increase in cases of human infection caused by a particular serovar but within a short time scale the outbreak comes to an end. There are many examples of this phenomenon. For example, between 1977–1981 there was a national outbreak of *S. enterica* serovar Hadar infection in England and Wales (Anonymous, 1989) that was associated with contaminated turkey meat. The animals had become infected as a result of being fed *S. enterica*-positive imported fishmeal. The outbreak was identified by surveillance, the source of infection was removed and the cases ended.

4.8 EGGS AND EGG PRODUCTS AS VEHICLES OF INFECTION AND THE *S. ENTERICA* SEROVAR ENTERITIDIS PANDEMIC

Eggs have long been associated with food poisoning (Harbour *et al.*, 1977). *S. enterica* was first found in eggs in the UK in the late 1920s and early 1930s (Scott, 1930). At this time there were a number of outbreaks of infection linked to the consumption of raw egg (Wilson, 1945). In the 1940s, further outbreaks of *S. enterica* infection in humans in the UK were associated with the consumption of spray-dried eggs imported from the USA (Solowey *et al.*, 1946; Winter *et al.*, 1946). Nearly all egg-related *S. enterica* cases at this time involved the consumption of eggs, which had become contaminated by bacteria coming in contact with the shell (Cantor and McFarlane, 1948). This may have been due to penetration of organisms through the eggshell or contamination of bulk egg during breaking (Borland, 1975). As a consequence of this, bulk egg in the UK and USA was pasteurised, and egg in the USA washed to remove fecal contaminants from the shell before packing (Winter *et al.*, 1946). These measures succeeded in bringing the problem under control. The egg pasteurization legislation that was introduced in the UK in 1963 to combat the above problems was remarkably successful and for about 20 years after its introduction eggs were only infrequently implicated as vehicles for human salmonellosis, in UK and elsewhere. It is believed that during this long period, egg contamination, if it occurred at all, was largely confined to the outer shell and was a manifestation of intestinal carriage of non-invasive *S. enterica* in commercial laying hens. *S. enterica* present on eggshells may gain access to egg contents when they are broken out and may grow in egg contents at ambient temperature, particularly when the yolk and albumen have been homogenized. Contamination of this kind can be important and is believed to have been a major contributory factor in large outbreaks of human *S. enterica* infection in the USA in 1962/63 (Ager *et al.*, 1967; Thatcher and Montford, 1962).

In the mid-1980s, a remarkable and rapid change occurred in the international epidemiology of human salmonellosis. There was a marked increase in the number of infections caused by different PTs of *S. enterica* serovar Enteritidis. Case-control studies in the USA (St Louis *et al.*, 1988) and England and Wales (Cowden *et al.*, 1989) revealed that the principal vehicle of infection was either undercooked or raw eggs, or dishes derived from them. In Europe, contaminated poultry meat was also important (Cowden *et al.*, 1989), but eggs represented the greater international public health threat.

4.8.1 The beginning of the *S. enterica* serovar Enteritidis pandemic

Until the early 1970s, *S. enterica* serovar Gallinarum and serovar Pullorum were common in the UK poultry flock. These caused invasive fowl typhoid and Pullorum disease respectively in hens, but only rarely caused illness in humans (Bullis, 1977). A campaign of vaccination and the culling of sero-positive hens resulted in the virtual eradication of these diseases in the UK and US by the mid 1970s (Baumler *et al.*, 2000). It has been postulated that the removal of these organisms left a niche that was filled by the antigenically similar *S. enterica* serovar Enteritidis (Baumler *et al.*, 2000). Up until this time, *S. enterica* serovar Enteritidis had been found occasionally in poultry and eggs but had caused only a small number of cases in humans. The organism may have found its way into poultry from the rodent population associated with hen houses, where it is known to have existed since the early

1930s (Edwards and Bruner, 1943), having been introduced as a rodenticide (Riemann *et al.*, 2000).

The fact that *S. enterica* serovar Enteritidis is able to infect poultry without causing observable disease, particularly in laying hens, probably assisted its spread (Cooper *et al.*, 1989), as this made detection of the disease more difficult. As *S. enterica* serovar Enteritidis infection of chickens increased, so did infections in humans in the UK, Europe and North America. It has been postulated that in the UK this may have been due to the introduction of infected layer breeder hens from Europe (Ward *et al.*, 2000) or to the declining number of hens with immunity to *S. enterica* serovar Pullorum, which might also protect against infection with serovar Enteritidis (Baumler *et al.*, 2000). It has also been suggested that the rapid spread of *S. enterica* serovar Enteritidis through Europe and the US could indicate the emergence and expansion of a new, more virulent, strain of the bacterium due to the recent acquisition of the ability to enter and persist in poultry (Rabsch *et al.*, 2001). In Australia, *S. enterica* serovar Enteritidis is present at levels comparable to those pre-pandemic in Europe. It is probable that this strain is a different clone to that now prevalent in Europe (Cox, 1995), supporting the idea that the pandemic has been caused by the expansion of a new strain of the bacterium. It is likely that all of these factors, in combination, played a role in the development of the international epidemic.

4.8.2 Vehicles of infection and the contamination of egg contents

S. enterica serovar Enteritidis can be isolated from eggshells (Humphrey *et al.*, 1989c) and can migrate through the shell and associated membranes and contaminate egg contents. This will occur with a higher frequency if eggshells are damaged (Vadehra *et al.*, 1969) or particularly when a freshly laid egg is exposed to *S. enterica*-positive feces (Padron, 1990; Sparks and Board, 1985). The 'new' feature of the international pandemic of *S.* Enteritidis infection, however, was the fact that the bacterium could be isolated with regularity from the contents of clean, dry, intact, commercially produced eggs. This new behaviour for this bacterium is the single most important factor for the spread and continued international prevalence of *S. enterica* serovar Enteritidis PTs, especially PT4. All available scientific evidence suggests that contamination of egg contents is a consequence of the infection of reproductive tissue (Hoop and Pospischil, 1993). The bacterium is almost always present in egg contents in pure culture (Humphrey, 1994) and studies on outbreak-associated eggs have shown that there is no association between the presence of *S. enterica* serovar Enteritidis on eggshells and egg contents contamination (Humphrey *et al.*, 1989c, 1991; Mawer *et al.*, 1989). Further evidence for the relative lack of involvement of shell/fecal contamination is the finding that the bacterium can be isolated from the reproductive tissues of naturally infected hens even in the absence of intestinal carriage (Bygrave and Gallagher, 1989; Lister, 1988).

The upper regions of the reproductive tract are the principal site of infection with European strains of *S. enterica* serovar Enteritidis (Humphrey *et al.*, 1996). The bacteria are principally localized in the albumen-secreting region, where they can persist for the life of the bird, causing no harm to the host. There may be differences in this respect with serovar Enteritidis strains from the USA, where infection of the ovary appears to be more common (Gast and Holt 2000). The molecular bases for the differences in the site of infection within the reproductive tract are not known. Some preliminary data suggest that tissue persistence may be related to expression of RpoS and LPS structure (Guard-Petter, 2001; Humphrey *et al.*, 1996).

The presence of *S. enterica* serovar Enteritidis in ovaries and oviducts does not always mean that egg contents will become *S. enterica*-positive. Studies of artificially infected hens have shown that, whilst many reproductive tissue samples can be *S. enterica*-positive, serovar Enteritidis is only rarely recovered from the contents of eggs laid by hens infected orally (Humphrey *et al.*, 1996; Keller *et al.*, 1997). Contamination rates can be much higher if birds are infected with aerosols, particularly with *S. enterica* serovar Typhimurium (Leach *et al.*, 1999). The observed prevalence of contamination of commercial eggs is very variable. In a study of eggs from 22 naturally infected commercial flocks associated with outbreaks of PT4 infection, 56/8,700 (0.6% eggs) were contents-positive. Rates for flocks ranged from 0.1 to 10% (Humphrey, 1994). A small study by Paul and Batchelor (1988) in the UK found that, of ten eggs from a batch implicated in a family outbreak of PT4 infection, five were contents-positive. In contrast, in Spain, Perales and Audicana (1989) found that only 0.1% of eggs from flocks implicated in outbreaks were contents-positive. In general, however, the examination of eggs from infected flocks yields more contaminated eggs than this, as recent examination of eggs imported into the UK has shown (Anonymous, 2004b). A national survey of eggs from retail outlets in the UK in 1995/6 found that about 1 in 600 eggs had *S. enterica*-positive shells but only approximately 1 in 6,000 had bacteria in their contents (Anonymous, 2001). A survey in 2003 found that the contamination rates had fallen significantly and no eggs were contents-positive (Anonymous, 2004b). Investigations by Chapman *et al.*, (1988), in an outbreak of *S. enterica* serovar Typhimurium infection, did not

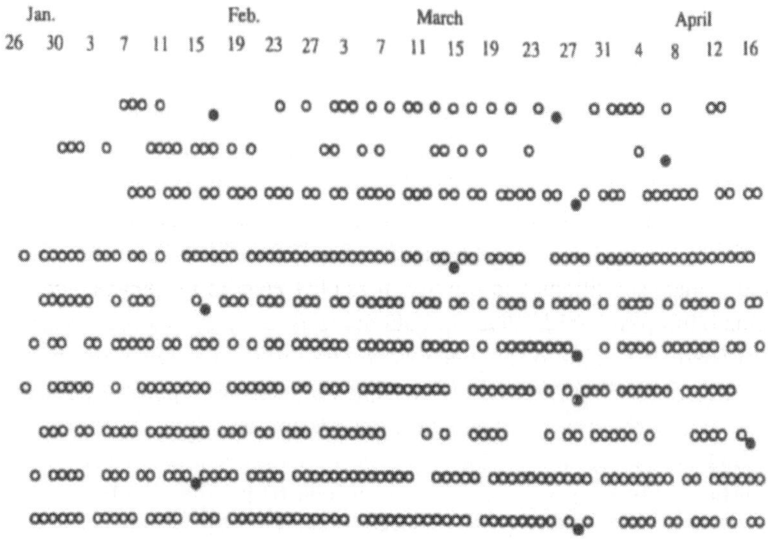

Figure 4.2. Distribution of *S. enterica*-positive eggs, laid by naturally infected commercial hens. Eggs with contaminated contents are denoted by the closed circles.

find *S. enterica* in the contents of 1,000 eggs taken from naturally infected flocks. A possible explanation for this is the finding (Keller *et al.*, 1997) that *S. enterica* serovar Typhimurium may survive less well in the contents of forming eggs than *S. enterica* serovar Enteritidis. It should be pointed out, however, that a full understanding has yet to be gained concerning the mechanisms of egg contamination by *S. enterica*. A study where naturally infected hens were caged individually and each egg laid by each hen was examined over a 3-month period (Humphrey *et al.*, 1989c) found that 1% of 1,100 eggs were contents-positive for PT4. The study also revealed that a number of hens could lay contaminated eggs at, or around the same time (Figure 4.2.). The reasons for this are not yet understood, but similar clustering has been identified in the investigation of egg-associated outbreaks. It is also possible that egg contamination will be stress-related. Stress on chickens will cause changes to the chemistry of the oviduct, which might create an environment more suitable for the persistence and growth of *S. enterica*. Hormones like adrenaline can have an influence on iron uptake (Burton *et al.*, 2002) and thus might cause local increases in bacterial numbers and/or affect bacterial survival in the iron-limited egg albumen. Current vaccination strategies in egg production are focused almost exclusively on *S. enterica* serovar Enteritidis. Given that there are other serovars that are able to contaminate eggs, there is a public health need to understand this important event.

Almost all currently available evidence on the levels of *S. enterica* contamination in egg contents suggests that bacterial numbers are low in fresh eggs (Humphrey *et al.*, 1989a, b, c; Humphrey, 1991; Mawer *et al.*, 1989; Timoney *et al.*, 1989; Gast and Beard 1990). Growth is possible, however, following storage, particularly at temperatures above 20 °C. The examination of naturally contaminated eggs revealed a strong association between egg age and storage conditions and the numbers of *S. enterica* present in egg contents (Humphrey *et al.*, 1991). A national study, where eggs obtained from retail outlets in England and Wales were stored at 20–21 °C for approximately 5 weeks before examination, revealed that 50% of contents-positive eggs contained more than 10^4 *S. enterica* per g of egg contents (de Louvois, 1993).

The close relationship between the poultry-adapted serovar *S. enterica* serovar Gallinarum and serovar Enteritidis may mean that the latter possesses some of the factors that made serovar Gallinarum successful as a poultry pathogen (Baumler *et al.*, 2000). Both isolates belong to serogroup D1, indicating that the structure of their LPS is very similar. Both are part of the same clonal lineage (Li *et al.*, 1993) and both possess SEF14 fimbriae, found in only a few group D serovars (Turcotte and Woodward, 1993), which may be involved in reproductive tissue colonization. This factor is critically important in the production of eggs with *S. enterica*-positive contents. The rapid spread of *S. enterica* serovar Enteritidis PT4 throughout Europe and the USA suggests that this strain possesses factors that give it an advantage in both pathogenesis and environmental survival. No factor has been identified, however, that is unique to this strain. It is to be hoped that the recent (McClelland *et al.*, 2001) and current sequencing (www.sanger.ac.uk) of *S. enterica* genomes may provide insights into this and what makes *S. enterica* serovar Enteritidis uniquely successful in egg contamination.

4.9 THE INFECTIOUS DOSE OF *S. ENTERICA*

To be able to cause disease, *S. enterica* cells must be present in sufficient quantity to overcome host defences. The dose required to cause disease is influenced by the physiological state of the bacteria, the matrix in which the bacteria are ingested, the context of the meal and the degree of resistance of the host. For example, *S. enterica* pre-exposed to sub-lethal acid, as in mayonnaise or pre-exposed to high temperature, which would occur as a consequence of incomplete cooking, may mount stress responses that enhance their ability to survive gastric acidity. A high fat matrix is also protective for

Table 4.3. *Estimates of infective dose of* S. enterica *serovars derived from epidemiological evidence of outbreaks*

Food	Serovar	Infectious dose (cfu)
Chocolate	Eastbourne	<100
Chocolate	Napoli	10–100
Cheese	Typhimurium	1–10
Maize snack	Agona	2–45
Chocolate	Typhimurium	10
Paprika-flavoured potato chips	Saint Paul, Javiana and Rubislaw	<45
Peanut butter	Mbandaka	10–100

S. enterica cells. For these reasons, estimates of infective doses for *S. enterica* in humans have been highly variable. Whilst the typical infective dose for *S. enterica* is considered to be in the range of 10^6–10^8 colony forming units (cfu), epidemiological evidence from a number of outbreaks has demonstrated that in many cases the infective dose may be substantially lower (Table 4.3). Such data indicate the importance of the conditions that *S. enterica* spp. experience, pre-infection, particularly where these invoke stress responses. Almost all the work on the effect of environmental conditions on *S. enterica* has examined changes in bacterial behaviour and protein expression after short exposures to stresses such as either low pH or high temperature. In nature, however, *S. enterica* can experience prolonged exposure to a hostile environment. Recent studies have shown that *S. enterica* and *E. coli* O157:H7 (Mattick *et al.*, 2003) respond to prolonged exposure to conditions just outside their growth range in a quite dramatic and potentially very important way.

For example, it has long been believed that refrigeration and food treatments, which lower pH or a_w, are bacteriostatic for *S. enterica* but this is not always the case. It is now apparent that these conditions may inhibit septation and cell division but do not stop bacterial growth or chromosome replication. Thus, growth continues in terms of an increase of biomass and long (> 200 μm) multi-nucleate filaments are formed. This behaviour has been seen in *S. enterica* serovar Enteritidis and serovar Typhimurium, in certain foods, at a low temperature and low a_w. These bacteria also form long filaments at 4–8 °C and 44 °C and at pH values of 4.4 and 9.4.

Although it is possible that low numbers of *S. enterica* (Table 4.3) are capable of infecting previously healthy adults, the finding that these bacteria form filamentous structures in foods could indicate that outbreak investigations may have underestimated the infective dose present in certain foods. Thus there may be important public health implications for many food products and food storage conditions. Work with laboratory media has shown that if conditions become conducive to septation this process is rapidly completed and many daughter cells result.

4.10 CONCLUSIONS

Salmonella remain important international zoonotic pathogens, responsible for much morbidity and mortality. They are also important in costs to health care systems and employers and also because of direct losses to the international farming community in lost animals and reduced productivity. Control of *S. enterica* infections requires the identification of rational, science-based and cost-effective intervention measures, which will be effective at the farm level but also, where appropriate, cover other parts of the food chain. Control is made more difficult by the frequent changes in *S. enterica* populations seen in animals and in humans and the abilities of these bacteria to interact with and persist in the environment between animal hosts. Difficulties in control are exacerbated by the global trade in food and food animals, which may assist in the movement of *Salmonella* serovars. This could lead to the introduction of new bacterial strains and/or their genetic material into receiving countries. The scientific community now knows more about *Salmonella* than ever before. The genomes of certain *Salmonella* serovars. have been sequenced (McClelland *et al.*, 2001), which will allow investigation of bacterial behaviour by the examination of the interaction of gene products. At present, control in the food chain can be difficult, particularly in primary production, and it is to be hoped that the new information will allow the identification of novel intervention strategies in animal production. There is a need to ensure that consumers, who ultimately pay for the very sophisticated research that is now possible, gain real benefit from it in the form of safer food. It is also important that information is produced in a form that is readily accessible to those with a direct responsibility for food safety, namely farmers and food processors.

The ubiquity of *S. enterica* serovars and the frequency and rapidity with which populations can change mean that it will never be possible to eradicate these bacteria from the food chain. Improved biosecurity on-farm can do much to reduce the challenge to the food animal population and proper slaughter hygiene can reduce contamination levels on animal carcasses. It is also clearly important that food processes are applied that eradicate salmonellae. It is also clear that many *S. enterica* serovars are better able to survive challenges than previously thought. Outbreaks caused by processed foods like chocolate and snacks (Table 4.3) support this view, and when new processes are investigated, it is vital that they are challenged with strains with proven tolerance and which have been pre-exposed to conditions relevant to the product in question. For example, when people become infected with *Salmonella* it is highly likely that they will have consumed attached cells. It is surprising that so little work has been undertaken on the resistance of such populations to heat and acid, for example, and on the influence of attachment on infective dose.

REFERENCES

Adak, G. K., Long, S. M. and O'Brien, S. J. (2002). Trends in indigenous foodborne disease and deaths, England and Wales: 1992 to 2000. *Gut*, **51**, 832–41.

Ager, E. A., Nelson, K. E., Galton, M. M., Boring, J. R., III, and Jernigan, J. R. (1967). Two outbreaks of egg-borne salmonellosis and implications for their prevention. *JAMA*, **199**, 372–8.

Alban, L., Olsen, A. M., Nielsen, B., Sorensen, R. and Jessen, B. (2002). Qualitative and quantitative risk assessment for human salmonellosis due to multi-resistant *Salmonella typhimurium* DT104 from consumption of Danish dry-cured pork sausages. *Prev Vet Med*, **52**, 251–65.

Angulo, F. J. and Swerdlow, D. L. (1999). Epidemiology of human *Salmonella enterica* serovar Enteritidis in the United States. In *Salmonella enterica* serovar Enteritidis *in Humans and Animals*, ed. A. M. Saeed. Ames, Iowa: Iowa State University Press, pp. 33–42.

Anonymous. (1986). The Report of the Committee of Inquiry into an Outbreak of Food Poisoning at Stanley Royd Hospital. *Department of Health and Social Security*. London: HMSO.

(1989). Salmonella in eggs: PHLS evidence to Agriculture Committee. *PHLS Microbiology Digest*, **6**, 1–9.

(1997). *COST Action 97*. Lelystad: European Union.

(2001). *Advisory Committee on the Microbiological Safety of Food: Second Report on Salmonella in Eggs*. London: The Stationery Office.

(2003). UK-wide Survey of *Salmonella* and *Campylobacter* Contamination of Fresh and Frozen Chicken on Retail Sale. *Food Standards Agency*.

(2004a). Zoonoses report United Kingdom 2002 issued by Defra, the Department of Health, the Food Standards Agency, the Scottish Executive Environment & Rural Affairs Department (SEERAD), the Welsh Assembly Government, and the Department of Agriculture & Rural Development Northern

Ireland (DARDNI). *Defra, 2004 A4 63pp., figures PB9248 Free (ISBN 0855210559)*

(2004b). *Salmonella enteritidis* outbreak in central London linked to Spanish eggs. *CDR Weekly*, **14**.

Bailey, J. S., Stern, N. J., Fedorka-Cray, P. *et al.* (2001). Sources and movement of *Salmonella* through integrated poultry operations: a multistate epidemiological investigation. *J Food Prot*, **64**, 1690–7.

Barrell, R. A. (1987). Isolations of salmonellas from humans and foods in the Manchester area: 1981–1985. *Epidemiol Infect*, **98**, 277–84.

Barrow, P. A., Simpson, J. M. and Lovell, M. A. (1988). Intestinal colonisation in the chicken by food-poisoning *Salmonella* serotypes; microbial characteristics associated with faecal excretion. *Avian Pathol*, **17**, 571–88.

Barrow, P. A., Simpson, J. M., Lovell, M. A. and Binns, M. M. (1987). Contribution of *Salmonella gallinarum* large plasmid toward virulence in fowl typhoid. *Infect Immun*, **55**, 388–92.

Baumler, A. J., Hargis, B. M. and Tsolis, R. M. (2000). Tracing the origins of *Salmonella* outbreaks. *Science*, **287**, 50–2.

Berends, B. R., Van Knapen, F., Mossel, D. A., Burt, S. A. and Snijders, J. M. (1998). Impact on human health of *Salmonella* spp. on pork in The Netherlands and the anticipated effects of some currently proposed control strategies. *Int J Food Microbiol*, **44**, 219–29.

Borland, E. D. (1975). *Salmonella* infection in poultry. *Vet Rec*, **97**, 406–8.

Bullis, K. L. (1977). The history of avian medicine in the U.S. II. Pullorum disease and fowl typhoid. *Avian Dis*, **21**, 422–9.

Burton, C. L., Chhabra, S. R., Swift, S. *et al.* (2002). The growth response of *Escherichia coli* to neurotransmitters and related catecholamine drugs requires a functional enterobactin biosynthesis and uptake system. *Infect Immun*, **70**, 5913–23.

Bygrave, A. C. and Gallagher, J. (1989). Transmission of *Salmonella enteritidis* in poultry. *Vet Rec*, **124**, 571.

Cantor, A. and McFarlane, V. H. (1948). *Salmonella* organisms on and in chicken eggs. *Br Poult Sci*, **27**, 350–5.

Chapman, P. A., Rhodes, P. and Rylands, W. (1988). *Salmonella typhimurium* phage type 141 infections in Sheffield during 1984 and 1985: association with hens' eggs. *Epidemiol Infect*, **101**, 75–82.

Cooper, G. L., Nicholas, R. A. and Bracewell, C. D. (1989). Serological and bacteriological investigations of chickens from flocks naturally infected with *Salmonella enteritidis*. *Vet Rec*, **125**, 567–72.

Corry, J. E., Allen, V. M., Hudson, W. R., Breslin, M. F. and Davies, R. H. (2002). Sources of *Salmonella* on broiler carcasses during transportation and processing: modes of contamination and methods of control. *J Appl Microbiol*, **92**, 424–32.

Cowden, J. M., Lynch, D., Joseph, C. A. *et al.* (1989). Case-control study of infections with *Salmonella*-enteritidis phage type-4 in England. *BMJ*, **299**, 771–3.

Cox, J. M. (1995). *Salmonella enteritidis* – the egg and I. *Aust Vet J*, **72**, 108–15.

Daniels, N. A., MacKinnon, L., Rowe, S. M. *et al.* (2002). Foodborne disease outbreaks in United States schools. *Pediatr Infect Dis J*, **21**, 623–8.

D'Aoust, J. Y. (1985). Infective dose of *Salmonella typhimurium* in cheddar cheese. *Am J Epidemiol*, **122**, 717–20.

de Louvois, J. (1993). *Salmonella* contamination of eggs. *Lancet*, **342**, 366–7.

Desenclos, J. C., Bouvet, P., Benz-Lemoine, E. *et al.* (1996). Large outbreak of *Salmonella enterica* serotype paratyphi B infection caused by a goats' milk cheese, France, 1993: a case finding and epidemiological study. *BMJ*, **312**, 91–4.

deWit, J. C., Broekhuizen, G. and Kampelmacher, E. H. (1979). Cross-contamination during the preparation of frozen chickens in the kitchen. *J Hyg (Lond.)*, **83**, 27–32.

Edwards, P. R. and Bruner, D. W. (1943). The occurrence and distribution of *Salmonella* types in the United States. *J Infect Dis*, **72**, 58–67.

Firstenberg-Eden, R. (1981). Attachment of bacteria to meat surface: a review. *J Food Prot*, **44**, 6002–7.

Gast, R. K. and Beard, C. W. (1990). Production of *Salmonella enteritidis*-contaminated eggs by experimentally infected hens. *Avian Dis*, **34**, 438–46.

Gast, R. K. and Holt, P. S. (2000). Deposition of phage type 4 and 13a *Salmonella enteritidis* strains in the yolk and albumen of eggs laid by experimentally infected hens. *Avian Dis*, **44**, 706–10.

Gay, J. M., Rice, D. H. and Steiger, J. H. (1994). Prevalence of faecal *Salmonella* shedding by cull dairy-cattle marketed in Washington-State. *J Food Prot*, **57**, 195–7.

Gill, O. N., Bartlett, C. L. R., Sockett, P. N. *et al.* (1983). Outbreak of *Salmonella*-Napoli infection caused by contaminated chocolate bars. *Lancet*, **1**, 574–7.

Grau, F. H. and Smith, M. G. (1974). *Salmonella* contamination of sheep and mutton carcasses related to pre-slaughter holding conditions. *J Appl Bacteriol*, **37**, 111–16.

Guard-Petter, J. (2001). The chicken, the egg and *Salmonella enteritidis*. *Environ Microbiol*, **3**, 421–30.

Haeghebaert, S., Duche, L., Gilles, C. *et al.* (2001). Minced beef and human salmonellosis: review of the investigation of three outbreaks in France. *Euro Surveill*, **6**, 21–6.

Harbour, H. E., Abell, J. M., Cavanagh, P. *et al.* (1977). *Salmonella: The Food Poisoner.* London: British Association for the Advancement of Science.

Helms, M., Vastrup, P., Gerner-Smidt, P. and Molbak, K. (2003). Short and long term mortality associated with foodborne bacterial gastrointestinal infections: registry based study. *BMJ*, **326**, 357–9.

Hoop, R. K. and Pospischil, A. (1993). Bacteriological, serological, histological and immunohistochemical findings in laying hens with naturally acquired *Salmonella enteritidis* phage type-4 infection. *Vet Rec*, **133**, 391–3.

Humphrey, T. J. (1991). Food poisoning – a change in patterns? *Veterinary Annual*, **31**, 32–7.

Humphrey, T. J. (1994). Contamination of egg shell and contents with *Salmonella enteritidis*: a review. *Int J Food Microbiol*, **21**, 31–40.

Humphrey, T. J. and Lanning, D. G. (1987). *Salmonella* and *Campylobacter* contamination of broiler chickens and scald tank water: the influence of water pH. *J Appl Bacteriol*, **63**, 21–5.

Humphrey, T. J., Baskerville, A., Chart, H. and Rowe, B. (1989a). Infection of egg-laying hens with *Salmonella* enteritidis PT4 by oral inoculation. *Vet Rec*, **18**, 531–2.

Humphrey, T. J., Baskerville, A., Mawer, S., Rowe, B. and Hopper, S. (1989b). *Salmonella enteritidis* phage type 4 from the contents of intact eggs: a study involving naturally infected hens. *Epidemiol Infect*, **103**, 415–23.

Humphrey T. J., Cruickshank, J. G. and Rowe, B. (1989c). *Salmonella enteritidis* phage type-4 and hens' eggs. *Lancet*, **1**, 281.

Humphrey, T. J., Martin, K. W. and Whitehead, A. (1994). Contamination of hands and work surfaces with *Salmonella enteritidis* PT4 during the preparation of egg dishes. *Epidemiol Infect*, **113**, 403–9.

Humphrey, T. J., Slater, E., McAlpine, K., Rowbury, R. J. and Gilbert, R. J. (1995). *Salmonella enteritidis* phage type 4 isolates more tolerant of heat, acid or hydrogen peroxide also survive longer on surfaces. *Appl Environ Microbiol*, **61**, 3161–4.

Humphrey, T. J., Whitehead, A., Gawler, A. H., Henley, A. and Rowe, B. (1991). Numbers of *Salmonella enteritidis* in the contents of naturally contaminated hens' eggs. *Epidemiol Infect*, **106**, 489–96.

Humphrey, T. J., Wilde, S. J., Rowbury, R. J. (1997). Heat tolerance of *Salmonella typhimurium* DT104 isolates attached to muscle tissue. *Lett Appl Microbiol*, **25**, 265–8.

Humphrey, T. J., Williams, A., McAlpine, K. *et al.* (1996). Isolates of *Salmonella* enterica Enteritidis PT4 with enhanced heat and acid tolerance are more virulent in mice and more invasive in chickens. *Epidemiol Infect*, **117**, 79–8.

Keller, L. H., Schifferli, D. M., Benson, C. E., Aslam, S. and Eckroade, R. J. (1997). Invasion of chicken reproductive tissues and forming eggs is not unique to *Salmonella enteritidis*. *Avian Dis*, **41**, 535–9.

Kennedy, M., Villar, R., Vugia, D. J. *et al.*; Emerging Infections Program FoodNet Working Group (2004). Hospitalizations and deaths due to *Salmonella infections*, FoodNet, 1996–1999. *Clin Infect Dis*, **15** (38 suppl. 3), S142–8.

Killalea, D., Ward, L. R., Roberts, D. *et al.* (1996). International epidemiological and microbiological study of outbreak of *Salmonella agona* infection from a ready to eat savoury snack.1. England and Wales and the United States. *BMJ*, **313**, 1105–7.

Leach, S. A., Williams, A., Davies, A. C. *et al.* (1999). Aerosol route enhances the contamination of intact eggs and muscle of experimentally infected laying hens by *Salmonella typhimurium* DT104. *FEMS Microbiol Lett*, **171**, 203–7.

Li, J., Smith, N. H., Nelson, K. *et al.* (1993). Evolutionary origin and radiation of the avian-adapted non-motile salmonellae. *J Med Microbiol*, **38**, 129–39.

Liebana, E., Garcia-Migura, L., Clouting, C. *et al.* (2003). Molecular fingerprinting evidence of the contribution of wildlife vectors in the maintenance of *Salmonella enteritidis* infection in layer farms. *J Appl Microbiol*, **94**, 1024–9.

Lillard, H. S. (1973). Contamination of blood systems and edible parts of poultry with *Clostridium perfringens* during water scalding. *J Food Sci*, **38**, 131–4.

Lister, S. A. (1988). *Salmonella enteritidis* infection in broilers and broiler breeders. *Vet Rec*, **123**, 50.

Luby, S. P., Jones, J. L. and Horan, J. M. (1993). A large *Salmonellosis* outbreak associated with a frequently penalized restaurant. *Epidemiol Infect*, **110**, 31–9.

Maguire, H., Cowden, J., Jacob, M. *et al.* (1992). An outbreak of *Salmonella dublin* infection in England and Wales associated with a soft unpasteurized cows' milk cheese. *Epidemiol Infect*, **109**, 389–96.

Mattick, K. L., Phillips, L. E., Jorgensen, F., Lappin-Scott, H. M. and Humphrey, T. J. (2003). Filament formation by *Salmonella* spp. inoculated into liquid food matrices at refrigeration temperatures, and growth patterns when warmed. *J Food Prot*, **66**, 215–19.

Mawer, S. L., Spain, G. E. and Rowe, B. (1989). *Salmonella enteritidis* phage type 4 and hens' eggs. *Lancet*, 280–1.

McBride, G. B., Skura, B. J., Yada, R. Y. and Bowmer, E. J. (1980). Relationship between incidence of *Salmonella* contamination among pre-scalded, eviscerated and post-chilled chickens in a poultry-processing plant. *J Food Prot*, **43**, 538–42.

McClelland, M., Sanderson, K. E., Spieth, J. *et al.* (2001). Complete genome sequence of *Salmonella enterica* serovar Typhimurium LT2. *Nature*, **413**, 852–6.

McIlroy, S. G., McCracken, R. M., Neill, S. D. and O'Brien, J. J. (1989). Control, prevention and eradication of *Salmonella* enteritidis infection in broiler and broiler breeder flocks. *Vet Rec*, **125**, 545–8.

Mead, P. S., Slutsker, L., Dietz, V. *et al.* (1999). Food-related illness and death in the United States. *Emerg Infect Dis*, **5**, 381–5.

Mulder, R. W., Dorresteijn, L. W. and Van der Broek, J. (1978). Cross-contamination during scalding and plucking of broilers. *Br Poult Sci*, **19**, 61–70.

Nietfeld, J. C., Yeary, T. J., Basaraba, R. J. and Schauenstein, K. (1999). Norepinephrine stimulates in vitro growth but does not increase pathogenicity of *Salmonella choleraesuis* in an *in vivo* model. *Adv Exp Med Biol*, **473**, 249–60.

Notermans, S., Kampelmacher, E. H. and van Schothorst, M. (1975). Studies on sampling methods used in control of hygiene in poultry processing. *J Appl Bacteriol*, **39**, 55–61.

Padron, M. (1990). *Salmonella typhimurium* penetration through the eggshell of hatching eggs. *Avian Dis*, **34**, 463–5.

Pang, T., Bhutta, Z. A., Finlay, B. B. and Altwegg, M. (1995). Typhoid-fever and other salmonellosis – a continuing challenge. *Trends Microbiol*, **3**, 253–5.

Patrick, M. E., Adcock, P. M., Gomez, T. M. *et al.* (2004). *Salmonella enteritidis* infections, United States, 1985–1999. *Emerg Infect Dis*, **10**, 1–7.

Paul, J. and Batchelor, B. (1988). *Salmonella enteritidis* phage type 4 and hens' eggs. *Lancet*, **2**, 1421.

Perales, I. and Audicana, A. (1989). The role of hens' eggs in outbreaks of salmonellosis in north Spain. *Int J Food Microbiol*, **8**, 175–80.

Rabsch, W., Tschape, H. and Baumler, A. J. (2001). Non-typhoidal salmonellosis: emerging problems. *Microbes Infect*, **3**, 237–47.

Rampling, A., Anderson, J. R., Upson, R. *et al.* (1989). *Salmonella enteritidis* phage-type-4 infection of broiler-chickens – a hazard to public-health. *Lancet*, **2**, 436–8.

Rees, J. R., Pannier, M. A., McNees, A. *et al.* (2004). Persistent diarrhea, arthritis, and other complications of enteric infections: a pilot survey based on California FoodNet surveillance, 1998–1999. *Clin Infect Dis*, **38** (suppl. 3), S311–17.

Refregier-Petton, J., Kemp, G. K., Nebout, J. M., Allo, J. C. and Salvat, G. (2003). Post treatment effects of a SANOVA immersion treatment on turkey carcases and subsequent influence on recontamination and cross contamination of breast fillet meat during turkey processing. *Br Poult Sci*, **44**, 790–1.

Rice, D. H., Besser, T. E. and Hancock, D. D. (1997). Epidemiology and virulence assessment of *Salmonella dublin*. *Vet Microbiol*, **56**, 111–24.

Riemann, H., Kass, P. and Cliver, D. (2000). *Salmonella enteritidis* epidemic. *Science*, **287**, 1754–5.

Roberts, D. (1986). Factors contributing to outbreaks of food-borne infection and intoxication in England and Wales 1970–1982. In 2nd *World Congress Food-borne Infections and Intoxication, Berlin*, **1**, 157–9.

Roberts, J. A., Cumberland, P., Sockett, P. N. *et al.* and Infectious Intestinal Disease Study Executive (2003). The study of infectious intestinal disease in England: socio-economic impact. *Epidemiol Infect*, **130**, 1–11.

Rowe, B., Hutchinson, D. N., Gilbert, R. J. *et al.* (1987). *Salmonella ealing* infections associated with consumption of infant dried milk. *Lancet*, **2**, 900–3.

Ryan, C. A., Nickels, M. K., Hargrett-Bean, N. T. *et al.* (1987). Massive outbreak of antimicrobial-resistant salmonellosis traced to pasteurized milk. *JAMA*, **258**, 3269–74.

Scott, W. M. (1930). Food poisoning due to eggs. *BMJ*, **12**, 56–8.

Selander, R. K., Beltran, P., Smith, N. H. *et al.* (1990). Evolutionary genetic relationships of clones of *Salmonella* serovars that cause human typhoid and other enteric fevers. *Infect Immun*, **58**, 2262–75.

Shelobolina, E. S., Sullivan, S. A., O'Neill, K. R., Nevin, K. P. and Lovley, D. R. (2004). Isolation, characterization, and U(VI)-reducing potential of a facultatively anaerobic, acid-resistant bacterium from Low-pH, nitrate- and U(VI)-contaminated subsurface sediment and description of *Salmonella subterranea* sp. nov. *Appl Environ Microbiol*, **70**, 2959–65.

Slader, J., Domingue, G., Jorgensen, F. *et al.* (2002). Impact of transport crate reuse and of catching and processing on *Campylobacter* and *Salmonella* contamination of broiler chickens. *Appl Environ Microbiol*, **68**, 713–19.

Solowey, M., Spaulding, E. H. and Goresline, H. E. (1946). An investigation of a source of mode of entry of *Salmonella* organisms in spray-dried whole-egg powder. *Food Research*, **11**, 380–90.

Sparks, N. H. C. and Board, R. G. (1985). Bacterial penetration of the recently oviposited shell of hens eggs. *Aust Vet J*, **62**, 169–70.

St Louis, M. E., Morse, D. L., Potter, M. E. *et al.* (1988). The emergence of grade A eggs as a major source of *Salmonella enteritidis* infections: new implications for the control of salmonellosis. *JAMA*, **259**, 2103–7.

Stevens, A., Joseph, C., Bruce, J. *et al.* (1989). A large outbreak of *Salmonella enteritidis* phage type-4 associated with eggs from overseas. *Epidemiol Infect*, **103**, 425–33.

Thatcher, F. S. and Montford, J. (1962). Egg products as a source of *Salmonellae* in processed foods. *Can J Public Health*, **53**, 61–9.

Threlfall, E. J., Hall, M. L. and Rowe, B. (1992). *Salmonella* bacteraemia in England and Wales, 1981–1990. *J Clin Pathol*, **45**, 34–6.

Timoney, J. F., Shivaprasad, H. L., Baker, R. C. and Rowe, B. (1989). Egg transmission after infection of hens with *Salmonella enteritidis* phage type 4. *Vet Rec*, **125**, 600–1.

Turcotte, C. and Woodward, M. J. (1993). Cloning, DNA nucleotide sequence and distribution of the gene encoding the SEF14 fimbrial antigen of *Salmonella enteritidis*. *J Gen Microbiol*, **139**, 1477–85.

Vadehra, D. V., Baker, R. C. and Naylor, H. B. (1969). *Salmonella* infection of cracked eggs. *Br Poult Sci*, **48**, 954–7.

van Duynhoven, Y. T., Widdowson, M. A., de Jager, C. M. *et al.* (2002). *Salmonella enterica* serotype Enteritidis phage type 4b outbreak associated with bean sprouts. *Emerg Infect Dis*, **8**, 440–3.

Voetsch, A. C., Van Gilder, T. J., Angulo, F. J. *et al.* Emerging Infections Program FoodNet Working Group (2004). FoodNet estimate of the burden of illness caused by nontyphoidal *Salmonella* infections in the United States. *Clin Infect Dis*, **38** (suppl. 3), S127–34.

Vought, K. J. and Tatini, S. R. (1998). *Salmonella enteritidis* contamination of ice cream associated with a 1994 multistate outbreak. *J Food Prot*, **61**, 5–10.

Wall, P. G., Morgan, D., Lamden, K. *et al.* (1994). A case control study of infection with an epidemic strain of multiresistant *Salmonella typhimurium* DT104 in England and Wales. *CDR Review*, **4**, R130–5.

Ward, L. R., Threlfall, E. J., Smith, H. R. and O' Brien, S. J. (2000). *Salmonella enteritidis* epidemic. *Science*, **287**, 1753–4.

Wheeler, J. G., Sethi, D., Cowden, J. M. *et al.* (1999). Study of infectious intestinal disease in England: rates in the community, presenting to general practice, and reported to national surveillance. *BMJ*, **318**, 1046–50.

Williams, E. F. and Spencer, R. (1973). Abattoir practices and their effect on the incidence of Salmonellae in meat. In B. C. Hobbs and J. H. B. Christian, eds., *The Microbiological Safety of Food*. London: Academic Press, pp. 41–6.

Williams, J. E. (1981). *Salmonella* in poultry feeds – a worldwide review. Part 1. *World's Poult Sci J*, **37**, 6–19.

Wilson, I. G., Wilson, T. S. and Weatherup, S. T. (1996). *Salmonella* in retail poultry in Northern Ireland. *CDR Review*, **6**, R64–6.

Wilson, J. E. (1945). Infected egg shells as a means of spread of salmonellosis in chicks and ducklings. *Vet Rec*, **57**, 411–13.

Winter, A. R., Stewart, G. F., McFarlane, V. H. and Solowey, M. (1946). Pasteurisation of liquid egg products III. Destruction of *Salmonella* in liquid whole egg. *Am J Public Health*, **36**, 451–60.

Wray, C. and Wray, A. (2000). *Salmonella in domestic animals.* (Wallingford, Oxfordshire: CABI Publishing).

Wray, C., Todd, N., Mclaren, I. M. and Beedell, Y. E. (1991). The epidemiology of *Salmonella* in calves – the role of markets and vehicles. *Epidemiol Infect*, **107**, 521–5.

CHAPTER 5

The *Salmonella* genome: a global view

Anne L. Bishop, Gordon Dougan and Stephen Baker

5.1 INTRODUCTION

Genome sequences of different salmonellae are available or are close to completion providing a rich data set to support studies on these micoorganisms. Comparative sequence analysis has been used to redefine the relationships between different *Salmonella* species and serovars and the first functional genomic analyses have been completed. In the near future genomic studies will facilitate a redefinition of the *Salmonella* genus from an evolutionary perspective and we can expect novel typing systems, diagnostic approaches and possibly therapies to emerge.

5.2 FULL GENOME SEQUENCES FACILITATE THE STUDY OF *SALMONELLA*

The availability of full genome sequences for several *Salmonella* serovars has radically advanced the fields of functional and comparative *Salmonella* genomics. The genomic era brings an opportunity to analyze more comprehensively the phylogenetic relationships between *Salmonella*, the evolution of pathogenicity and the genetic variability within natural populations – comparative genomics. The precise genetic makeup of the bacterium combined with host factors are thought to account for the observed differences in the disease spectra and host specificities for different salmonellae. The recent rapid expansion of bacterial genome sequence information has enhanced our ability to investigate the activities of the genes involved on the bacterial side of this equation – functional genomics. There is hope that these genetic insights

'Salmonella' Infections: Clinical, Immunological and Molecular Aspects, ed. Pietro Mastroeni and Duncan Maskell. Published by Cambridge University Press.

may contribute not only to a clearer understanding of *Salmonella* pathogenicity and epidemiology, but also to the design of better vaccines, diagnostic kits and surveillance tools. Two technical fields, discussed below, have come to the forefront of microbiological studies in response to the availability of full genome information, those being bioinformatics and microarray technology.

5.3 COMPARATIVE GENOMICS: OLD AND NEW TECHNIQUES

The Kauffmann-White scheme classifies *Salmonella* into serotypes on the basis of variation in the somatic lipopolysaccharide (O) and flagella (H) antigens (Kauffmann 1957; Popoff *et al.*, 2003). This is a convenient way to categorize isolates, but surface antigens alone cannot provide information about the overall genetic relatedness of strains. In some cases, strains in different serovars are actually genetically very similar, conversely strains within one serovar can have a very different genetic makeup, *Salmonella enterica* subspecies I serovar Paratyphi B (*S*. Paratyphi B) being a clear example (Selander *et al.*, 1990a; Prager *et al.*, 2003). These differences between strains suggest that horizontal transfer and acquisition of genes that mediate cell-surface antigens has been a significant influence in the evolution of *Salmonella* (Beltran *et al.*, 1988). On the other hand, serovars such as Typhi are comparatively similar in their genetic content (Reeves *et al.*, 1989; Selander *et al.*, 1990b). Even so, regions of variation between different strains of *S. enterica* serovar Typhi do exist (Boyd *et al.*, 2003). A clear phenotypic example of this is that certain Indonesian isolates of *S. enterica* serovar Typhi express an unusual H antigen (Franco *et al.*, 1992).

Phage typing, based upon susceptibility to a panel of bacteriophages (Felix, 1956), and biotyping (Duguid *et al.*, 1975) are historically important epidemiological tools for the categorization of *Salmonella*, but like serotyping they do not reflect the overall bacterial genotype. Serotyping and phage typing are often combined with other relatively simple molecular techniques. These include: a) pulse-field gel electrophoresis (PFGE), based on the analysis of restriction enzyme digested DNA fragments (Nair *et al.*, Koay *et al.*, 1997); b) amplified fragment length polymorphism (AFLP), which is a PCR-based modification of PFGE that gives more discrimination between strains (Nair *et al.*, 2000); c) IS200 typing or ribotyping, which analyze the multi-copy IS200 elements or rRNA genes respectively in a *Salmonella* genome using either restriction digests and Southern blotting (sometimes termed Restriction Fragment Length Polymorphism – RFLP analysis) or PCR-based techniques (Olsen *et al.*, Threlfall *et al.*, 1998; Millemann *et al.*, 2000). Multilocus Enzyme Electrophoresis (MLEE), which is based on separating and then detecting the activity and isoelectric point of enzymes, is a useful tool for global epidemiology (Reeves *et al.*, 1989; Selander *et al.*, 1990b). However, MLEE data can be difficult to re-produce from one laboratory to another and does not have the level of discernment required for detailed studies of *Salmonella* evolution. A more reproducible and discerning method to study genetic variation is Multilocus Sequence Typing (MLST), based on the sequencing of selected genes (Boyd *et al.*, 1996; Kidgell *et al.*, 2002).

Comparisons based upon full genome sequences, carried out *in silico* or using microarray technology, allow the analysis of variations across the whole genome in great detail, and thus provide a clear advantage over other techniques used for the study of genetic variation between *Salmonella* species and strains. However, it must not be forgotten that genome sequencing is still comparatively expensive and time-consuming. Microarray techniques provide information about genes common to, or absent from, a strain of unknown genetic content compared with a sequenced strain. However, microarrays are not sensitive enough to detect point mutations compared to the reference strain, also genes present in the test strain, but not in the sequenced strain, can not be detected. In this case, techniques such as subtractive hybridization can be used due to their ability to identify regions of DNA present in one strain (termed the tester) and not in another (termed the driver), even when genome sequence is not available (Agron *et al.*, 2002). The identification of serovar-specific sequences is sometimes of prime interest, for example when developing diagnostic tests. For this reason, in the absence of a complete genome sequence for *S. enterica* serovar Enteritidis, suppression subtraction hybridization, which uses PCR to enrich for restriction fragments present in one strain and absent in another, was used to identify novel *S. enterica* serovar Enteritidis-specific DNA sequences (Agron *et al.*, 2001). Subtractive hybridization does not give immediate information about genes within the DNA identified unless this is subsequently cloned and sequenced.

5.4 *IN SILICO* TOOLS FOR COMPARATIVE GENOMICS

Vast amounts of information can potentially be obtained from the multiple *Salmonella* genome sequences now available. Along with this capacity comes the practical challenge of how to analyze the data. Computer-based tools are improving rapidly and these are essential aids in the practical application of genome comparisons and the analysis of microarray data. User-friendly bioinformatic tools, now freely available to non-profit making institutions, facilitate the handling and analysis of whole genomes *in silico*. For instance, Artemis (a genome viewer and annotation tool) and Artemis Comparison Tool (ACT, for comparing multiple genome sequences) are both written in java script and are available to run on UNIX, GNU/Linux, Macintosh and MS Windows systems (www.sanger.ac.uk/Software/). Three more genome comparison visualization tools (*Enteric, Menteric* and *Maj*), that allow comparative views of a reference genome (*E. coli* K-12 or O157: H7, *S. enterica* serovar Typhimurium LT2, or *S. enterica* serovar Typhi CT18) with sequences from several related bacteria, are available at http://globin.cse.psu.edu/enterix/ (Florea *et al.*, 2003). Many software packages for genome analysis are available at www.tigr.org/software/, including tools for genome annotation such as Glimmer (Delcher *et al.*, 1999) and comparison such as MUMmer (Delcher *et al.*, 1999). Websites co-ordinating genome analysis software, full genome information and partial genome information, such as ColiBase for the analysis of enteric bacteria (http://colibase.bham.ac.uk/), provide further assistance in the rapid interrogation of data from un-finished genome projects.

5.5 MICROARRAY TECHNOLOGY AS A TOOL FOR COMPARATIVE GENOMICS

The availability of whole genome sequences and the ability to attach DNA fragments to solid supports has led to the development of DNA microarray technology (Shalon *et al.*, 1996). Essentially relying on chemical interactions similar to the ones that facilitate both Northern and Southern blotting, DNA microarrays support both whole genome DNA comparisons (genomotyping) and entire cDNA transcriptome comparisons (expression profiling) (Ye *et al.*, 2001). DNA microarrays have been used successfully with *Salmonella* serovars to perform both expression profiling under different conditions and also genome comparisons (Detweiler *et al.*, 2001; Chan *et al.*, 2003; Eriksson *et al.*, 2003; Porwollik *et al.*, 2002). Custom microarrays are usually designed to display all or the majority of predicted coding sequences from a finished genome, either in the form of small oligonucleotides or as specific amplified PCR products. The DNA sequences serve as specific probes that capture complementary DNA strands during hybridization. The most common form of microarray-based analysis is a competitive two-color (green/red, Cy3/Cy5) reaction, whereby the reference DNA or cDNA is labeled in one color and experimental DNA or cDNA is labeled in another. The labeling tends to take the form of a dye incorporation reaction using the Klenow fragment of DNA polymerase I, whereby the Cy labeled nucleotide is incorporated into the synthesis of the complementary strand. Bioinformatic tools have been developed to aid both the handling and the analysis of the huge amount of data produced during a microarray experiment. Commercial tools such as Genespring (http://www.silicongenetics.com/) are widely available and user friendly. Free software for data analysis such as Cluster, treeview and SAM are also available (http://genome-www5.stanford.edu/). There are many factors that may affect the reproducibility and significance of a microarray experiment. Consistence in sample preparation is essential, especially for expression profiling, to ensure reproducible results across the dataset. Variables such as the choice of the reference DNA or cDNA and the mathematical normalization procedure used to analyze data all influence the final data profile. With genomes from several species of *Salmonella* becoming available in the public databases it is now possible to combine all of the genes from one *Salmonella* genome with additional genes from related genomes on one microarray, termed a non-redundant microarray, which can be updated as new sequence becomes available (Porwollik *et al.*, 2003).

5.6 SEQUENCED *SALMONELLA* GENOMES AS TOOLS FOR COMPARATIVE GENOMICS

There are currently over 2,400 recognized *Salmonella* serovars, so sequencing all of them would obviously be a lengthy and expensive task using current sequencing technology. *Salmonella* genome sequencing projects have therefore concentrated upon serovars that are either of importance to human disease or are representative of a particular branch of the *Salmonella* genus. Table 5.1 lists *Salmonella* sequence projects that have been completed or are in progress and the sequencing centers that are undertaking these projects. Up to date information about all of the *Salmonella* genome sequencing projects can be found at www.salmonella.org/genomics/ and the progress of specific projects can be monitored on the websites for each sequencing center (see Table 5.1).

S. enterica serovar Paratyphi A and *S. enterica* serovar Typhi isolates were sequenced because the serovars they represent normally cause systemic diseases that are restricted and adapted to humans and are a public health problem (Parry *et al.*, 2002). The two fully sequenced *S. enterica* serovar Typhi clinical isolates were chosen for quite different reasons. The CT18 strain was chosen mainly because it is a recent clinical isolate that harbors two plasmids, one of which confers multi-drug resistance (Parkhill *et al.*, 2001); the Ty2 strain was chosen (Deng *et al.*, 2003) because it is a commonly used laboratory isolate and has been the background strain for the production of the live attenuated *S. enterica* serovar Typhi Ty21a vaccine strain. Serovars that cause systemic diseases that are adapted to other animal hosts are also represented in the *Salmonella* genome sequence list. These include *S. enterica* serovar

Table 5.1. Salmonella *sequencing projects*

Species	Subspecies, Serovar, Strain	Sequencing centre	Website/contact	Status (2004)
S. bongori	12419	The Wellcome Trust Sanger Institute	www.sanger.ac.uk/Projects/Salmonella/	In Annotation
S. enterica	subspecies IIIa/ arizonae serotype 62: z 4, z 23:- strain RSK2980	Genome Sequencing Centre, University of Washington	www.genome.wustl.edu/	Shotgun completed
S. enterica	Subspecies IIIb/ diarizonae serotype 61:1,v:1,5,(7) strain 01–0005	Genome Sequencing Centre, University of Washington	www.genome.wustl.edu/	In Shotgun
Subspecies I				
S. enterica	Choleraesuis CGSC67	Chang Gung University Taoyuan, Taiwan	chchiu5adm.cgmh.org.tw	Finishing/gap closure
S. enterica	Dublin	University of Illinois	www.salmonella.org	In Shotgun
S. enterica	Enteritidis LK5	University of Illinois	www.salmonella.org	In Shotgun
S. enterica	Enteritidis PT4	The Wellcome Trust Sanger Institute	www.sanger.ac.uk/Projects/Salmonella/	Finishing/gap closure
S. enterica	Gallinarum 287/91	The Wellcome Trust Sanger Institute	www.sanger.ac.uk/Projects/Salmonella/	Finishing/gap closure

S. enterica	Paratyphi A ATCC9150	Genome Sequencing Centre, University of Washington	ww.genome.wustl.edu/	In Annotation
S. enterica	Paratyphi B SPB7	Genome Sequencing Centre, University of Washington	ww.genome.wustl.edu/	Shotgun completed
S. enterica	Paratyphi C	University of Calgary, Alberta, Canada	slliu5ucalgary.ca	In Shotgun
S. enterica	Pullorum	University of Illinois	www.salmonella.org	In Shotgun
S. enterica	Typhi CT18	The Wellcome Trust Sanger Institute	www.sanger.ac.uk/Projects/Salmonella/	Finished
S. enterica	Typhi Ty2	University of Wisconsin	www.genome.wisc.edu/	Finished
S. enterica	Typhimurium DT104	The Wellcome Trust Sanger Institute	www.sanger.ac.uk/Projects/Salmonella/	Finishing/gap closure
S. enterica	Typhimurium LT2	Genome Sequencing Centre, University of Washington	www.genome.wustl.edu/	Finished
S. enterica	Typhimurium SL1344	The Wellcome Trust Sanger Institute	www.sanger.ac.uk/Projects/Salmonella/	Finishing/gap closure

Dublin that is adapted to cattle, and *S. enterica* serovar Galinarum and serovar Pullorum that are adapted to chickens where they cause typhoid-like disease or dysentery respectively. *S. enterica* serovar Typhimurium and serovar Enteritidis are commonly associated with food borne gastroenteritis in humans and can also cause septicemia in immuno-compromised individuals. In addition, *S. enterica* serovar Typhimurium causes a systemic disease in mice, which is often used as a model for typhoid. The LT2 and SL1344 strains of *S. enterica* serovar Typhimurium have been widely used as laboratory strains. *S. bongori* 12419 provides a representative of the only other species that, together with *S. enterica*, makes up the *Salmonella* genus. Enteric bacteria other than *Salmonella* have also been sequenced, for instance strains of *Shigella* (Wei *et al.*, 2003; Jin *et al.*, 2002), and strains of non-pathogenic (Blattner *et al.*, 1997) and pathogenic (Hayashi *et al.*, 2001; Perna *et al.*, 2001) *Escherichia coli*, which provides additional sequence information against which to compare the *Salmonella* genomes.

5.7 *IN SILICO* ANALYSIS OF *SALMONELLA* GENOMES AND COMPARISONS BETWEEN GENOME SEQUENCES

5.7.1 General

Genome information can be used to gain insights into the evolution of the *Salmonella* genus, to identify stable regions conserved between different *Salmonella* species and serovars, and to identify regions that appear to be specific for individual serovars. Prior to the advent of full genome sequence information, whole genome hybridization studies indicated that *S. enterica* subspecies I serovars were 85–100% related to *S. enterica* serovar Typhimurium and that other subspecies were still 70–80% related to *S. enterica* serovar Typhimurium (Crosa *et al.*, 1973). *In silico* whole genome comparisons can now be used to assess globally and accurately the relatedness of different salmonellae. DNA comparisons between the genomes of *S. enterica* serovar Typhimurium LT2 and *E. coli* K12 demonstrate a median homology between predicted open reading frames of 80%. Similar comparisons between the genomes of *S. enterica* serovar Typhimurium LT2 and serovar Typhi CT18 show a median homology of 98% (McClelland *et al.*, 2001). Detailed *in silico* comparisons have been made between the fully sequenced non-pathogenic laboratory *E. coli* strain K12 (Blattner *et al.*, 1997), *S. enterica* serovar Typhimurium LT2 (McClelland *et al.*, 2001), *S. enterica* serovar Typhi CT18 (Parkhill *et al.*, 2001)) and *S. enterica* serovar Paratyphi A ATCC9150 (McClelland *et al.*, 2000; McClelland, Sanderson *et al.*, 2004). Genome comparisons have also been used specifically to identify tRNA-associated regions that vary between *E. coli* K12, *E. coli* O157: H7 (Perna *et al.*, 2001), *S. enterica* serovar Typhi CT18 and *S. enterica* serovar Typhimurium LT2, prior to further analysis using Southern blotting (Hansen-Wester and Hensel, 2002). Sequencing of the second *S. enterica* serovar Typhi strain, Ty2, enabled an inter-isolate comparison to be made (Deng *et al.*, 2003), which showed that *S. enterica* serovar Typhi strains are very similar in genetic content. These *in silico* comparative genomics studies, combined with analyses using methods such as Southern blotting, plasmid typing and subtractive hybridization, have revealed a number of interesting features of *Salmonella* genomes discussed in the following sections. One interesting observation arising from the completed annotated *Salmonella* genome sequences is the very large numbers of previously unidentified genes to which no predicted function can currently be assigned (hypothetical proteins) (Parkhill *et al.*, 2001; McClelland *et al.*, 2001). Some of these hypothetical proteins lie within defined regions involved in pathogenicity (pathogenicity islands) and some have orthologues in other bacterial genomes. These hypothetical proteins will no doubt be the subject of future phenotypic studies.

5.7.2 Large-scale genomic rearrangements

Salmonella diverged from *E. coli* over 100 million years ago (Ochman and Wilson, 1987). However, the two genera maintain a remarkable synteny between their genomes. The common genetic backbone shared by enteric bacteria is most likely maintained through the selective pressure of inhabiting a similar ecological niche. This conservation of gene order in enteric bacteria is all the more surprising when one considers that genomic duplications and inversions are frequent (10^{-3} to 10^{-5}) in *S. enterica* serovar Typhimurium and *E. coli* following laboratory culture (Haack and Roth, 1995; Hill and Harnish, 1981). These rearrangement types must be selected against in the wild, as they are rarely detected in wild type isolates in the field. However, more frequent large inversions have occurred in the CT18 and Ty2 strains of *S. enterica* serovar Typhi, as well as in the *E. coli* laboratory strain K12, compared with *S. enterica* serovar Typhimurium LT2. These inversions appear to have been mediated by recombination between IS elements (Alokam *et al.*, 2002). The different orientation of a region of the *S. enterica* serovar Typhimurium LT2 genome compared with *E. coli* K12 or *S. enterica* serovar

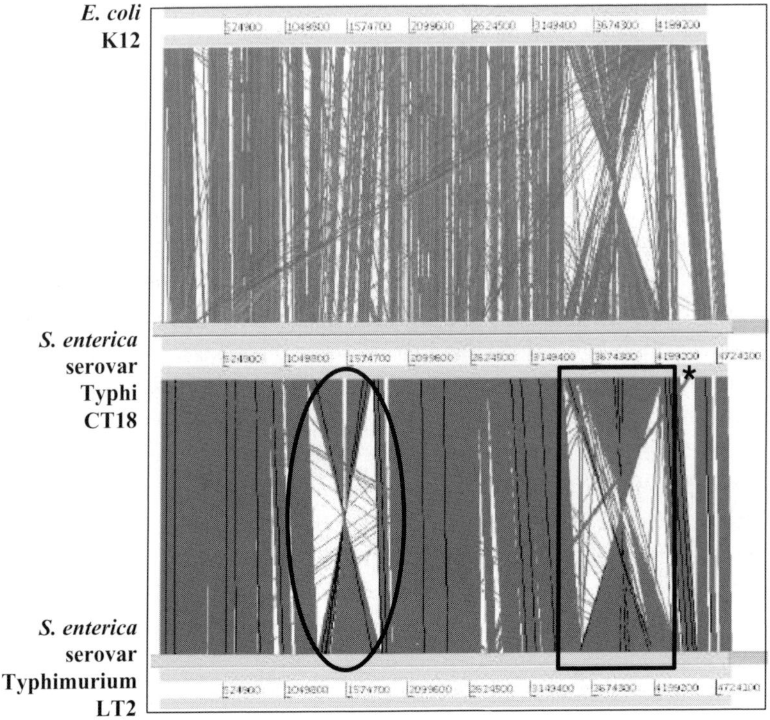

Figure 5.1. Image generated using the Artemis Comparison Tool, comparing the entire genomes of *E. coli* K12, *S. enterica* serovar Typhi CT18 and *S. enterica* serovar Typhimurium LT2. Lines between the genomes indicate regions of homology. Circled is a large inversion caused by recombination of IS200 elements (Alokam *et al.*, 2002). Boxed are large inversions due to recombination of rRNA genes in *S. enterica* serovar Typhi compared with *E. coli* K12 and *S. enterica* serovar Typhimurium LT2 (Liu and Sanderson, 1995). A star denotes the position of *Salmonella* pathogenicity island 7, which is present in *S. enterica* serovar Typhi, but not in *E. coli* K12 or *S. enterica* serovar Typhimurium LT2 (Pickard *et al.*, 2003).

Typhi CT18 can be seen in Figure 5.1 (circled region). The cross of lines, which indicate homology, in the ACT-generated genome comparison image shown in Figure 5.1 represents a region of DNA that has opposite orientation in two different genomes. A spectrum of large rearrangements around rRNA genes can be detected using I-CeuI restriction digestion and separation of DNA fragments by PFGE (Liu *et al.*, 1993). These rearrangements appear to be tolerated by wild-type isolates of *S. enterica* serovar Typhi, and other host-restricted salmonellae. However, such rearrangements are not frequently observed in isolates of *Salmonella* with broad host spectra, such as *S. enterica* serovar Typhimurium (Liu and Sanderson, 1995; Liu and Sanderson, 1998). The rRNA gene-mediated rearrangements in the *S. enterica* serovar Typhi CT18 genome compared with *S. enterica* serovar Typhimurium LT2 and *E. coli* K12 can also be visualized using ACT, as shown in Figure 5.1 (boxed region). At present there are no fully sequenced strains of *S. enterica* serovar Pullorum, although a project is underway (Table 5.1), but a physical genome map has been constructed for strain 4930. This map reveals major inversions of I-CeuI fragments relative to *S. enterica* serovar Typhimurium, which are suggested to have occurred as a means of stabilizing the genome in response to the insertion of a large (157kb) island (Liu *et al.*, 2002). It is possible that in other serovars major chromosomal inversions or rearrangements were also triggered by the need to rebalance the genome after large DNA insertion events. For example, the insertion of the 134kb large SPI-7 pathogenicity island into the genome of *S. enterica* serovar Typhi could have triggered the recombination events noted above (Parkhill *et al.*, 2001; Pickard *et al.*, 2003).

5.7.3 Pseudogenes

Pseudogenes are genes that appear to have acquired point mutations, deletions, insertions or frame shifts/stop codons, and consequently may be inactive or have altered functionality. There are a surprisingly large number of genes with the appearance of pseudogenes in *S. enterica* serovar Typhi (Parkhill *et al.*, 2001). Orthologues of some of these pseudogenes are known to contribute to the pathogenesis of other *Salmonella* serovars. For example, *S. enterica* serovar Typhi pseudogenes are found in a large number of fimbrial genes (Townsend *et al.*, 2001), in genes coding for type III secretion system effectors such as *sopD2* (Brumell *et al.*, 2003), *sopE2* (Bakshi *et al.*, 2000) and *sopA* (Wood *et al.*, 2000), in the gene encoding a putative *S. enterica* serovar Typhimurium host range factor *slrP* (Tsolis *et al.*, 1999) and in the *shdA* gene, involved in the shedding response to *S. enterica* serovar Typhimurium (Kingsley *et al.*, 2000). A number of genes involved in coenzyme B_{12} (cobalamin)-dependent utilization of 1,2-propanediol and ethanolamine (Roth *et al.*, 1996) are also pseudogenes in *S. enterica* serovar Typhi, which suggests that cobalamin-dependent metabolic pathways are not required for the lifecycle of this host-restricted pathogen. In contrast to *S. enterica* serovar Typhi, most bacteria harbor relatively few pseudogenes (Lawrence *et al.*, 2001), although there are exceptions such as *Mycobacterium leprae*, which is another host-restricted pathogen that has acquired numerous pseudogenes (Cole *et al.*, 2001). The acquisition of large numbers of pseudogenes has been suggested to correlate with a sheltered intracellular lifestyle, and possibly restricted host range (Lawrence *et al.*, 2001).

5.7.4 Salmonella pathogenicity islands

Salmonella pathogenicity islands (SPI) are large regions of DNA often associated with virulence traits and have general features suggesting that they have been horizontally transferred into the genome. The presence or absence of *S. enterica* serovar Typhimurium-associated pathogenicity islands in different *Salmonella* species or serovars has been assessed using microarrays (Porwollik *et al.*, 2002; Chan *et al.*, 2003). Prior to the availability of microarray technology, Southern blots or sequencing of genes from specific pathogenicity islands were relied upon to gain an insight into whether similar horizontally transferred regions had been acquired by different salmonellae. In many cases the initial detection of a SPI was through a specific phenotype, where mutations in these regions were found to lead to reduced pathogenic phenotypes and/or the genes from these regions were found to confer pathogenic phenotypes on otherwise non-pathogenic bacteria (Galan and Curtiss, 1989; Shea *et al.*, 1996). With the availability of genome sequences, potential horizontally transferred DNA can now be identified more readily and several SPIs have been designated and annotated even before phenotypic data has shown an actual link with pathogenicity. Ten potential SPIs have been identified in *S. enterica* serovar Typhi CT18 (Parkhill *et al.*, 2001), only some of which have been clearly implicated in pathogenicity. SPI-1 to 5 have been relatively well characterized in *S. enterica* serovar Typhimurium (Marcus *et al.*, 2000). SPI-1 and SPI-2 are the most studied, each of which codes for a type III secretion system (TTSS) that has been implicated in the infection process (Galan, 2001). Some of the SPIs, such as SPI-1 (Ochman and Groisman, 1996), are highly conserved across the *Salmonella* genus, whereas others, such as SPI-7 (Pickard *et al.*, 2003) appear to have been more recently acquired and are detected in only a few serovars. The majority of *Salmonella*, even within subspecies I, do not harbor SPI-7 (Pickard *et al.*, 2003) (Figure 5.2). Occasionally an *S. enterica* serovar Typhi isolate is detected that is missing SPI-7, such as SARB64 (marked with an arrow in Figure 5.2) (Boyd *et al.*, 2003; Pickard *et al.*, 2003), suggesting that this particular pathogenicity island is relatively unstable. SPI often have mosaic structures, which suggest that they have been built up through multiple insertion events. For instance, SPI-2 seems to have been acquired in two sections, one of which contains genes

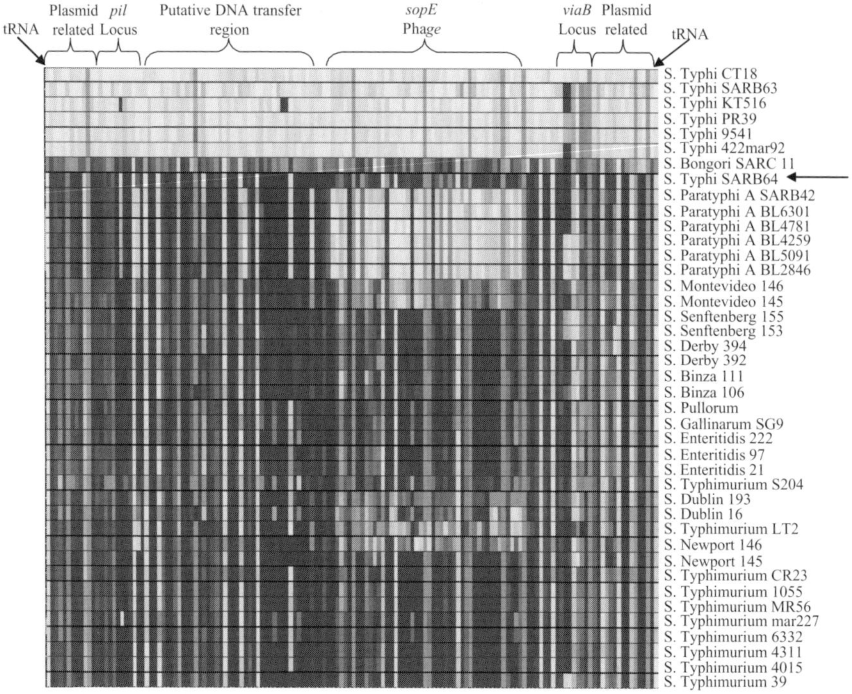

Figure 5.2. GeneSpring-generated image depicting *S. enterica* serovar Typhi CT18 microarray data for *Salmonella* pathogenicity island 7 (SPI-7) across 40 different *Salmonella* strains (Baker, Dougan *et al.*, unpublished data). Each column represents a specific gene within SPI-7 (STY4521 to STY4680). Each row is the result of challenging the microarray with a different *Salmonella* isolate (listed on right). Genes considered absent/divergent are shown in blue (see Figure 5.2 in color plate section), while genes that are present/conserved are shown in yellow. Orange regions are where hybridisation with test DNA is higher than reference DNA from *S. enterica* serovar Typhi CT18. The majority of the SPI-7 region is absent (blue) in salmonellae other than *S. enterica* serovar Typhi. The SopE phage is present (yellow) in isolates of both *S. enterica* serovar Typhi and serovar Paratyphi A. An unusual *S. enterica* serovar Typhi strain, SARB64, that is missing SPI-7 is marked with an arrow. (See colour plate section.)

that encode the TTSS and the other, at the 3′ end of the island, which harbors five *ttr* genes (involved in anaerobic tetrathionate reduction) plus seven genes encoding hypothetical proteins (Hensel *et al.*, 1999). The *ttr* region is common to *S. enterica* and *S. bongori*, so may have been acquired before the two *Salmonella* species diverged, whereas the larger TTSS-containing insertion appears to have been acquired later (Ochman and Groisman, 1996; Hensel *et al.*, 1999). Other examples of mosaic pathogenicity islands, such as SPI-7, are discussed in Chapter 6.

5.8 MOBILE GENETIC ELEMENTS: PLASMIDS AND BACTERIOPHAGES

5.8.1 General

In addition to the pathogenicity islands of *Salmonella*, major sites of variation between the genomes of *E. coli* and *Salmonella*, and between *Salmonella* serovars are found in the form of mobile genetic elements such as plasmids and prophages. Subtractive hybridization studies and genome sequence comparisons of DNA from *S. enterica* serovars Typhi and Typhimurium revealed that genes from F-type plasmids and lambdoid phages are major points of diversity between these two *Salmonella* serovars (Emmerth *et al.*, 1999).

5.8.2 Plasmids

Many salmonellae contain autonomously replicating plasmids, some of which can be transferred between bacteria by conjugation. Some of these plasmids are involved in virulence or harbor drug-resistance genes, while others are cryptic, with no ascribed function. Some *Salmonella* plasmids have been sequenced. These include the two *S. enterica* serovar Typhi CT18 plasmids pHCM1 and pHCM2 (Parkhill *et al.*, 2001) and the pSLT virulence plasmid of *S. enterica* serovar Typhimurium LT2 (McClelland *et al.*, 2001). In addition, the 50kb virulence plasmid pKDSC50 present in *S. enterica* serovar Choleraesuis has been fully sequenced (Haneda *et al.*, 2001) and genetic maps and partial sequence data are available for other plasmids such as the *S. enterica* serovar Enteritidis 60kb virulence plasmid (Rodriguez-Pena *et al.*, 1997).

F-type virulence plasmids of 50–100kb are present in many isolates of different *Salmonella* serovars, including *S. enterica* serovar Typhimurium, serovar Choleraesuis, serovar Enteritidis and serovar Dublin, whereas serovar Typhi does not harbor this type of virulence plasmid. Most plasmids are reported to be mobilizable, but only some have retained their conjugative ability. For instance the 90kb plasmid of *S. enterica* serovar Typhimurium strain LT2 is self-transmissible, whereas the plasmid of strain SL1344 is not (Ahmer *et al.*, 1999). Heteroduplex analysis using DNA probes from a plasmid of *S. enterica* serovar Choleraesuis indicates that the majority of virulence plasmids are closely related (Montenegro *et al.*, 1991). The virulence plasmids have a common ~8kb DNA region encoding the *spv* (*Salmonella* plasmid virulence) genes (Gulig *et al.*, 1993), which in subspecies II, IIIa and VII are located on the chromosome (Boyd and Hartl, 1998). SpvB can ADP-ribosylate actin causing destabilization of the cytoskeleton of eukaryotic cells (Lesnick *et al.*, 2001), which may account for the *spv*-dependent killing of human macrophages and epithelial cell lines by *S. enterica* serovar Typhimurium (Libby *et al.*, 2000; Paesold *et al.*, 2002). These plasmids are required for the expression of a full virulence phenotype in vivo. For instance the persistence of *S. enterica* serovar Enteritidis in chicks is reduced in strains cured of the virulence plasmid (Virlogeux-Payant *et al.*, 2003). Persistence and enhanced virulence can be attributed to the *spv* genes. This is indicated by the fact that the ~8kb region containing all of the *spv* genes can restore virulence of plasmid-cured *S. enterica* serovar Typhimurium in mice (Gulig *et al.*, 1993). Furthermore, *spvBC* are sufficient to confer full virulence in mice to *S. enterica* serovar Typhimurium lacking the rest of the plasmid (Matsui *et al.*, 2001) and a *spvR* mutant of *S. enterica* serovar is highly attenuated in calves (Libby *et al.*, 1997). F-plasmid genes include also the full *pef* fimbrial operon (Baumler *et al.*, 1996) and *rck* complement resistance gene (Heffernan *et al.*, 1992). Genes related to the *E. coli* K88 fimbrial genes *faeH* and *faeI* can influence virulence of *S. enterica* serovar Galinarum, and are also found in virulence plasmids of *S. enterica* serovar Pullorum and serovar Dublin, but not in those of *S. enterica* serovar Typhimurium, serovar Enteriditids or serovar Choleraesuis (Rychlik *et al.*, 1998).

Drug resistance in *Salmonella* and the plasmids that often carry resistance genes are discussed in detail in Chapter 2. Some of these drug-resistance plasmids have been sequenced, such as the *Salmonella* prototype 180kb H incompatibility group IncHI1 plasmid R27 (Sherburne *et al.*, 2000) and the ~200 kb IncHI1 plasmid pHCM1 isolated from CT18 (Wain *et al.*, 2003). An unusual self-transferable virulence plasmid (pUO-*St*VR2) isolated from a multidrug-resistant *S. enterica* serovar Typhimurium strain carries the usual *spv* gene complement as well as drug-resistance genes (Guerra *et al.*, 2002). *S. enterica* serovar Typhi CT18 harbors a second plasmid of 106kb, pHCM2, whose function is currently unknown (Kidgell *et al.*, 2002). Cryptic plasmids have been detected in other serovars, including the small ~3kb *S. enterica* serovar Typhimurium plasmid pIMVS1, which was suggested to correlate with pathogenicity in humans, but did not carry any obvious genes that could account for this phenotype (Astill *et al.*, 1993). Small ColE1-type plasmids of 3–5.6kb have been found in *S. enterica* serovar Enteritidis and one of these plasmids (pC) carries an active restriction modification system, which could explain the high resistance of pC-carrying *S. enterica* serovar Enteritidis strains to phage infection (Gregorova *et al.*, 2002).

5.8.3 Bacteriophages

Bacteriophages that mediate the transduction of genetic material are readily isolated from the intestine and the environment. Many bacteriophages are capable of lysogeny, whereby they can integrate into the chromosome of their host. These events can be significant if the bacteriophage carry additional genes (cargo genes) that are not essential for phage proliferation (Boyd and Brussow, 2002). The most significant manifestation of this is "lysogenic conversion," whereby a non-pathogenic bacterial strain is converted into a pathogenic strain via the addition of pathogenic determinants carried by the bacteriophage (Canchaya *et al.*, 2003). One of the most surprising revelations of the recent *Salmonella* genome sequencing projects was the relatively high proportion of bacteriophage genes within the chromosome, thus suggesting a vital role for bacteriophages in the evolution of *Salmonella* species (Parkhill *et al.*, 2001; McClelland *et al.*, 2001). *Salmonella* prophage analysis has focused on the sequenced strains of *S. enterica* serovar Typhi CT18 and *S. enterica* serovar Typhimurium LT2. Both these genomes contain bacteriophage DNA mainly in two distinct forms. The first form consists of small clusters of less than five genes, which often include transposases and are evidence of a previous bacteriophage excision event in an ancestor bacterium. The second form of bacteriophage DNA consists either of larger gene clusters thought to encode full-length prophage, or of large phage remnants. In the case of S. *enterica* serovar Typhi CT18 these gene clusters represent approximately 4% of the genome (Parkhill *et al.*, 2001). The genome sequence of *S. enterica* serovar Typhimurium LT2 contains six regions that represent sizable prophage or prophage remnants. The two lambdoid-like prophages Gifsy 1 and Gifsy 2 affect the virulence of their bacterial hosts. In fact, the curing of Gifsy-2 reduces the virulence of *S. enterica* serovar Typhimurium for mice by up to 100 fold (Figueroa-Bossi, Bossi, 1999; Figueroa-Bossi *et al.*, 2001). Non-phage genes carried within Gifsy phages include *sodCI,* encoding a superoxide dismutase, linked to resistance to the macrophage oxidative burst (Uzzau *et al.*, 2002). Gifsy-1 encodes *gipA,* implicated in bacterial colonization of the small intestine, and *sseI,* which is a type III secreted effector protein associated with SPI-2 (Figueroa-Bossi *et al.*, 2001). The *S. enterica* serovar Typhi CT18 sequence harbors seven phage associated regions that exhibit strong similarities to the P2, P4, Lambda and Mu families (Thomson *et al.*, 2004). The most intriguing of these sequences is the *sopE* phage, which shares significant homology with bacteriophage P2 and carries the *sopE* gene. The *sopE* gene is located within a hyper-variable cassette of the tail fiber genes on the phage genome (Mirold *et al.*, 2001). Further analysis of this region suggests that this particular site has a role similar to that of the cos site and has the ability to carry a variety of genes that are unnecessary for phage proliferation (Thomson *et al.*, 2004). A recent microarray study comparing the phage content of *S. enterica* serovar Typhi CT18 with other serovar Typhi isolates and members of other serovars indicates great diversity of bacteriophage genes within the *Salmonella* genomes. Some of the prophage regions were specific to strain CT18, suggesting inter-serovar variation and more subtle intra-serovar variation within individual serovar Typhi strains (Thomson *et al.*, 2004). From the currently sequenced *Salmonella* genomes it is clear that the number and repertoire of prophages represents a significant source of genetic variation, and may have played a crucial role in evolution and specialization.

5.9 FIMBRIAL AND PILUS GENES ARE HIGHLY VARIABLE BETWEEN *SALMONELLA* GENOMES

Fimbriae and pili extend from the surface of *Salmonella*, where they can interact with host cells or other bacteria (Humphries *et al.*, 2001). The repertoire of fimbrial operons present in different *Salmonella* is highly variable (Baumler *et al.*, 1997; Townsend *et al.*, 2001). Subtractive hybridization showed that *S. enterica* serovar Typhi does not possess the *stf* (*S.* enterica serovar Typhimurium fimbriae) fimbrial operon that is present in other serovars, including *S. enterica* serovar Typhimurium (Emmerth *et al.*, 1999). Genome sequence data show that many fimbrial operons present in *S. enterica* serovar Typhimurium are entirely missing or inactivated by pseudogenes in *S. enterica* serovar Typhi (Townsend *et al.*, 2001). A number of fimbrial or pilus operons are associated with mobile genetic elements or horizontally transferred pathogenicity islands. These include the *pef* operon carried by F-type virulence plasmids (Baumler *et al.*, 1997), *tcf* (present in *S. enterica* serovar Typhi and not in serovar Typhimurium) and *saf* in the SPI-6 region (Folkesson *et al.*, 1999), *sef* in the SPI-10 region of some *S. enterica* subspecies I serovars (Clouthier *et al.*, 1993; Baumler *et al.*, 1997) and the type IV pilus in the SPI-7 region of *S. enterica* serovar Typhi (Pickard *et al.*, 2003). These genes appear to be recent acquisitions, being present only in a limited number of *S. enterica* subspecies I serovars. The *lpf* (long polar fimbriae) and *fim* (prototype *S. enterica* serovar Typhimurium mannose-sensitive fimbriae – then termed type 1) fimbrial operons may have entered the *Salmonella* genome earlier, as they are present in *S. bongori* as well as *S. enterica* (Swenson *et al.*, 1991; Baumler *et al.*, 1997). The *lpf* operon appears to have been subsequently lost by a number of *Salmonella* serovars including serovar Typhi (Baumler *et al.*, 1997). The *agf* operon (aggregative fimbriae, also known as mannose-insensitive or thin curled fimbriae) codes for nucleating fimbriae, rather than chaperone-usher fimbriae, and is the most ancient *Salmonella* fimbrial operon, common to some *E. coli* strains and all *Salmonella* (Doran *et al.*, 1993; Baumler *et al.*, 1997).

5.10 ANALYSIS OF *SALMONELLA* GENOMES BASED ON MICROARRAY TECHNOLOGY

The advent of genome sequencing revealed the full extent of the genetic diversity between different samonellae. This facilitated the construction of *Salmonella* genome-based DNA microarrays that underpin the study of total chromosomal genetic diversity without the necessity for a genome sequence. Microarrays have been used successfully to establish the range of genetic diversity within *Salmonella* populations, have aided the knowledge of the relationship between serovars, and importantly have added insights into the conundrum of *Salmonella* host adaptation. Although this powerful technique provides a large amount of quantitative data about the presence or absence of genes, a more precise picture about strain variation will still require complete genome sequences. Current microarray comparison studies are centered on using the *S. enterica* serovar Typhimurium LT2 and *S. enterica* serovar Typhi CT18 genomes on separate or combined arrays. Initial studies have focused on phylogenetic tree construction based on homologue content relative to the known genome sequence (Chan *et al.*, 2003; Porwollik *et al.*, 2002). These studies have provided novel information about gene content and have shown that in the course of evolution of the serovars there has been a high level of gene gain, loss and also rapid divergence. The data sets produced allow clustering of the serovars by gene content, as well as revealing the presence or absence of horizontally transferred DNA, including genomic islands and bacteriophages. Variation between isolates of the same serovar has also been observed, including differences between *S. enterica* serovar Typhimurium LT2 and a monophasic multidrug-resistant *S. enterica* serovar Typhimurium lacking the *fljB* locus (Garaizar *et al.*, 2002). Even variations within *S. enterica* serovar Typhi which is widely considered to be a "clonal" organism have been revealed using microarrays (Boyd *et al.*, 2003). These differences tend to be in specific regions of the genome associated with small gene islets, in pathogenicity islands and in bacteriophages and demonstrate that the chromosome of *S. enterica* serovar Typhi may have certain regions of plasticity that act as an unstable genomic resource. In summary, microarrays will not abolish the need for genome sequencing or true phylogenic studies, but they are a powerful aid in studying gene content and population organization.

5.11 GENOME SEQUENCES FACILITATE FUNCTIONAL GENOMICS

Genome information can be used for the study of bacterial pathogenicity. Transcriptome analyses with microarrays take a snap-shot of the full bacterial mRNA expression profile under particular conditions. This technique can facilitate the rapid identification of bacterial genes whose transcription is increased or decreased under particular circumstances, including exposure to host immune defenses or conditions of environmental stress (Conway and Schoolnik, 2003). For instance, gene expression responses of *S. enterica* serovar Typhimurium to intracellular survival inside a mouse macrophage cell line have been studied (Eriksson *et al.*, 2003). In vitro responses of *S. enterica* serovar Typhimurium to low-shear stress, as a model for low gravity, have also been assessed using microarrays (Wilson *et al.*, 2002). Comparisons made between genes expressed by a wild type *S. enterica* serovar Typhimurium strain with those expressed by an isogenic *csrA* mutant have enabled the identification of novel genes under the control of this global regulator (Lawhon *et al.*, 2003).

Genome information also facilitates directed mutagenesis, by aiding the rational design of mutagenesis strategies and removing the need to first clone and sequence the gene to be mutated. Specific primers to target each predicted open reading frame in the genomes of *S. enterica* serovar Typhi CT18, *S. enterica* serovar Typhimurium LT2 and *E. coli* strains, have been systematically identified for the purpose of mutagenesis studies (http://falkow.stanford.edu/whatwedo/wanner/). These primers are designed to be combined with the Red Recombinase mutagenesis system, which allows PCR products to be recombined into the genome requiring only short (at least 35bp) regions of homology with the gene to be mutated (Datsenko and Wanner, 2000). Using this system, genes of interest identified through genome comparisons can be rapidly mutated and subjected to functional analysis.

Microarrays can be used in combination with signature-tagged mutagenesis to facilitate rapid random screening in a system sometimes termed TRASH (transposon site hybridization). Basically, a pool of bacteria containing mutations inserted by signature-tagged transposon mutagenesis techniques are put through an in vitro or in vivo screen and DNA is isolated from bacteria before and after the screen. The DNA is digested and ligated to adaptors, regions around the transposons are amplified, labeled (DNA from before the screen being labeled with a different dye to that from after the screen) and hybridized to a microarray containing all of the genes from the test bacterium. Genes that have been lost from the test pool compared to the control pool are liable to be important for bacterial survival under the assay conditions chosen. This type of system for mutagenesis and screening has been successfully used for the identification of genes essential to growth under different culture conditions with *E. coli* (Badarinarayana *et al.*, 2001) and *Mycobacterium bovis* (Sassetti *et al.*, 2001). The TRASH system has the major advantage, over classic signature-tagged mutagenesis, of avoiding the arduous task of isolating and identifying each mutant in the mutagenized pool. The clearest disadvantage of this system compared with signature-tagged mutagenesis is that at the end of the experiment the individual mutant clones are not available for additional analysis.

Host responses to *Salmonella* infection are also being studied using microarrays. For example, one study has looked at the effects of infection with wild type *S. enterica* serovar Typhimurium compared with a *phoP* mutant on gene expression by a human macrophage cell line (Detweiler *et al.*, 2001). Another study has looked at the role of flagellin in the responses of an intestinal epithelial cell line to different *Salmonella* (Zeng *et al.*, 2003). Future aims include combining host cell and bacterial mRNA isolations in a single experiment with microarray analysis for each organism to gain simultaneous host and bacterial response information. The complete sequencing of additional host genomes will make these types of analyses more feasible and will enable us to gain a detailed global insight into the interplay between host and pathogen. It should not be forgotten that proteomics provides an important complement to genomic and transcriptional analysis, and the study of *Salmonella* proteomics will certainly be aided by the presence of genome information (Jungblut, 2001).

5.12 CONCLUSIONS

Genome sequencing has enhanced studies in the evolution of the *Salmonella* genus, and will facilitate the development of diagnostic and surveillance tools. Genome information will also facilitate more rapid, higher throughput functional screens for the identification of genes involved in bacterial pathogenesis. Genome sequence information is moving *Salmonella* research into an era where whole genome-based genetic data or expression data can be generated relatively rapidly and easily. New challenges now lie in making the most effective use of these capabilities.

5.13 ACKNOWLEDGEMENTS

Recent work on *Salmonella* genetics in G. Dougan's laboratory has been supported by The Wellcome Trust and the BBSRC.

REFERENCES

Agron, P. G., Macht, M., Radnedge, L. *et al.* (2002). Use of subtractive hybridization for comprehensive surveys of prokaryotic genome differences. *FEMS Microbiol Lett,* **211**, 175–82.

Agron, P. G., Walker, R. L., Kinde, H. *et al.* (2001). Identification by subtractive hybridization of sequences specific for *Salmonella enterica* serovar Enteritidis. *Appl Environ Microbiol,* **67**, 4984–91.

Ahmer, B. M., Tran, M. and Heffron, F. (1999). The virulence plasmid of *Salmonella typhimurium* is self-transmissible. *J Bacteriol,* **181**, 1364–8.

Alokam, S., Liu, S. L., Said, K. and Sanderson, K. E. (2002). Inversions over the terminus region in *Salmonella* and *Escherichia coli*: IS200s as the sites of homologous recombination inverting the chromosome of *Salmonella enterica* serovar Typhi. *J Bacteriol,* **184**, 6190–7.

Astill, D. S., Manning, P. A. and Heuzenroeder, M. W. (1993). Characterization of the small cryptic plasmid, pIMVS1, of *Salmonella enterica* ser. Typhimurium. *Plasmid,* **30**, 258–67.

Badarinarayana, V., Estep, P. W., III, Shendure, J. *et al.* (2001). Selection analyses of insertional mutants using subgenic-resolution arrays. *Nat Biotechnol,* **19**, 1060–5.

Bakshi, C. S., Singh, V. P., Wood, M. W. *et al.* (2000). Identification of SopE2, a *Salmonella* secreted protein which is highly homologous to SopE and involved in bacterial invasion of epithelial cells. *J Bacteriol,* **182**, 2341–4.

Baumler, A. J., Gilde, A. J., Tsolis, R. M. *et al.* (1997). Contribution of horizontal gene transfer and deletion events to development of distinctive patterns of fimbrial operons during evolution of *Salmonella* serotypes. *J Bacteriol,* **179**, 317–22.

Baumler, A. J., Tsolis, R. M., Bowe, F. A. *et al.* (1996). The *pef* fimbrial operon of *Salmonella typhimurium* mediates adhesion to murine small intestine and is necessary for fluid accumulation in the infant mouse. *Infect Immun,* **64**, 61–8.

Beltran, P., Musser, J. M., Helmuth, R. *et al.* (1988). Toward a population genetic analysis of *Salmonella*: genetic diversity and relationships among strains of serotypes *S. choleraesuis, S. derby, S. dublin, S. enteritidis, S. heidelberg, S. infantis, S. newport,* and *S. typhimurium. Proc Natl Acad Sci USA,* **85**, 7753–7.

Blattner, F. R., Plunkett, G., III, Bloch, C. A. *et al.* (1997). The complete genome sequence of *Escherichia coli* K-12. *Science*, **277**, 1453–74.

Boyd, E. F. and Brussow, H. (2002). Common themes among bacteriophage-encoded virulence factors and diversity among the bacteriophages involved. *Trends Microbiol*, **10**, 521–9.

Boyd, E. F. and Hartl, D. L. (1998). *Salmonella* virulence plasmid. Modular acquisition of the *spv* virulence region by an F-plasmid in *Salmonella enterica* subspecies I and insertion into the chromosome of subspecies II, IIIa, IV and VII isolates. *Genetics*, **149**, 1183–90.

Boyd, E. F., Porwollik, S., Blackmer, F. and McClelland, M. (2003). Differences in gene content among *Salmonella enterica* serovar Typhi isolates. *J Clin Microbiol*, **41**, 3823–8.

Boyd, E. F., Wang, F. S., Whittam, T. S. and Selander, R. K. (1996). Molecular genetic relationships of the salmonellae. *Appl Environ Microbiol*, **62**, 804–8.

Brumell, J. H., Kujat-Choy, S., Brown, N. F. *et al.* (2003). SopD2 is a novel type III secreted effector of *Salmonella typhimurium* that targets late endocytic compartments upon delivery into host cells. *Traffic*, **4**, 36–48.

Canchaya, C., Fournous, G., Chibani-Chennoufi, S., Dillmann, M. L. and Brussow, H. (2003). Phage as agents of lateral gene transfer. *Curr Opin Microbiol*, **6**, 417–24.

Chan, K., Baker, S., Kim, C. C. *et al.* (2003). Genomic comparison of *Salmonella enterica* serovars and *Salmonella bongori* by use of a *S. enterica* serovar *Typhimurium* DNA microarray. *J Bacteriol*, **185**, 553–63.

Clouthier, S. C., Muller, K. H., Doran, J. L., Collinson, S. K. and Kay, W. W. (1993). Characterization of three fimbrial genes, *sefABC*, of *Salmonella enteritidis*. *J Bacteriol*, **175**, 2523–33.

Cole, S. T., Eiglmeier, K., Parkhill, J. *et al.* (2001). Massive gene decay in the leprosy bacillus. *Nature*, **409**, 1007–11.

Conway, T. and Schoolnik, G. K. (2003). Microarray expression profiling: capturing a genome-wide portrait of the transcriptome. *Mol Microbiol*, **47**, 879–89.

Crosa, J. H., Brenner, D. J., Ewing, W. H. and Falkow, S. (1973). Molecular relationships among the salmonelleae. *J Bacteriol*, **115**, 307–15.

Datsenko, K. A. and Wanner, B. L. (2000). One-step inactivation of chromosomal genes in *Escherichia coli* K-12 using PCR products. *Proc Natl Acad Sci USA*, **97**, 6640–5.

Delcher, A. L., Harmon, D., Kasif, S., White, O. and Salzberg, S. L. (1999a). Improved microbial gene identification with GLIMMER. *Nucleic Acids Res*, **27**, 4636–41.

Delcher, A. L., Kasif, S., Fleischmann, R. D. *et al.* (1999b). Alignment of whole genomes. *Nucleic Acids Res*, **27**, 2369–76.

Deng, W., Liou, S. R., Plunkett, G., III *et al.* (2003). Comparative genomics of *Salmonella enterica* serovar Typhi strains Ty2 and CT18. *J Bacteriol*, **185**, 2330–7.

Detweiler, C. S., Cunanan, D. B. and Falkow, S. (2001). Host microarray analysis reveals a role for the *Salmonella* response regulator *phoP* in human macrophage cell death. *Proc Natl Acad Sci USA*, **98**, 5850–5.

Doran, J. L., Collinson, S. K., Burian, J. *et al.* (1993). DNA-based diagnostic tests for *Salmonella* species targeting *agfA*, the structural gene for thin, aggregative fimbriae. *J Clin Microbiol*, **31**, 2263–73.

Duguid, J. P., Anderson, E. S., Alfredsson, G. A., Barker, R. and Old, D. C. (1975). A new biotyping scheme for *Salmonella typhimurium* and its phylogenetic significance. *J Med Microbiol*, **8**, 149–66.

Emmerth, M., Goebel, W., Miller, S. I. and Hueck, C. J. (1999). Genomic subtraction identifies *Salmonella typhimurium* prophages, F-related plasmid sequences, and a novel fimbrial operon, *stf*, which are absent in *Salmonella typhi*. *J Bacteriol*, **181**, 5652–61.

Eriksson, S., Lucchini, S., Thompson, A., Rhen, M. and Hinton, J. C. (2003). Unravelling the biology of macrophage infection by gene expression profiling of intracellular *Salmonella enterica*. *Mol Microbiol*, **47**, 103–18.

Felix, A. (1956). Phage typing of *Salmonella typhimurium*: its place in epidemiological and epizootiological investigations. *J Gen Microbiol*, **14**, 208–22.

Figueroa-Bossi, N. and Bossi, L. (1999). Inducible prophages contribute to *Salmonella* virulence in mice. *Mol Microbiol*, **33**, 167–76.

Figueroa-Bossi, N., Uzzau, S., Maloriol, D. and Bossi, L. (2001). Variable assortment of prophages provides a transferable repertoire of pathogenic determinants in *Salmonella*. *Mol Microbiol*, **39**, 260–71.

Florea, L., McClelland, M., Riemer, C., Schwartz, S. and Miller, W. (2003). EnteriX 2003: visualization tools for genome alignments of *Enterobacteriaceae*. *Nucleic Acids Res*, **31**, 3527–32.

Folkesson, A., Advani, A., Sukupolvi, S. *et al.* (1999). Multiple insertions of fimbrial operons correlate with the evolution of *Salmonella* serovars responsible for human disease. *Mol Microbiol*, **33**, 612–22.

Franco, A., Gonzalez, C., Levine, O. S. *et al.* (1992). Further consideration of the clonal nature of *Salmonella typhi*: evaluation of molecular and clinical characteristics of strains from Indonesia and Peru. *J Clin Microbiol*, **30**, 2187–90.

Galan, J. E. (2001). *Salmonella* interactions with host cells: type III secretion at work. *Annu Rev Cell Dev Biol*, **17**, 53–86.

Galan, J. E. and Curtiss, R. (1989). Cloning and molecular characterization of genes whose products allow *Salmonella typhimurium* to penetrate tissue culture cells. *Proc Natl Acad Sci USA*, **86**, 6383–7.

Garaizar, J., Porwollik, S., Echeita, A. *et al.* (2002). DNA microarray-based typing of an atypical monophasic *Salmonella enterica* serovar. *J Clin Microbiol*, **40**, 2074–8.

Gregorova, D., Pravcova, M., Karpiskova, R. and Rychlik, I. (2002). Plasmid pC present in *Salmonella enterica* serovar Enteritidis PT14b strains encodes a restriction modification system. *FEMS Microbiol Lett*, **214**, 195–8.

Guerra, B., Soto, S., Helmuth, R. and Mendoza, M. C. (2002). Characterization of a self-transferable plasmid from *Salmonella enterica* serotype *Typhimurium* clinical isolates carrying two integron-borne gene cassettes together with virulence and drug resistance genes. *Antimicrob Agents Chemother*, **46**, 2977–81.

Gulig, P. A., Danbara, H., Guiney, D. G. *et al.* (1993). Molecular analysis of *spv* virulence genes of the *Salmonella* virulence plasmids. *Mol Microbiol*, **7**, 825–30.

Haack, K. R. and Roth, J. R. (1995). Recombination between chromosomal IS200 elements supports frequent duplication formation in *Salmonella typhimurium*. *Genetics*, **141**, 1245–52.

Haneda, T., Okada, N., Nakazawa, N., Kawakami, T. and Danbara, H. (2001). Complete DNA sequence and comparative analysis of the 50-kilobase virulence plasmid of *Salmonella enterica* serovar Choleraesuis. *Infect Immun*, **69**, 2612–20.

Hansen-Wester, I. and Hensel, M. (2002). Genome-based identification of chromosomal regions specific for *Salmonella* spp. *Infect Immun*, **70**, 2351–60.

Hayashi, T., Makino, K., Ohnishi, M. *et al.* (2001). Complete genome sequence of enterohemorrhagic *Escherichia coli* O157: H7 and genomic comparison with a laboratory strain K-12. *DNA Res*, **8**, 11–22.

Heffernan, E. J., Harwood, J., Fierer, J. and Guiney, D. (1992). The *Salmonella typhimurium* virulence plasmid complement resistance gene *rck* is homologous to a family of virulence-related outer membrane protein genes, including *pagC* and *ail*. *J Bacteriol*, **174**, 84–91.

Hensel, M., Nikolaus, T. and Egelseer, C. (1999). Molecular and functional analysis indicates a mosaic structure of *Salmonella* pathogenicity island 2. *Mol Microbiol*, **31**, 489–98.

Hill, C. W. and Harnish, B. W. (1981). Inversions between ribosomal RNA genes of *Escherichia coli*. *Proc Natl Acad Sci USA*, **78**, 7069–72.

Humphries, A. D., Townsend, S. M., Kingsley, R. A. *et al.* (2001). Role of fimbriae as antigens and intestinal colonization factors of *Salmonella* serovars. *FEMS Microbiol Lett*, **201**, 121–5.

Jin, Q., Yuan, Z., Xu, J. *et al.* (2002). Genome sequence of *Shigella flexneri* 2a: insights into pathogenicity through comparison with genomes of *Escherichia coli* K12 and O157. *Nucleic Acids Res*, **30**, 4432–41.

Jungblut, P. R. (2001). Proteome analysis of bacterial pathogens. *Microbes Infect*, **3**, 831–40.

Kauffmann, F. (1957). [The Kauffmann-White schema; diagnostic *Salmonella* antigen schema.] *Ergeb Mikrobiol Immunitatsforsch Exp Ther*, **30**, 160–216.

Kidgell, C., Pickard, D., Wain, J. *et al.* (2002). Characterisation and distribution of a cryptic *Salmonella typhi* plasmid pHCM2. *Plasmid*, **47**, 159–71.

Kingsley, R. A., van Amsterdam, K., Kramer, N. and Baumler, A. J. (2000). The *shdA* gene is restricted to serotypes of *Salmonella enterica* subspecies I and contributes to efficient and prolonged fecal shedding. *Infect Immun*, **68**, 2720–7.

Koay, A. S., Jegathesan, M., Rohani, M. Y. and Cheong, Y. M. (1997). Pulsed-field gel electrophoresis as an epidemiologic tool in the investigation of laboratory acquired *Salmonella typhi* infection. *Southeast Asian J Trop Med Public Health*, **28**, 82–4.

Lawhon, S. D., Frye, J. G., Suyemoto, M. *et al.* (2003). Global regulation by CsrA in *Salmonella typhimurium*. *Mol Microbiol*, **48**, 1633–45.

Lawrence, J. G., Hendrix, R. W. and Casjens, S. (2001). Where are the pseudogenes in bacterial genomes? *Trends Microbiol*, **9**, 535–40.

Lesnick, M. L., Reiner, N. E., Fierer, J. and Guiney, D. G. (2001). The *Salmonella spvB* virulence gene encodes an enzyme that ADP-ribosylates actin and destabilizes the cytoskeleton of eukaryotic cells. *Mol Microbiol*, **39**, 1464–70.

Libby, S. J., Adams, L. G., Ficht, T. A. *et al.* (1997). The *spv* genes on the *Salmonella dublin* virulence plasmid are required for severe enteritis and systemic infection in the natural host. *Infect Immun*, **65**, 1786–92.

Libby, S. J., Lesnick, M., Hasegawa, P., Weidenhammer, E. and Guiney, D. G. (2000). The *Salmonella* virulence plasmid *spv* genes are required for cytopathology in human monocyte-derived macrophages. *Cell Microbiol*, **2**, 49–58.

Liu, G. R., Rahn, A., Liu, W. Q., Sanderson, K. E., Johnston, R. N. and Liu, S. L. (2002). The evolving genome of *Salmonella enterica* serovar *Pullorum*. *J Bacteriol*, **184**, 2626–33.

Liu, S. L. and Sanderson, K. E. (1995). Rearrangements in the genome of the bacterium *Salmonella typhi*. *Proc Natl Acad Sci USA*, **92**, 1018–22.

(1998). Homologous recombination between rrn operons rearranges the chromosome in host-specialized species of *Salmonella*. *FEMS Microbiol Lett*, **164**, 275–81.

Liu, S. L., Hessel, A. and Sanderson, K. E. (1993). The XbaI-BlnI-CeuI genomic cleavage map of *Salmonella typhimurium* LT2 determined by double digestion, end labelling, and pulsed-field gel electrophoresis. *J Bacteriol*, **175**, 4104–20.

Marcus, S. L., Brumell, J. H., Pfeifer, C. G. and Finlay, B. B. (2000). *Salmonella* pathogenicity islands: big virulence in small packages. *Microbes Infect*, **2**, 145–56.

Matsui, H., Bacot, C. M., Garlington, W. A. *et al.* (2001). Virulence plasmid-borne *spvB* and *spvC* genes can replace the 90-kilobase plasmid in conferring virulence to *Salmonella enterica* serovar Typhimurium in subcutaneously inoculated mice. *J Bacteriol*, **183**, 4652–8.

McClelland, M., Florea, L., Sanderson, K. *et al.* (2000). Comparison of the *Escherichia coli* K-12 genome with sampled genomes of a *Klebsiella pneumoniae* and three *Salmonella enterica* serovars, Typhimurium, Typhi and Paratyphi. *Nucleic Acids Res*, **28**, 4974–86.

McClelland, M., Sanderson, K. E., Spieth, J. *et al.* (2001). Complete genome sequence of *Salmonella enterica* serovar Typhimurium LT2. *Nature*, **413**, 852–6.

McClelland, M., Sanderson, K. and Clifton, S. W. (2004). Comparison of genome degradation in Paratyphi A and Typhi, human-restricted serovars of *Salmonella enterica* that cause typhoid. *Nat. Genet.*, **36**, 1268–74.

Millemann, Y., Gaubert, S., Remy, D. and Colmin, C. (2000). Evaluation of IS200-PCR and comparison with other molecular markers to trace *Salmonella enterica* subsp. enterica serotype typhimurium bovine isolates from farm to meat. *J Clin Microbiol*, **38**, 2204–9.

Mirold, S., Rabsch, W., Tschape, H. and Hardt, W. D. (2001). Transfer of the *Salmonella* type III effector sopE between unrelated phage families. *J Mol Biol*, **312**, 7–16.

Montenegro, M. A., Morelli, G. and Helmuth, R. (1991). Heteroduplex analysis of *Salmonella* virulence plasmids and their prevalence in isolates of defined sources. *Microb Pathog*, **11**, 391–7.

Nair, S., Schreiber, E., Thong, K. L., Pang, T. and Altwegg, M. (2000). Genotypic characterization of *Salmonella typhi* by amplified fragment length polymorphism fingerprinting provides increased discrimination as compared to pulsed-field gel electrophoresis and ribotyping. *J Microbiol Methods*, **41**, 35–43.

Ochman, H. and Groisman, E. A. (1996). Distribution of pathogenicity islands in *Salmonella* spp. *Infect Immun*, **64**, 5410–12.

Ochman, H. and Wilson, A. C. (1987). Evolution in bacteria: evidence for a universal substitution rate in cellular genomes. *J Mol Evol*, **26**, 74–86.

Paesold, G., Guiney, D. G., Eckmann, L. and Kagnoff, M. F. (2002). Genes in the *Salmonella* pathogenicity island 2 and the *Salmonella* virulence plasmid are essential for *Salmonella*-induced apoptosis in intestinal epithelial cells. *Cell Microbiol*, **4**, 771–81.

Parkhill, J., Dougan, G., James, K. D. *et al.* (2001). Complete genome sequence of a multiple drug resistant *Salmonella enterica* serovar Typhi CT18. *Nature*, **413**, 848–52.

Parry, C. M., Hien, T. T., Dougan, G., White, N. J. and Farrar, J. J. (2002). Typhoid fever. *N Engl J Med*, **347**, 1770–82.

Perna, N. T., Plunkett, G., III, Burland, V. *et al.* (2001). Genome sequence of enterohaemorrhagic *Escherichia coli* O157:H7. *Nature*, **409**, 529–33.

Pickard, D., Wain, J., Baker, S. *et al.* (2003). Composition, acquisition, and distribution of the Vi exopolysaccharide-encoding *Salmonella enterica* pathogenicity island SPI-7. *J Bacteriol*, **185**, 5055–65.

Popoff, M. Y., Bockemuhl, J. and Gheesling, L. L. (2003). Supplement 2001 (no. 45) to the Kauffmann-White scheme. *Res Microbiol*, **154**, 173–4.

Porwollik, S., Frye, J., Florea, L. D., Blackmer, F. and McClelland, M. (2003). A non-redundant microarray of genes for two related bacteria. *Nucleic Acids Res*, **31**, 1869–76.

Porwollik, S., Wong, R. M. and McClelland, M. (2002). Evolutionary genomics of *Salmonella*: gene acquisitions revealed by microarray analysis. *Proc Natl Acad Sci USA*, **99**, 8956–61.

Prager, R., Rabsch, W., Streckel, W. *et al.* (2003). Molecular properties of *Salmonella enterica* serotype paratyphi B distinguish between its systemic and its enteric pathovars. *J Clin Microbiol*, **41**, 4270–8.

Reeves, M. W., Evins, G. M., Heiba, A. A., Plikaytis, B. D. and Farmer, J. J., III. (1989). Clonal nature of *Salmonella typhi* and its genetic relatedness to other salmonellae as shown by multilocus enzyme electrophoresis, and proposal of *Salmonella bongori* comb. nov. *J Clin Microbiol*, **27**, 313–20.

Rodriguez-Pena, J. M., Buisan, M., Ibanez, M. and Rotger, R. (1997). Genetic map of the virulence plasmid of *Salmonella enteritidis* and nucleotide sequence of its replicons. *Gene*, **188**, 53–61.

Roth, J. R., Lawrence, J. G. and Bobik, T. A. (1996). Cobalamin (coenzyme B12): synthesis and biological significance. *Annu Rev Microbiol*, **50**, 137–81.

Rychlik, I., Lovell, M. A. and Barrow, P. A. (1998). The presence of genes homologous to the K88 genes *faeH* and *faeI* on the virulence plasmid of *Salmonella gallinarum*. *FEMS Microbiol Lett*, **159**, 255–60.

Sassetti, C. M., Boyd, D. H. and Rubin, E. J. (2001). Comprehensive identification of conditionally essential genes in mycobacteria. *Proc Natl Acad Sci USA*, **98**, 12712–7.

Selander, R. K., Beltran, P., Smith, N. H. *et al.* (1990a). Genetic population structure, clonal phylogeny, and pathogenicity of *Salmonella paratyphi* B. *Infect Immun*, **58**, 1891–901.

Selander, R. K., Beltran, P., Smith, N. H. *et al.* (1990b). Evolutionary genetic relationships of clones of *Salmonella* serovars that cause human typhoid and other enteric fevers. *Infect Immun*, **58**, 2262–75.

Shalon, D., Smith, S. J. and Brown, P. O. (1996). A DNA microarray system for analyzing complex DNA samples using two-color fluorescent probe hybridization. *Genome Res*, **6**, 639–45.

Shea, J. E., Hensel, M., Gleeson, C. and Holden, D. W. (1996). Identification of a virulence locus encoding a second type III secretion system in *Salmonella typhimurium*. *Proc Natl Acad Sci USA*, **93**, 2593–7.

Sherburne, C. K., Lawley, T. D., Gilmour, M. W. *et al.* (2000). The complete DNA sequence and analysis of R27, a large IncHI plasmid from *Salmonella typhi* that is temperature sensitive for transfer. *Nucleic Acids Res*, **28**, 2177–86.

Swenson, D. L., Clegg, S. and Old, D. C. (1991). The frequency of *fim* genes among *Salmonella* serovars. *Microb Pathog*, **10**, 487–92.

Thomson, N., Baker, S., Pickard, D. *et al.* (2004). The role of prophage-like elements in the diversity of *Salmonella enterica* serovars. *J Mol Biol*, **339**, 279–300.

Threlfall, E. J., Ward, L. R., Hampton, M. D. *et al.* (1998). Molecular fingerprinting defines a strain of *Salmonella enterica* serotype Anatum responsible for an international outbreak associated with formula-dried milk. *Epidemiol Infect*, **121**, 289–93.

Townsend, S. M., Kramer, N. E., Edwards, R. *et al.* (2001). *Salmonella* enterica serovar Typhi possesses a unique repertoire of fimbrial gene sequences. *Infect Immun*, **69**, 2894–901.

Tsolis, R. M., Townsend, S. M., Miao, E. A. *et al.* (1999). Identification of a putative *Salmonella enterica* serotype Typhimurium host range factor with homology to IpaH and YopM by signature-tagged mutagenesis. *Infect Immun*, **67**, 6385–93.

Uzzau, S., Bossi, L. and Figueroa-Bossi, N. (2002). Differential accumulation of *Salmonella* [Cu, Zn] superoxide dismutases SodCI and SodCII in intracellular bacteria: correlation with their relative contribution to pathogenicity. *Mol Microbiol*, **46**, 147–56.

Virlogeux-Payant, I., Mompart, F., Velge, P., Bottreau, E. and Pardon, P. (2003). Low persistence of a large-plasmid-cured variant of *Salmonella enteritidis* in ceca of chicks. *Avian Dis*, **47**, 163–8.

Wain, J., Diem Nga, L. T., Kidgell, C. *et al.* (2003). Molecular analysis of incHI1 antimicrobial resistance plasmids from *Salmonella* serovar Typhi strains associated with typhoid fever. *Antimicrob Agents Chemother*, **47**, 2732–9.

Wei, J., Goldberg, M. B., Burland, V. *et al.* (2003). Complete genome sequence and comparative genomics of *Shigella flexneri* serotype 2a strain 2457T. *Infect Immun*, **71**, 2775–86.

Wilson, J. W., Ramamurthy, R., Porwollik, S. *et al.* (2002). Microarray analysis identifies *Salmonella* genes belonging to the low-shear modeled microgravity regulon. *Proc Natl Acad Sci USA*, **99**, 13807–12.

Wood, M. W., Jones, M. A., Watson, P. R. *et al.* (2000). The secreted effector protein of *Salmonella dublin*, SopA, is translocated into eukaryotic cells and influences the induction of enteritis. *Cell Microbiol*, **2**, 293–303.

Ye, R. W., Wang, T., Bedzyk, L. and Croker, K. M. (2001). Applications of DNA microarrays in microbial systems. *J Microbiol Methods*, **47**, 257–72.

Zeng, H., Carlson, A. Q., Guo, Y. *et al.* (2003). Flagellin is the major proinflammatory determinant of enteropathogenic *Salmonella*. *J Immunol*, **171**, 3668–74.

CHAPTER 6

Pathogenicity islands and virulence of *Salmonella enterica*

Michael Hensel

6.1 INTRODUCTION

6.1.1 Horizontal gene transfer and bacterial virulence

Bacteria show a remarkable ability to adapt rapidly to new habitats. This observation also applies to pathogenic bacteria that have evolved strategies to colonize various anatomical niches of their multi-cellular hosts. The acquisition of genetic material by a process termed horizontal gene transfer is considered to be a driving force for the rapid evolution of bacteria as pathogens. Extrachromosomal DNA such as plasmids conferring resistance to antibiotics were the first horizontally transferred DNA elements to be identified, but later it became obvious that there are also mechanisms that allow the horizontal transfer of chromosomal DNA elements. The observation that certain virulence functions are clustered in distinct regions of the chromosome and that these regions are genetically unstable and were deleted with high frequencies gave the first clue about the existence of the new form of genetic elements. The term 'Pathogenicity Island' (PAI) was first introduced in 1983 by Hacker and colleagues who observed genetic instability of genes associated with the haemolytic activity of uropathogenic strains of *Escherichia coli* (Hacker *et al.*, 1983). PAI were initially defined as large, unstable regions of the chromosome. With the identification of a large number of additional PAIs in various groups of bacterial pathogens, further common features were found.

'Salmonella' Infections: Clinical, Immunological and Molecular Aspects, ed. Pietro Mastroeni and Duncan Maskell. Published by Cambridge University Press.

6.1.2 Definition of pathogenicity islands

The following common characteristics were defined for PAIs:

- They are large, distinct genetic entities on the bacterial chromosome.
- They harbor one or more virulence genes.
- They are often genetically unstable. This feature correlates with the presence of genetic elements involved in DNA-mobility, such as direct repeats, integrases, transposases and bacteriophage genes.
- The base composition of their DNA is often different from that of the core genome. For example, the base composition expressed as the percentage of guanine and cytosine residues (% G + C) is a characteristic specific of individual bacterial species. A DNA fragment with a G + C content that is higher or lower than the average base composition of the chromosome, often indicates the horizontal acquisition of this region from another species.
- They are frequently associated with genes encoding tRNAs. tRNA genes also function as integration points to certain lysogenic bacteriophages. It has been proposed that tRNA genes can serve as 'anchoring points' for horizontally acquired DNA due to their conserved structure in different bacterial species.

These definitions for PAIs are not stringent and the majority of PAIs only meet a subset of the listed criteria.

6.1.3 Pathogenicity islands as driving forces in evolution of bacterial virulence

If genes for a virulence function are scattered over the chromosome, several horizontal gene transfer events have to take place before the virulence function is successfully transferred. In contrast, the clustering of complex, multigenic virulence traits within PAIs increases the chance of distribution. Acquisition of a PAI by horizontal gene transfer enables a bacterium to adopt complex virulence functions from other pathogens very rapidly. Such acquisitions allow pathogens to extend their host range and to colonize new niches in a host organism.

6.2 PATHOGENICITY ISLANDS OF *SALMONELLA*

Important virulence traits of *Salmonella enterica* are the interaction with enterocytes resulting in diarrhoea, the invasion of non-phagocytic cells, and the ability to survive phagocytosis and to proliferate within eukaryotic host cells. These virulence functions are complex multigenic traits. Analyses of the genetic bases of virulence phenotypes led to the identification of large clusters of virulence genes in *S. enterica* that had various characteristics of PAIs and were termed SPI for '*Salmonella* Pathogenicity Island'. Further SPIs were identified by comparative genomics but for several of these loci their role in virulence has not been established.

The PAIs of *Salmonella* are described in detail in the following sections and the main features of these PAIs are summarized in Table 6.1. The nomenclature of SPI-6 to SPI-10 follows the designation used by Parkhill *et al.* for *S. enterica* serovar Typhi (Parkhill *et al.*, 2001a). It should be noted that alternative designations for some loci exist, and that some of the loci are restricted to certain serotypes as detailed in Table 6.1.

6.3 *SALMONELLA* PATHOGENICITY ISLAND 1

6.3.1 General

Genes within SPI-1 were first identified during the characterization of invasion-deficient strains of *S. enterica* using a transposon mutagenesis approach (Galan and Curtiss, 1989). Subsequent analyses revealed the requirement for a large number of genes for the invasion phenotype. These genes clustered within centisome 63 of the *Salmonella* chromosome in a distinct invasion locus that had several characteristics of a PAI (Mills *et al.*, 1995). This locus was later referred to as *Salmonella* Pathogenicity Island 1 (SPI-1).

6.3.2 Evolution of SPI-1

SPI-1 is present in *S. bongori* and all subspecies of *S. enterica*. Sequence comparison of SPI-1 genes led to the conclusion that this gene cluster is conserved between *Salmonella* species and invasion genes within the *spa/mxi* locus on the virulence plasmid of *Shigella* species (Groisman and Ochman, 1993). There is no similarity in structure and function to the invasion mechanisms utilized by *Yersinia* or *Listeria*. A recent study identified a gene cluster in *Burkholderia pseudomallei* that is highly similar to the SPI-1 locus of *S. enterica* (Stevens *et al.*, 2002). The presence of SPI-1 in other species has not been reported.

There are no genetic elements directly associated with DNA instability or mobility, indicating that the locus has evolved into a rather stable element of the chromosome of all clinical isolates of *Salmonella* analyzed so far. However,

Table 6.1. *Characteristics of* Salmonella *pathogenicity islands*

Designation (alternative)	Size in kb	Base comp % G + C (range)	Insertion point	Distribution	Variability (stability)	Virulence functions
SPI-1	39.8	47	*flhA-mutS*	*Salmonella* spp.	conserved	TTSS, iron uptake
SPI-2	39.7	44.6	tRNA *valV*	*S. enterica*	conserved	TTSS
SPI-3	17.3	39.8–49.3	tRNA *selC*	*Salmonella* spp.	variable	Mg^{2+} uptake
SPI-4	23.4	44.8	(tRNA like)	*Salmonella* spp.	conserved	unknown
SPI-5	7.6	43.6	tRNA *serT*	*Salmonella* spp.	variable	TTSS effectors
SPI-6 (SCI)	59	51.5	tRNA *aspV*	subsp. I, parts in IIIB, IV, VII	?	fimbriae
SPI-7 (MPI)	133	44–53	tRNA *pheU*	subsp. I serovars	unstable	Vi antigen, pilus assembly, *sopE*
SPI-8	6.8	38.1	tRNA *pheV*	sv. Typhi	?	unknown
SPI-9	16.3	56.7	prophage	subsp. I serovars	?	putative toxin, unknown
SPI-10	32.8	46.6	tRNA *leuX*	subsp. I serovars	?	Sef fimbriae
SGI-I	43	48.4	*thdF–yidY*	subsp. I serovars	variable	5 antibiotic resistance genes
HPI	?	?	tRNA *ansT* (*ychF*)	subspecies IIIa, IIIb, IV	?	high affinity iron uptake

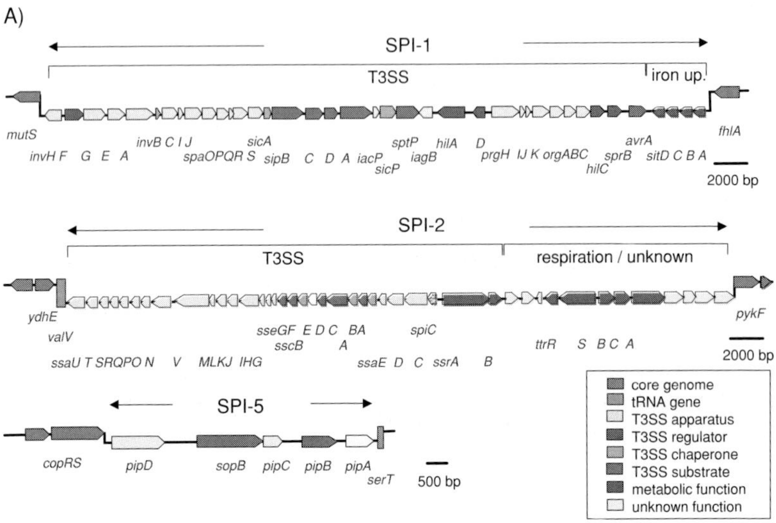

Figure 6.1. Topologies of representative *Salmonella* Pathogenicity Islands.
Genetic organization, gene functions and the role in virulence are shown. Gene designations of relevant loci are indicated. Incomplete elements (IS, tRNA) are indicated by parenthesis.
A) SPI encoding type III secretion systems (TTSS) and cognate effector proteins. SPI-1 and SPI-2 encode TTSS for the secretion and translocation of virulence proteins into eukaryotic target cells. Both loci encode the TTSS apparatus, regulatory systems and a subset of substrate proteins, *i.e.* secreted proteins involved in translocation and translocated effector proteins. While the SPI-1-encoded TTSS is involved in host cell invasion and enteropathogenesis, the SPI-2-encoded TTSS is required for the intracellular proliferation of *S. enterica* and systemic pathogenesis. SPI-1 and SPI-2 both contain portions that are not related to TTSS function but have functions in iron uptake (*sit* gene in SPI-1) or anaerobic respiration (*ttr* genes in SPI-2). Maps of SPI-1 and SPI2 were adapted from Hansen-Wester and Hensel (2001). The SPI-5 map was adapted from Wood *et al.* (1998).
B) SPI-3, SPI-7 and SGI-1 harbor DNA sequences and genes involved in DNA mobility. SPI-3 encodes the high affinity Mg^{2+} uptake system MgtCB, required for intracellular proliferation. A central portion of SPI-3 (*rmb*A to *marT*) is flanked by IS element-like DNA sequences. SGI-1 encodes five antibiotic resistances. Various genes with similarity to integrases, transposases and conjugational transfer systems are present, as well as cryptic prophage. SGI-1 is flanked by direct repeats, and further direct repeats flank the *floR* gene. SPI-7 is composed of various regions with different functions. The *pil* gene cluster encodes a type IVB pilus assembly system and the *viaB* region is required for biosynthesis of the Vi antigen capsule. Furthermore, the *sopE* phage harboring the *sopE* gene for an effector protein of the SPI-1-encoded TTSS is present within SPI-7. Maps of SPI-3 (Blanc-Potard *et al.*, 1999), SGI-1 (Boyd *et al.*, 2001) and SPI-7 (Pickard *et al.*, 2003) were adapted from the original publications. (See colour plate section.)

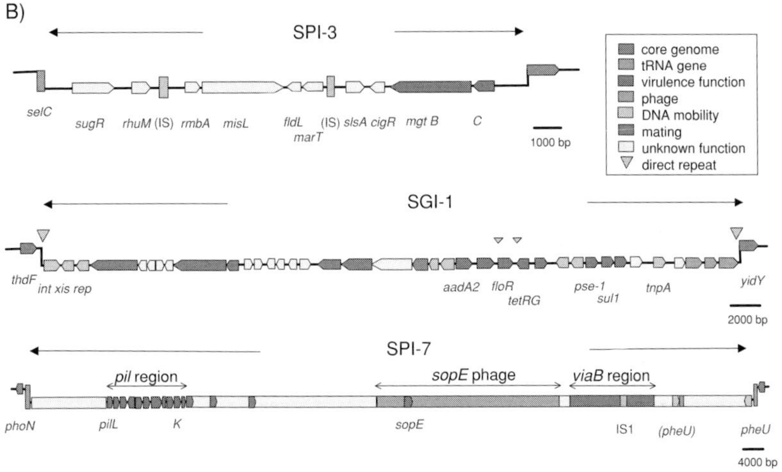

Figure 6.1. (*cont.*)

the deletion of major parts of SPI-1 has been reported for environmental isolates of *S. enterica* serovars Senftenberg and Litchfield (Ginocchio *et al.*, 1997).

6.3.3 Structure of SPI-1

SPI-1 is about 40 kb in size and forms a *Salmonella*-specific insertion between genes that are consecutive in *E. coli* K-12 (Mills *et al.*, 1995). This PAI is not associated with a tRNA gene, and the base composition of SPI-1 of 47% G + C is lower than the average G + C content of the core genome of about 52%. Functional analyses indicated that SPI-1 is composed of at least two distinct elements. One element consists of about 30 genes required for the assembly and function of the type III secretion system (TTSS). TTSS are macromolecular complexes found in Gram-negative bacteria that mediate the contact-dependent translocation of proteins into a eukaryotic target cell (Hueck, 1998). The SPI-1-encoded TTSS forms needle-like surface appendages (Kubori *et al.*, 1998) that mediate the delivery of proteins by extracellular *Salmonella* into host cells. Another element, consisting of a small cluster of *sitABCD* genes at a flank of SPI-1, is not involved in invasion and encodes an iron uptake system (Zhou *et al.*, 1999).

6.3.4 Function and effector proteins of SPI-1

SPI-1 is required for the invasion of non-phagocytic cells, an important virulence trait of *S. enterica* (Galan, 2001). Molecular analyses indicate that the invasion genes of *Salmonella* and the *mxi/spa* invasion genes of *Shigella* are closely related. Invasion of epithelial cells and M cells in the Peyer's patches of the gut is one of the mechanisms used by *S. enterica* to gain access to the systemic compartment. Mutations in SPI-1 genes result in reduced invasiveness of *Salmonella* in cell culture and in animal models (Galan, 2001). During invasion of the gut, the SPI-1-encoded SipB protein triggers the activation of intracellular Caspase-1 within resident macrophages (Hersh *et al.*, 1999). Caspase-1, a member of a family of cysteine proteases, induces apoptosis in the infected macrophages resulting in escape of *Salmonella* from these cells. Caspase-1 also directly cleaves the pro-inflammatory cytokines IL1β and IL18 to produce bioactive cytokines that enhance local inflammation and infiltration of polymorphonuclear phagocytes (PMN). SipB-mediated activation of Caspase-1 results in enhanced colonization of the Peyer's patches and mesenteric lymph nodes (Monack *et al.*, 2001). However, an alternative SPI-1-independent mechanism of invasion has been described where *Salmonella* does not interact with M cells but is engulfed by dendritic cells that open the tight junctions between epithelial cells and sample bacteria at the mucosal surface (Rescigno *et al.*, 2001). *Salmonella* is then transported from the gastrointestinal tract to the bloodstream by CD18-expressing phagocytes (Vazquez-Torres *et al.*, 1999).

Translocation into the host cell has been demonstrated for a large number of effector proteins of the SPI-1-encoded TTSS. SipA, SipB, SipC, SptP and AvrA are encoded by genes within the SPI-1 locus and whilst *sipABC* and *sptP* are conserved in *Salmonella* species, *avrA* is located at a flank of SPI-1 and is hyper-variable. The effector proteins SopA, SopB, SopD, SopE and SopE2 are encoded by independent loci outside of SPI-1. The *sopB*, *sopD* and *sopE2* loci again are conserved throughout *Salmonella* species, whilst the distribution of the phage harboring *sopE* is variable (Mirold *et al.*, 2001).

One subset of the effector proteins modifies signal transduction pathways resulting in the temporal reorganization of the actin cytoskeleton of the host cell. The structures and functions of SptP, SopE and SopE2 have been analyzed in detail. These proteins interfere with the function of host cell proteins that form a family of small GTPases (e.g. Cdc42, Rac-1 and Rho) and regulate the formation of F-actin filaments and the dynamics of the cytoskeleton. In detail, SopE and SopE2 function as GTP-exchange factors, resulting in activation of Cdc42 leading to the formation of actin filaments at the site of translocation of these effector proteins (Hardt *et al.*, 1998). The localized modification of the cytoskeleton is followed by dramatic changes in the host cell surface that appear as membrane ruffles. As a consequence non-phagocytic cells, in this case epithelial cells, internalize larger particles such as bacteria in a process termed macropinocytosis. It has also been demonstrated that the effector protein SptP has an antagonistic effect by its function as a GTPase-activating factor. SptP can mediate the inactivation of Cdc42 and Rac-1 resulting in the termination of actin polymerization and membrane ruffling (Fu and Galan, 1999). Translocation of SPI-1 effector proteins into macrophages induces a very rapid form of apoptosis. As mentioned above, the SPI-1-encoded protein SipB is involved in macrophage apoptosis by activation of Caspase-1, a function similar to that performed by IpaB of *Shigella* (Hersh *et al.*, 1999). This also results in the release of pro-inflammatory cytokines like IL8 and IL18. Caspase-1 induces apoptosis probably by activating other Caspases (Knodler and Finlay, 2001; Monack *et al.*, 2000).

It has been demonstrated that the function of SPI-1 is also related to the diarrhea associated with *Salmonella* infection (Wallis and Galyov, 2000). The second subset of effector proteins (SopA, SopB, and SopD) translocated by the SPI-1 system is required for this phenotype. SopB is an inositol phosphate phosphatase and its enzymatic activity results in activation of chloride channels in the membranes of epithelial target cells, finally leading to the secretion of chloride and loss of fluid into the intestinal lumen (Norris *et al.*, 1998). The functions of SopA and SopD are not fully understood but both effectors also contribute to the diarrhea phenotype in the bovine model of *Salmonella* enteritis (Jones *et al.*, 1998; Zhang *et al.*, 2000).

6.3.5 Regulation of SPI-1 gene expression

Regulation of the expression of invasion genes of SPI-1 is a complex process involving local as well as global regulatory systems (Lucas and Lee, 2000). HilA, a SPI-1-encoded transcriptional regulator of the AraC class has a central function in SPI-1 expression and controls InvF, a SPI-1-encoded transcriptional activator. HilC and HilD are further SPI-1-encoded regulators that modulate HilA activity. Global regulatory systems such as the PhoPQ, OmpR/EnvZ and BarA/SirA two-component systems, as well as the flagella assembly system affect expression of SPI-1 genes. Collectively these regulatory systems integrate environmental stimuli that are encountered by *Salmonella* in the intestinal lumen such as low oxygen tension, moderately basic pH, high osmolarity and presence of short chain fatty acids, resulting in the expression of invasion genes.

6.4 *SALMONELLA* PATHOGENICITY ISLAND 2

6.4.1 General

SPI-2 is essential for the ability of *Salmonella* to cause systemic infections and to proliferate within host organs. This virulence phenotype is linked to the ability of *S. enterica* to survive in phagocytic cells and to replicate within vacuoles in a variety of eukaryotic cells.

Genes within SPI-2 were first identified by application of signature-tagged mutagenesis to *S. enterica*, in order to perform parallel in vivo screening for attenuated mutant strains in a mouse model of infection (Hensel *et al.*, 1995). A fragment of SPI-2 was also identified by selection of *Salmonella*-specific genomic regions (Ochman *et al.*, 1996).

6.4.2 Evolution of SPI-2

SPI-1 is present in *S. enterica* as well as in *S. bongori*, a phylogenetically older group of *Salmonella* that is rarely associated with human disease. In contrast, SPI-2 has only been detected in *S. enterica*, thus probably forming a more recent acquisition (Hensel *et al.*, 1997; Ochman and Groisman, 1996). The acquisition of SPI-2 is considered as an evolutionary step towards the systemic colonization of warm-blooded hosts (Bäumler, 1997; Groisman and Ochman, 1997). Naturally occurring deletions of SPI-2 or portions of SPI-2 have not been reported. There are no genes within SPI-2 with putative functions in DNA mobility or transfer. The stability of SPI-2 and the presence of various loci encoding effector proteins outside of it suggest that it has become part of the core genome and that adaptation is mediated by acquisition and loss of the additional loci.

No direct homologues of SPI-2 with known functions in virulence have been reported for other bacteria. However in *Yersinia pestis* a chromosomal locus was identified with high levels of sequence similarity to SPI-2 genes (Parkhill *et al.*, 2001b). Interestingly, the *Y. pestis ysa* locus lacks homologues of the *sse* genes, that encode translocon proteins SseBCD and translocated effectors SseFG. In light of these observations, it is questionable whether a functional TTSS is encoded by the *ysa* locus. Comparisons between TTSS of various pathogens placed the SPI-2-encoded system in a group containing the TTSS of enteropathogenic *E. coli* (EPEC), but showed it to be only a distant relation of the SPI-1-encoded TTSS (Foultier *et al.*, 2002; Hensel *et al.*, 1997). This observation suggests that the SPI-1 and SPI-2 encoded TTSS resulted from independent events of horizontal gene transfer and not from duplication of one gene cluster.

6.4.3 Structure of SPI-2

The SPI-2 locus is 40 kb in size and is inserted adjacent to the tRNA *valV* gene (Hensel *et al.*, 1997). SPI-2 is composed of at least two distinct elements (Hensel *et al.*, 1999b). A portion of 25 kb with a G + C content of 43% is only present in *S. enterica* and is essential for systemic pathogenesis. This element encodes a second TTSS in *S. enterica* that is activated when the bacteria are in an intracellular environment. Another portion of SPI-2 encoding the tetrathionate reductase (Ttr) is 15 kb in size and was detected in *S. bongori* and *S. enterica*. This element has a G + C content of 54% (Hensel *et al.*, 1999a). The *ttr* gene cluster was detected in genome sequences of bacteria including *Haemophilus somnus, Pasteurella multocida, Bordetella bronchiseptica, Vibrio parahaemolyticus* and several others.

6.4.4 SPI-2-encoded functions

SPI-2 encodes a second TTSS in *S. enterica* needed for the translocation of effector proteins by intracellular bacteria into their host cell. All structural components of the second TTSS, the two-component regulatory system SsrAB and a subset of effector proteins are encoded by this locus. Mutant strains deficient in the SPI-2-encoded TTSS or the SsrAB system are highly attenuated in virulence and unable to proliferate in systemic sites (Shea *et al.*, 1999; Shea *et al.*, 1996). The use of such strains as vaccines against typhoid fever or as carrier strains for recombinant vaccines has been evaluated (Hindle *et al.*, 2002; Medina *et al.*, 1999).

There are several cellular phenotypes related to SPI-2. The function of the SPI-2-encoded TTSS is required to protect the pathogens within the *Salmonella*-containing vacuole (SCV) against the effector functions of innate immunity. It has been reported that SPI-2 prevents co-localization of the phagocyte oxidase (Vazquez-Torres *et al.*, 2000) and the inducible nitric oxide synthase to the SCV (Chakravortty *et al.*, 2002). Both functions might be related to the modification of host cell trafficking by SPI-2 (Uchiya *et al.*, 1999). As a consequence, intracellular *Salmonella* are protected against damage by reactive nitrogen species (RNS) and reactive oxygen species (ROS) and against the potent antimicrobial activity of peroxynitrite, which is generated by reaction of RNS and ROS. These defense mechanisms represent a specific adaptation to the intracellular environment especially within phagocytic cells. In *Salmonella*-infected epithelial cells, the formation of tubular aggregates of host cell endosomes was observed (Garcia-del Portillo *et al.*, 1993), and this phenotype is dependent on the translocation of the effector protein SifA by the SPI-2-encoded TTSS (Beuzon *et al.*, 2000). Mutant strains deficient in SifA fail to maintain the SCV and are released into the cytoplasm of the host cell (Beuzon *et al.*, 2000). Whilst the rapid apoptosis of macrophages is triggered by SPI-1-dependent invasion by *Salmonella*, a delayed Caspase-1-dependent form of apoptosis was reported to be dependent on SPI-2 function (Monack *et al.*, 2001).

The function of other genes within the SPI-2 locus is not related to systemic pathogenesis. The *ttr* gene cluster encodes three structural subunits of the tetrathionate reductase of anaerobic respiration and the cognate TtrSR two-component regulatory system. The role of additional ORFs in SPI-2 is not known but these genes are also dispensable for systemic pathogenesis (Hensel *et al.*, 1999b).

6.4.5 Effector proteins of the SPI2-encoded TTSS

SPI-2 encodes three secreted proteins SseBCD that function as translocators for the delivery of effector proteins into infected host cells. Translocated effectors encoded by SPI-2 are SseF and SseG (Kuhle and Hensel, 2002). For the SPI-2-encoded SpiC, translocation and a function in interference with cellular trafficking were reported (Uchiya *et al.*, 1999). However, there is also evidence that SpiC is a functional component of the TTSS itself (Freeman *et al.*, 2002; Yu *et al.*, 2002).

Seven loci outside of the SPI-2 locus encoding the *Salmonella* translocated effector (STE) proteins were identified (Miao and Miller, 2000). In contrast to TTSS effectors of SPI-1 and of other pathogens, STE proteins SifA, SifB, SspH1, SspH2, SlrP, SseI and SseJ share a conserved N-terminal domain of 150 amino acids. Other effector proteins encoded outside SPI-2 are PipB (encoded by SPI-5), PipB2 and SopD2. These effectors, as well as the SPI-2-encoded effectors lack the conserved domain.

The relative functions of the various effectors in the pathogenesis of *S. enterica* are only partially understood (Waterman and Holden, 2003). Inactivation of most effector genes does not result in strong attenuation of systemic virulence, as has been observed for null mutations in the SPI-2-encoded TTSS. However for the STE protein SifA a role in virulence was observed. SifA is important for the formation of endosomal aggregates in infected host cells and is required to maintain the integrity of the phagosomal membrane of the SCV during intracellular proliferation (Beuzon *et al.*, 2000).

6.4.6 SPI-1 and SPI-2 – fixed PAI and variable effector loci

A common characteristic of both SPI-1 and SPI-2 is the observation that only a subset of the translocated effector proteins is encoded by genes within the island. In fact, most of the effectors are encoded by distinct loci scattered over the chromosome (Miao and Miller, 2000; Mirold *et al.*, 2001). Some of these loci are associated with bacteriophage genes (Hansen-Wester *et al.*, 2002; Mirold *et al.*, 2001), or are located within SPI-5, indicating that these elements have been acquired by separate horizontal gene transfer events. SPI-1 and SPI-2 have evolved into stable regions of the *Salmonella* genome and each locus encodes a TTSS and a small number of translocated effector proteins. The majority of effector proteins however is encoded by separate loci outside of SPI-1 or SPI-2. The frequent association of the effector loci with cryptic, as well as functional bacteriophages indicates that these effector genes form a highly dynamic and mobile pool of virulence factors (Figueroa-Bossi *et al.*, 2001). It is possible that the combination of effector genes in different serovars of *Salmonella* species contributes to host specificity of the various serotypes as well as to disease outcome.

6.4.7 Regulation of SPI-2 gene expression

Expression of the SPI-2-encoded TTSS is directly controlled by SsrAB, a two-component system encoded by SPI-2 (Deiwick *et al.*, 1999). Expression of *ssrAB* is modulated by OmpR/EnvZ, another two-component system with global regulatory functions (Lee *et al.*, 2000). Effects of environmental stimuli on SPI-2 expression in vitro suggest that the global regulatory system PhoPQ also modulates SPI-2 expression (Deiwick *et al.*, 1999). Expression of SPI-2 genes is activated by the intracellular environment (Cirillo *et al.*, 1998; Deiwick *et al.*, 1999). Although the signals received by SsrA are not known, the definition of in vitro culture conditions activating SPI-2 expression suggests that limited availability of components such as Mg^{2+} and inorganic phosphate activate SPI-2 expression (Deiwick *et al.*, 1999). These components appear to be limited within the SCV. The role of acidic pH as a signal for SPI-2 induction is controversial. Acidification of the SCV is observed after phagocytosis, but acid pH appears to be a signal for the functional assembly of the SPI-2-encoded TTSS rather than for inducing expression (Rappl *et al.*, 2003).

Expression of structural *ttr* genes is independent from the function of SsrAB but under the control of a further SPI-2-encoded two-component system termed TtrRS (Hensel *et al.*, 1999a). The function of a further ORF in

SPI-2 encoding a putative DNA-binding protein is independent from SPI-2 regulation but has not been studied in detail.

6.5 *SALMONELLA* PATHOGENICITY ISLAND 3

6.5.1 General

Pathogenicity islands of uropathogenic and enteropathogenic *E. coli* (Schmidt and Hensel, 2004), as well as a retronphage of *E. coli* are located adjacent to the tRNA *selC* gene, leading to the hypothesis that this locus served as an insertion point for horizontally acquired DNA elements. These observations led to the screening of the area of the genome adjacent to the *selC* locus of *S. enterica* serovar Typhimurium for the presence of features indicative of a PAI. This analysis resulted in the identification of a *Salmonella*-specific insertion adjacent to *selC* that was termed SPI-3 (Blanc-Potard and Groisman, 1997). The SPI-3 locus is required for the intracellular proliferation of *S. enterica* and for systemic pathogenicity in the murine model of salmonellosis.

6.5.2 Evolution and structure of SPI-3

SPI-3 is conserved between *S. enterica* serovars Typhi and Typhimurium. Analysis of the distribution of SPI-3 in various salmonellae revealed extensive variations in its structure, ranging from deletions to insertions of additional gene clusters (Amavisit *et al.*, 2003; Blanc-Potard *et al.*, 1999). These variations were mainly located in the portion of SPI-3 adjacent to *selC*, whilst a portion containing *mgtCB* and various other genes appeared to be conserved in various *S. enterica* isolates and in *S. bongori* (Amavisit *et al.*, 2003). The variability of SPI-3 in *Salmonella* and the observation that different PAIs are inserted at this locus in various pathogens indicate that *selC* is a hot spot for the integration of foreign DNA elements.

The average base composition of SPI-3 is 47.5% G + C but extensive variations of the base composition have been observed (Blanc-Potard *et al.*, 1999). There are several indications for a composite structure of SPI-3. Two fragments of insertion sequence (IS) elements are located in the central region of SPI-3.

6.5.3 SPI-3-encoded functions

A prerequisite for survival and replication of *Salmonella* in the SCV is adaptation to the microbicidal and nutrient-poor environment found there, which limits concentrations of purines, pyrimidines, particular amino acids and Mg^{2+}. The main virulence factor encoded by SPI-3 is the high affinity magnesium transport system MgtCB, which is important for intracellular survival of *Salmonella* (Blanc-Potard and Groisman, 1997). Mutant strains deficient in the MgtCB system are deficient in intracellular proliferation and systemic virulence. MgtB and MgtC are located in the cytoplasmic membrane. Whereas MgtB is a magnesium transporter (Snavely *et al.*, 1991), the function of MgtC has not yet been clarified (Moncrief and Maguire, 1998). A further putative virulence gene is *misL*, encoding a type V secretion system (autotransporter) with similarity to VirG of *Shigella flexneri* and the AIDA-1 adhesin of enteropathogenic *E. coli* (Blanc-Potard *et al.*, 1999). The role in virulence of *misL* and other genes within SPI-3 has not yet been elucidated.

6.6 *SALMONELLA* PATHOGENICITY ISLAND 4

6.6.1 Structure and evolution of SPI-4

Comparison of *E. coli* and *S. enterica* genomic DNA by hybridization analysis identified a *Salmonella*-specific region that was designated SPI-4 (Wong *et al.*, 1998).

SPI-4 is an insertion of 23 kb with a low G + C content of 44.8% and is located adjacent to a tRNA-like sequence (Wong *et al.*, 1998). SPI-4 appears to be conserved among various serovars of *S. enterica* (Amavisit *et al.*, 2003; Wong *et al.*, 1998). However, comparison of genome sequences revealed differences in the organization of SPI-4 in *S. enterica* serovar Typhi and serovar Typhimurium (Parkhill *et al.*, 2001a).

6.6.2 SPI-4-encoded functions

A systematic mutational and functional analysis of SPI-4 has not been reported so far. Putative virulence factors in SPI-4 are genes with sequence similarity to genes encoding type I secretion systems for toxins. Furthermore, transposon mutagenesis identified several loci whose functions are required for survival in macrophages (Bäumler *et al.*, 1996). One of these loci termed *ims98* is located within SPI-4 (Wong *et al.*, 1998).

6.7 *SALMONELLA* PATHOGENICITY ISLAND 5

6.7.1 Structure and evolution of SPI-5

The SPI-5 locus was identified during the characterization of genes encoding Sops, the SPI-1 effectors introduced above. The *sopB* gene is located within a small *Salmonella*-specific gene cluster insertion with specific characteristics of a PAI (Wood *et al.*, 1998).

The SPI-5 locus is 7.6 kb in size, has a G + C content of 43.6% compared to 52% for the core genome of *S. enterica*, and is located adjacent to the gene encoding *serT* tRNA. SPI-5 has a mosaic structure probably consisting of elements separately acquired during evolution. In fact, the *sopB* gene is present in *S. bongori* and all subspecies of *S. enterica*, whilst a portion of SPI-5 harboring *pipAB* is absent from *S. bongori* and *S. enterica* subspecies II (Knodler *et al.*, 2002). There is also a difference in base composition in the different portions of SPI-5, supporting the independent acquisition of two elements at the *serT* tRNA gene.

6.7.2 SPI-5-encoded functions

SPI-5 encodes effector proteins for both of the TTSS encoded by SPI-1 and SPI-2. SopB is translocated by the SPI-1-encoded TTSS. SopB is an inositol phosphatase involved in triggering fluid secretion, resulting in diarrhea (Norris *et al.*, 1998). In contrast, PipB is a translocated effector of the SPI-2-encoded TTSS (Knodler *et al.*, 2002). The contribution of the SPI-2 effector protein PipB in virulence is less well understood. As with many other effectors of the SPI-2-encoded TTSS, PipB is not essential for virulence (Knodler *et al.*, 2002).

6.7.3 Regulation of SPI-5 gene expression

Regulation of effector genes in SPI-5 mirrors that of the cognate TTSS. Expression of *pipB* is dependent on the function of the SsrAB system encoded by SPI-2. Expression profiles of SPI-5 genes are similar to those of SPI-2 genes, activation occurring under intracellular conditions or in vitro when the bacteria are exposed to limiting amounts of Mg^{2+}. In contrast, *sopB* expression is dependent on HilA function, the SPI-1-encoded central regulatory protein (Ahmer *et al.*, 1999).

6.8 *SALMONELLA* PATHOGENICITY ISLAND 6 (OR *SALMONELLA* CENTISOME 7 GENOMIC ISLAND)

6.8.1 Structure and evolution of SPI-6

The SPI-6 locus has been postulated as a PAI based on the genome sequence of *S. enterica* serovar Typhi (Parkhill *et al.*, 2001a). The locus has also been abbreviated as SCI which stands for '*Salmonella enterica* centisome 7 genomic island' in *S. enterica* serovar Typhimurium (Folkesson *et al.*, 2002).

SPI-6/SCI is inserted adjacent to the *aspV* tRNA gene. Sizes of 47 kb and 59 kb were determined for SPI-6 loci of *S. enterica* serovar Typhimurium and serovar Typhi, respectively. The SPI-6 locus was detected in *S. enterica* subspecies I, and the presence of portions of SPI-6 at the *aspV* tRNA locus of isolates of subspecies IIIb, IV and VII was taken as an indication for the mosaic structure of this PAI. There is partial synteny with PAI OI#7 of enterohaemorrhagic *E. coli* that is also associated with the *aspV* tRNA gene. Further homologues of SPI-6 genes were identified in the genomes of *P. aeruginosa* and *Y. pestis*, but their functions in these pathogens is unknown.

6.8.2 SPI-6-encoded functions

The functions of few genes in SPI-6/SCI are known. SPI-6 contains genes such as the *saf* fimbrial operon, encoding fimbriae that could potentially contribute to virulence in *S. enterica*. However, a mutant strain with a deletion of the entire SPI-6/SCI locus had reduced ability to invade epithelial cells but retained virulence after oral and intraperitoneal infection of mice (Folkesson *et al.*, 2002).

6.9 *SALMONELLA* PATHOGENICITY ISLAND 7 (OR MAJOR PATHOGENICITY ISLAND)

6.9.1 Structure and evolution of SPI-7

SPI-7 is the largest SPI present exclusively in *S. enterica* serovar Typhi, serovar Dublin and serovar Paratyphi C and absent in other *S. enterica* serovars. The SPI-7 locus has also been referred to as the 'Major Pathogenicity Island' of *S. enterica* serovar Typhi (Zhang *et al.*, 1997).

SPI-7 is 133 kb in size and is adjacent to the *pheU* tRNA gene (Hansen-Wester and Hensel, 2002). The genetic organization of SPI-7 is rather complex and indicates that this locus is composed of individual horizontally acquired elements. The presence of *pil*, *tra* and *sam* genes indicates that portions of SPI-7 originated from a conjugative plasmid or conjugative transposon. A recent analysis demonstrated that a portion of SPI-7 is also present in several other bacteria including the plant pathogen *Xanthomonas axonopodis* and *Pseudomonas aeruginosa* SG17M where the locus is referred to as PAGI-3 (Pickard *et al.*, 2003). Furthermore, loss of the Vi capsule (whose biosynthesis depends on genes within SPI-7) can be observed in isolates of *S. enterica* serovar Typhi, suggesting instability of the SPI-7 locus. The extensive synteny between SPI-7 of *S. enterica* and related loci in plant pathogens and *P. aeruginosa* was considered as an indication for the acquisition of the locus by contact of *Salmonella* with environmental bacteria (Pickard *et al.*, 2003).

The *sopE* phage is also present in a subset of *S. enterica* subspecies I isolates that lack SPI-7 and it was demonstrated that the *sopE* phage can be activated and transfer the *sopE* gene to other isolates (Mirold *et al.*, 1999). The absence of the *sopE* phage in the SPI-7 locus of *S. enterica* serovar Dublin and serovar Paratyphi C indicates that the insertion of *sopE* in the SPI-7 locus of *S. enterica* serovar Typhi is rather recent.

6.9.2 SPI-7-encoded functions

The SPI-7 locus contains genetic elements encoding virulence factors that were identified individually before the definition of a PAI. These genetic elements include the following: the *viaB* gene locus for the biosynthesis of the Vi exopolysaccharide capsule; the *sopE* prophage harboring the *sopE* gene which encodes an effector protein of the SPI-1 system (Mirold *et al.*, 1999); a gene cluster which encodes type IVB pili that contribute to invasion of epithelial cells by *S. enterica* serovar Typhi (Zhang *et al.*, 2000).

6.10 *SALMONELLA* PATHOGENICITY ISLANDS 8 TO 10

Based on the genome sequence *of S. enterica* serovar Typhi, the presence of three additional SPIs has been postulated (Parkhill *et al.*, 2001a). These loci have several characteristics of PAIs but experimental evidence is required to elucidate the contribution of SPI-8, SPI-9 and SPI-10 to pathogenesis.

The SPI-8 locus consists of 6.8 kb and is adjacent to the *pheV* tRNA gene (Parkhill *et al.*, 2001a). Putative virulence factors contained within SPI-8 are genes encoding bacteriocins. The presence of a gene encoding for an integrase suggests the potential for mobility of this genetic element. SPI-8 appears to be specific for *S. enterica* serovar Typhi, however the distribution of SPI-8 has not yet been investigated in detail.

SPI-9 is a locus of 16.2 kb adjacent to a lysogenic bacteriophage in the chromosome of *S. enterica* serovar Typhi (Parkhill *et al.*, 2001a). Putative virulence factors encoded by SPI-9 are a type I secretion system and a large RTX-like protein. This locus is also present on the chromosome of *S. enterica* serovar Typhimurium where the ORF for the RTX-like toxin appears to be a pseudogene. Parts of SPI-9 and the adjacent bacteriophage genome are also present in the genome sequences of other *S. enterica* serovars and *S. bongori*, indicating a conserved distribution across several *Salmonella* species and serovars.

SPI-10 is a large insertion of 32.8 kb located at the *leuX* tRNA locus. There is also a bacteriophage present within SPI-10 (Parkhill *et al.*, 2001a). Known virulence factors encoded by SPI-10 are the Sef fimbriae. The distribution of Sef fimbriae is restricted to a subset of *S. enterica* serovars including serovar Typhi and serovar Enteritidis and is considered as one factor that determines host specificity (Townsend *et al.*, 2001). The role of the bacteriophage in distribution of SPI-10 has not been elucidated as yet.

6.11 *SALMONELLA* GENOMIC ISLAND 1

6.11.1 Evolution of SGI-1

The emergence of multidrug-resistant strains of *S. enterica* is a major problem associated with the clinical management of *Salmonella* infections in humans and in animals. Multidrug-resistant strains of *S. enterica* serovar Typhimurium DT 104, serovar Paratyphi B and serovar Agona contain a genomic fragment that has been termed *Salmonella* genomic island I (SGI-1) (Boyd *et al.*, 2001).

Genes associated with DNA mobility such as those encoding transposase, integrase and excisionase have been detected in SGI-1 and have sequence similarity to transposon genes (Boyd *et al.*, 2001). Variants of SGI-1 were identified in several *S. enterica* serovars at the same chromosomal location, indicating horizontal transfer and site-specific recombination. Recently a new variant of SGI-1 was identified in *S. enterica* serovar Albany (Doublet *et al.*, 2003). In this variant, the integron in SGI-1 containing the streptomycin resistance gene was replaced by an integron with the trimethoprim resistance gene. These observations indicate that chromosomal antibiotic resistance loci are recent acquisitions and are still in the process of genetic exchange and evolve as a mechanism of adaptation to antibiotic resistance.

6.11.2 SGI-1-encoded functions

Within SGI-1, genes conferring resistance to at least five antibiotics (multidrug resistance phenotype) are clustered in a region composed of two integrons. Resistance to ampicillin, chloramphenicol, streptomycin, sulphonamides and tetracycline are encoded in the SGI-1 variant of *S. enterica* serovar Typhimurium strain DT104. The streptomycin resistance gene is replaced by a trimethoprim resistance gene in the SGI-1 of *S. enterica* serovar Albany. Additional genetic elements in SGI-1 appear to be related to DNA mobility (transposases, DR, IS) or have unknown functions.

6.11.3 Structure of SGI-1

SGI-1 has a size of 43 kb, is flanked by DR and is not associated with a tRNA gene (Boyd *et al.*, 2001). The base composition of SGI-1 is variable, with regions of significantly lower and higher G + C-content than the rest of the genome. A cryptic retronphage was identified in SGI-1. It should be noted that multidrug resistance can also be mediated by plasmid-borne resistance genes, as observed by genome sequencing of *S. enterica* serovar Typhi harboring resistance plasmid pHCM1 (Parkhill *et al.*, 2001a). In contrast to plasmid-borne antibiotic resistance factors, the chromosomal SGI-1 appears to be stable in the absence of selective pressure.

6.12 HIGH PATHOGENICITY ISLAND

The High Pathogenicity Island (HPI) is a typical PAI that was initially identified and characterized in detail in *Y. pestis* and in highly pathogenic isolates of *Y. enterocolitica* and *Y. pseudotuberculosis* (Carniel, 2001). HPI contains genes that encode enzymes involved in the biosynthetic pathway for a siderophore, the cognate iron uptake system and regulators for the expression of structural genes under iron-limiting conditions. HPI was detected in a variety of other Gram-negative species and its presence appears to be related to the ability to cause septicaemic infections. Recently the presence of HPI in subspecies IIIa, IIIb and IV of *S. enterica* was reported (Oelschlaeger *et al.*, 2003). However, HPI is absent from human-adapted subspecies I isolates, and its role in the other subspecies remains to be analyzed. Similar to the situation in *Yersinia* species, the HPI in one group of *Salmonella* isolates is inserted adjacent to *asp* tRNA loci. There is also a second group of *Salmonella* isolates where HPI is not inserted at a tRNA locus (Oelschlaeger *et al.*, 2003).

6.13 OTHER SPI OF *SALMONELLA*?

Additional genomic regions of *S. enterica* have been identified that fulfill certain criteria of PAI. A genome-based analysis for *Salmonella*-specific insertions at tRNA genes identified SPI-7 and several other putative SPIs (Hansen-Wester and Hensel, 2002). However deletion of two of these loci did not affect systemic pathogenesis. The availability of genome sequences for *S. bongori* and various serovars of *S. enterica* subspecies I in the near future will allow the identification of additional SPI by comparative genomics.

6.14 CONCLUSIONS

6.14.1 General

PAIs are of major importance for the virulence of *S. enterica*. Hallmarks of *Salmonella* virulence, such as cell invasion, intracellular survival and the production of the Vi antigen capsule are encoded by SPIs. The large number of SPIs present in *Salmonella* indicate the adaptation of a prototrophic, free-living species to a complex pathogenic life-style. *Salmonella* serovars can occur as gastrointestinal commensals, but can also cause infections by penetration of the gastrointestinal mucosa. The bacteria need to adapt to extremely different habitats ranging from intracellular locations in various cell types to biofilms in gall bladder stones. During evolution of the genus *Salmonella*, different genetic elements were acquired and efficiently combined. For example, about 7.8% of the 4.8 Mb chromosome of *S. enterica* serovar Typhi consists of PAIs (Table 6.1). A similar complexity and high number of PAIs is found in only a few pathogens, namely certain pathotypes of *E. coli* and in *Staphylococcus aureus* (Schmidt and Hensel, 2004).

Comparison of structural features and evolutionary characteristics indicated that SPI extend over a range of ancient horizontal acquisitions, such as SPI-1 that became stable constituents of the *Salmonella* genome, to very variable elements such as SGI-1 that are still in the process of extensive alteration and horizontal distribution.

6.14.2 Common features of SPI

Salmonella pathogenicity islands 2 to 8, SPI-10 and HPI are associated with tRNA loci. This observation demonstrates the role of these loci as insertion points for DNA elements acquired by horizontal gene transfer and may indicate that certain loci function as hotspots for repetitive insertion of horizontally acquired DNA elements. Genome-based screening for further *Salmonella*-specific insertions at tRNA genes identified other putative PAIs with unknown functions in pathogenesis (Hansen-Wester and Hensel, 2002).

The majority of SPIs are stable constituents of the *Salmonella* chromosome. Genes associated with DNA mobility are absent from the majority of

SPIs and this observation correlates with their high stability. Most SPIs have become part of the core set of genes of *S. enterica* and encode species-specific traits. A smaller subset of SPIs is limited to certain subspecies or even specific serovars. These SPIs harbor genes associated with DNA mobility and are likely to be more recent acquisitions. It should be mentioned that the stability of several of the SPIs described here has not been analyzed systematically and it is possible that SPIs associated with bacteriophages, such as SPI-7 or SPI-10 are less stable than SPI-1 to SPI-5.

Within most of the SPI loci, significant structural and functional heterogeneities were observed. Examples are the *sit* genes (SPI-1) encoding an iron uptake system and the *ttr* genes (SPI-2) encoding an enzyme system for anaerobic respiration, whilst other portions of both SPIs encode TTSS. The organization of SPI-3 is highly variable in different subspecies of *Salmonella*. These genetic mosaics are indicative of repetitive insertion of foreign DNA elements of various origins into the same insertion site (e.g. at a tRNA gene). Obviously, this situation complicates the definition of function of SPI genes and the understanding of evolutionary events.

6.14.3 Perspectives and future research on SPI

The acquisition of most SPIs in *S. enterica* occurred millions of years ago but we also observe that the current acquisition of new SPIs leads to the emergence of new antibiotic-resistant epidemic strains. Whilst the roles in pathogenesis of some SPIs are well defined, the functions in virulence of many genes within SPIs are not understood. It is conceivable that several of these SPIs contribute to host specificity. This may also apply to the pool of effector genes of the SPI-1 and SPI-2 encoded TTSS that show a heterogeneous distribution in *Salmonella* species.

Although the contribution of SPIs to pathogenesis can be very different, several common motifs were identified between SPIs. In conclusion, the acquisition of a large number of PAI has had a pivotal role in the evolution of *S. enterica* to a highly successful pathogen. Clarification of the functions of SPIs and their interactions with other mobile genetic elements will broaden the understanding of the molecular mechanisms of virulence of *S. enterica*.

It can be assumed that SPIs with central virulence functions in the majority of *S. enterica* serotypes have already been identified. However, many other SPIs might be present that have a more limited distribution or functions that are restricted to specific hosts.

6.15 ACKNOWLEDGEMENTS

Work in the group of M. Hensel was supported by grants of the Deutsche Forschungsgemeinschaft, the BMBF and the European Commission.

REFERENCES

Ahmer, B. M., van Reeuwijk, J., Watson, P. R., Wallis, T. S. and Heffron, F. (1999). *Salmonella* SirA is a global regulator of genes mediating enteropathogenesis. *Mol Microbiol*, **31**, 971–82.

Amavisit, P., Lightfoot, D., Browning, G. F. and Markham, P. F. (2003). Variation between pathogenic serovars within *Salmonella* pathogenicity islands. *J Bacteriol*, **185**, 3624–35.

Bäumler, A. J. (1997). The record of horizontal gene transfer in *Salmonella*. *Trends Microbiol*, **5**, 318–22.

Bäumler, A. J., Tsolis, R. M., van der Velden, A. W. *et al.* (1996). Identification of a new iron regulated locus of *Salmonella typhi*. *Gene*, **183**, 207–13.

Beuzon, C. R., Meresse, S., Unsworth, K. E. *et al.* (2000). *Salmonella* maintains the integrity of its intracellular vacuole through the action of *sifA*. *EMBO J*, **19**, 3235–49.

Blanc-Potard, A. B. and Groisman, E. A. (1997). The *Salmonella selC* locus contains a pathogenicity island mediating intramacrophage survival. *EMBO J*, **16**, 5376–85.

Blanc-Potard, A. B., Solomon, F., Kayser, J. and Groisman, E. A. (1999). The SPI-3 pathogenicity island of *Salmonella enterica*. *J Bacteriol*, **181**, 998–1004.

Boyd, D., Peters, G. A., Cloeckaert, A. *et al.* (2001). Complete nucleotide sequence of a 43-kilobase genomic island associated with the multidrug resistance region of *Salmonella enterica* serovar Typhimurium DT104 and its identification in phage type DT120 and serovar Agona. *J Bacteriol*, **183**, 5725–32.

Carniel, E. (2001). The *Yersinia* high-pathogenicity island: an iron-uptake island. *Microbes Infect*, **3**, 561–9.

Chakravortty, D., Hansen-Wester, I. and Hensel, M. (2002). *Salmonella* pathogenicity island 2 mediates protection of intracellular *Salmonella* from reactive nitrogen intermediates. *J Exp Med*, **195**, 1155–66.

Cirillo, D. M., Valdivia, R. H., Monack, D. M. and Falkow, S. (1998). Macrophage-dependent induction of the *Salmonella* pathogenicity island 2 type III secretion system and its role in intracellular survival. *Mol Microbiol*, **30**, 175–88.

Deiwick, J., Nikolaus, T., Erdogan, S. and Hensel, M. (1999). Environmental regulation of *Salmonella* Pathogenicity Island 2 gene expression. *Mol Microbiol*, **31**, 1759–73.

Doublet, B., Lailler, R., Meunier, D. *et al.* (2003). Variant *Salmonella* genomic island 1 antibiotic resistance gene cluster in *Salmonella enterica* serovar Albany. *Emerg Infect Dis*, **9**, 585–91.

Figueroa-Bossi, N., Uzzau, S., Maloriol, D. and Bossi, L. (2001). Variable assortment of prophages provides a transferable repertoire of pathogenic determinants in *Salmonella*. *Mol Microbiol*, **39**, 260–71.

Folkesson, A., Lofdahl, S. and Normark, S. (2002). The *Salmonella enterica* subspecies I specific centisome 7 genomic island encodes novel protein families present in bacteria living in close contact with eukaryotic cells. *Res Microbiol*, **153**, 537–45.

Foultier, B., Troisfontaines, P., Muller, S., Opperdoes, F. R. and Cornelis, G. R. (2002). Characterization of the *ysa* pathogenicity locus in the chromosome of *Yersinia enterocolitica* and phylogeny analysis of type III secretion systems. *J Mol Evol*, **55**, 37–51.

Freeman, J. A., Rappl, C., Kuhle, V., Hensel, M. and Miller, S. I. (2002). SpiC is required for translocation of *Salmonella* Pathogenicity Island 2 effectors and secretion of translocon proteins SseB and SseC. *J Bacteriol*, **184**, 4971–80.

Fu, Y. and Galan, J. E. (1999). A *Salmonella* protein antagonizes Rac-1 and Cdc42 to mediate host-cell recovery after bacterial invasion. *Nature*, **401**, 293–7.

Galan, J. E. (2001). *Salmonella* interactions with host cells: type III secretion at work. *Annu Rev Cell Dev Biol*, **17**, 53–86.

Galan, J. E. and Curtiss, R. (1989). Cloning and molecular characterization of genes whose products allow *Salmonella typhimurium* to penetrate tissue culture cells. *Proc Natl Acad Sci USA*, **86**, 6383–7.

Garcia-del Portillo, F., Zwick, M. B., Leung, K. Y. and Finlay, B. B. (1993). *Salmonella* induces the formation of filamentous structures containing lysosomal membrane glycoproteins in epithelial cells. *Proc Natl Acad Sci USA*, **90**, 10544–8.

Ginocchio, C. C., Rahn, K., Clarke, R. C. and Galan, J. E. (1997). Naturally occurring deletions in the centisome 63 pathogenicity island of environmental isolates of *Salmonella* spp. *Infect Immun*, **65**, 1267–72.

Groisman, E. A. and Ochman, H. (1993). Cognate gene clusters govern invasion of host epithelial cells by *Salmonella typhimurium* and *Shigella flexneri*. *EMBO J*, **12**, 3779–87.

(1997). How *Salmonella* became a pathogen. *Trends Microbiol*, **5**, 343–9.

Hacker, J., Knapp, S. and Goebel, W. (1983). Spontaneous deletions and flanking regions of the chromosomally inherited hemolysin determinant of an *Escherichia coli* O6 strain. *J Bacteriol*, **154**, 1145–52.

Hansen-Wester, I. and Hensel, M. (2002). Genome-based identification of chromosomal regions specific for *Salmonella* spp. *Infect Immun*, **70**, 2351–60.

Hansen-Wester, I., Stecher, B. and Hensel, M. (2002). Analyses of the evolutionary distribution of *Salmonella* translocated effectors. *Infect Immun,* **70**, 1619–22.

Hardt, W. D., Chen, L. M., Schuebel, K. E., Bustelo, X. R. and Galan, J. E. (1998). *S. typhimurium* encodes an activator of Rho GTPases that induces membrane ruffling and nuclear responses in host cells. *Cell,* **93**, 815–26.

Hensel, M., Hinsley, A. P., Nikolaus, T., Sawers, G. and Berks, B. C. (1999a). The genetic basis of tetrathionate respiration in *Salmonella typhimurium. Mol Microbiol,* **32**, 275–88.

Hensel, M., Nikolaus, T. and Egelseer, C. (1999b). Molecular and functional analysis indicates a mosaic structure of *Salmonella* Pathogenicity Island 2. *Mol Microbiol,* **31**, 489–98.

Hensel, M., Shea, J. E., Bäumler, A. J. *et al.* (1997). Analysis of the boundaries of *Salmonella* pathogenicity island 2 and the corresponding chromosomal region of *Escherichia coli* K-12. *J Bacteriol,* **179**, 1105–11.

Hensel, M., Shea, J. E., Gleeson, C. *et al.* (1995). Simultaneous identification of bacterial virulence genes by negative selection. *Science,* **269**, 400–3.

Hersh, D., Monack, D. M., Smith, M. R. *et al.* (1999). The *Salmonella* invasin SipB induces macrophage apoptosis by binding to Caspase-1. *Proc Natl Acad Sci USA,* **96**, 2396–401.

Hindle, Z., Chatfield, S. N., Phillimore, J. *et al.* (2002). Characterization of *Salmonella enterica* derivatives harboring defined *aroC* and *Salmonella* Pathogenicity Island 2 type III secretion system (*ssaV*) mutations by immunization of healthy volunteers. *Infect Immun,* **70**, 3457–67.

Hueck, C. J. (1998). Type III protein secretion systems in bacterial pathogens of animals and plants. *Microbiol Mol Biol Rev,* **62**, 379–433.

Jones, M. A., Wood, M. W., Mullan, P. B. *et al.* (1998). Secreted effector proteins of *Salmonella dublin* act in concert to induce enteritis. *Infect Immun,* **66**, 5799–804.

Knodler, L. A. and B. B. Finlay (2001). *Salmonella* and apoptosis: to live or let die? *Microbes Infect,* **3**, 1321–6.

Knodler, L. A., Celli, J., Hardt, W. D. *et al.* (2002). *Salmonella* effectors within a single pathogenicity island are differentially expressed and translocated by separate type III secretion systems. *Mol Microbiol,* **43**, 1089–103.

Kubori, T., Matsushima, Y., Nakamura, D. *et al.* A., Galan, J. E. and Aizawa, S. I. (1998). Supramolecular structure of the *Salmonella typhimurium* type III protein secretion system. *Science,* **280**, 602–5.

Kuhle, V. and Hensel, M. (2002). SseF and SseG are translocated effectors of the type III secretion system of *Salmonella* pathogenicity island 2 that modulate aggregation of endosomal compartments. *Cell Microbiol,* **4**, 813–24.

Lee, A. K., Detweiler, C. S. and Falkow, S. (2000). OmpR regulates the two-component system SsrA-SsrB in *Salmonella* pathogenicity island 2. *J. Bacteriol*, **182**, 771–81.

Lucas, R. L. and Lee, C. A. (2000). Unravelling the mysteries of virulence gene regulation in *Salmonella typhimurium*. *Mol Microbiol*, **36**, 1024–33.

Medina, E., Paglia, P., Nikolaus, T. *et al.* (1999). Pathogenicity Island 2 mutants of *Salmonella typhimurium* are efficient carriers for heterologous antigens and enable modulation of immune responses. *Infect Immun*, **67**, 1093–9.

Miao, E. A. and Miller, S. I. (2000). A conserved amino acid sequence directing intracellular type III secretion by *Salmonella typhimurium*. *Proc Natl Acad Sci USA*, **97**, 7539–44.

Mills, D. M., Bajaj, V. and Lee, C. A. (1995). A 40 kb chromosomal fragment encoding *Salmonella typhimurium* invasion genes is absent from the corresponding region of the *Escherichia coli* K-12 chromosome. *Mol Microbiol*, **15**, 749–59.

Mirold, S., Ehrbar, K., Weissmüller, A. *et al.* (2001). *Salmonella* host cell invasion emerged by acquisition of a mosaic of separate genetic elements, including *Salmonella* Pathogenicity Island 1 (SPI-1), SPI-5, and *sopE2*. *J Bacteriol*, **183**, 2348–58.

Mirold, S., Rabsch, W., Rohde, M. *et al.* (1999). Isolation of a temperate bacteriophage encoding the type III effector protein SopE from an epidemic *Salmonella typhimurium* strain. *Proc Natl Acad Sci USA*, **96**, 9845–50.

Monack, D. M., Detweiler, C. S. and Falkow, S. (2001). *Salmonella* pathogenicity island 2-dependent macrophage death is mediated in part by the host cysteine protease Caspase-1. *Cell Microbiol*, **3**, 825–37.

Monack, D. M., Hersh, D., Ghori, N. *et al.* (2000). *Salmonella* exploits Caspase-1 to colonize Peyer's patches in a murine typhoid model. *J Exp Med*, **192**, 249–58.

Moncrief, M. B. and Maguire, M. E. (1998). Magnesium and the role of MgtC in growth of *Salmonella typhimurium*. *Infect Immun*, **66**, 3802–9.

Norris, F. A., Wilson, M. P., Wallis, T. S., Galyov, E. E. and Majerus, P. W. (1998). SopB, a protein required for virulence of *Salmonella dublin*, is an inositol phosphate phosphatase. *Proc Natl Acad Sci USA*, **95**, 14057–9.

Ochman, H. and Groisman, E. A. (1996). Distribution of pathogenicity islands in *Salmonella* spp. *Infect Immun*, **64**, 5410–12.

Ochman, H., Soncini, F. C., Solomon, F. and Groisman, E. A. (1996). Identification of a pathogenicity island required for *Salmonella* survival in host cells. *Proc Natl Acad Sci. USA*, **93**, 7800–4.

Oelschlaeger, T. A., Zhang, D., Schubert, S. *et al.* (2003). The High-Pathogenicity Island is absent in human pathogens of *Salmonella enterica* subspecies I but present in isolates of subspecies III and VI. *J Bacteriol*, **185**, 1107–11.

Parkhill, J., Dougan, G., James, K. D. (2001a). Complete genome sequence of a multiple drug resistant *Salmonella enterica* serovar Typhi CT18. *Nature,* **413**, 848–52.

Parkhill, J., Wren, B. W., Thomson, N. R. (2001b). Genome sequence of *Yersinia pestis,* the causative agent of plague. *Nature,* **413**, 523–7.

Pickard, D., Wain, J., Baker, S. *et al.* (2003). Composition, acquisition, and distribution of the Vi exopolysaccharide-encoding *Salmonella enterica* pathogenicity island SPI-7. *J Bacteriol,* **185**, 5055–65.

Rappl, C., Deiwick, J. and Hensel, M. (2003). Acidic pH is required for the functional assembly of the type III secretion system encoded by *Salmonella* pathogenicity island 2. *FEMS Microbiol. Lett,* **226**, 363–72.

Rescigno, M., Urbano, M., Valzasina, B. *et al.* (2001). Dendritic cells express tight junction proteins and penetrate gut epithelial monolayers to sample bacteria. *Nat Immunol,* **2**, 361–7.

Schmidt, H. and Hensel, M. (2004). Pathogenicity islands in bacterial pathogenesis. *Clin Microbiol. Rev,* **17**, 14–56.

Shea, J. E., Beuzon, C. R., Gleeson, C., Mundy, R. and Holden, D. W. (1999). Influence of the *Salmonella typhimurium* Pathogenicity Island 2 type III secretion system on bacterial growth in the mouse. *Infect Immun,* **67**, 213–19.

Shea, J. E., Hensel, M., Gleeson, C. and Holden, D. W. (1996). Identification of a virulence locus encoding a second type III secretion system in *Salmonella typhimurium. Proc Natl Acad Sci USA,* **93**, 2593–7.

Snavely, M. D., Miller, C. G. and Maguire, M. E. (1991). The *mgtB* Mg^{2+} transport locus of *Salmonella typhimurium* encodes a P-type ATPase. *J Biol Chem,* **266**, 815–23.

Stevens, M. P., Wood, M. W., Taylor, L. A. *et al.* (2002). An Inv/Mxi-Spa-like type III protein secretion system in *Burkholderia pseudomallei* modulates intracellular behaviour of the pathogen. *Mol Microbiol,* **46**, 649–59.

Townsend, S. M., Kramer, N. E., Edwards, R. *et al.* (2001). *Salmonella enterica* serovar Typhi possesses a unique repertoire of fimbrial gene sequences. *Infect Immun,* **69**, 2894–901.

Uchiya, K., Barbieri, M. A., Funato, K. *et al.* (1999). A *Salmonella* virulence protein that inhibits cellular trafficking. *EMBO J,* **18**, 3924–33.

Vazquez-Torres, A., Jones-Carson, J., Bäumler, A. J. *et al.* (1999). Extraintestinal dissemination of *Salmonella* by CD18-expressing phagocytes. *Nature,* **401**, 804–8.

Vazquez-Torres, A., Xu, Y., Jones-Carson, J. *et al.* (2000). *Salmonella* Pathogenicity Island 2-dependent evasion of the phagocyte NADPH oxidase. *Science,* **287**, 1655–8.

Wallis, T. S. and Galyov, E. E. (2000). Molecular basis of *Salmonella*-induced enteritis. *Mol Microbiol*, **36**, 997–1005.

Waterman, S. R. and Holden, D. W. (2003). Functions and effectors of the *Salmonella* pathogenicity island 2 type III secretion system. *Cell Microbiol*, **5**, 501–11.

Wong, K. K., McClelland, M., Stillwell, L. C. *et al.* (1998). Identification and sequence analysis of a 27-kilobase chromosomal fragment containing a *Salmonella* pathogenicity island located at 92 minutes on the chromosome map of *Salmonella enterica* serovar *typhimurium* LT2. *Infect Immun*, **66**, 3365–71.

Wood, M. W., Jones, M. A., Watson, P. R. *et al.* (1998). Identification of a pathogenicity island required for *Salmonella* enteropathogenicity. *Mol Microbiol*, **29**, 883–91.

Yu, X. J., Ruiz-Albert, J., Unsworth, K. E. *et al.* (2002). SpiC is required for secretion of *Salmonella* Pathogenicity Island 2 type III secretion system proteins. *Cell Microbiol*, **4**, 531–40.

Zhang, X. L., Morris, C. and Hackett, J. (1997). Molecular cloning, nucleotide sequence, and function of a site-specific recombinase encoded in the major 'pathogenicity island' of *Salmonella typhi*. *Gene*, **202**, 139–46.

Zhang, X. L., Tsui, I. S., Yip, C. M. *et al.* (2000). *Salmonella enterica* serovar *typhi* uses type IVB pili to enter human intestinal epithelial cells. *Infect Immun*, **68**, 3067–73.

Zhou, D., Hardt, W. D. and Galan, J. E. (1999). *Salmonella typhimurium* encodes a putative iron transport system within the centisome 63 pathogenicity island. *Infect Immun*, **67**, 1974–81.

CHAPTER 7

In vivo identification, expression and function of *Salmonella* virulence genes

Helene Andrews-Polymenis, Caleb W. Dorsey, Manuela Raffatellu and Andreas J. Bäumler

7.1 INTRODUCTION

Any determinant that enables a *Salmonella* serotype to enter a host, to find a unique niche to multiply, to avoid or subvert the host defenses, to cause disease and to be transmitted to the next susceptible host may be considered a virulence determinant. Essential genes required for growth in standard laboratory medium are usually not included under this broad definition of virulence genes. The total number of virulence genes present in the *Salmonella* genome can be estimated by screening a bank of mutants generated by random transposon mutagenesis using an animal model of infection. The genome of *Salmonella enterica* serovar Typhimurium strain LT2 contains 4552 intact open reading frames (McClelland *et al.*, 2001). Of these, approximately 490 genes are essential during growth in rich medium (Knuth *et al.*, 2004). Thus, only the function of about 4062 genes is assessed when *S. enterica* serovar Typhimurium transposon mutants are generated and analyzed. Analysis of 197 randomly generated transposon mutants of *S. enterica* serovar Typhimurium for virulence in mice upon intra gastric infection identified 8 mutants that were more than 1000 fold attenuated (Bowe *et al.*, 1998). Extrapolating to the actual number of intact genes present in the genome (4062 genes) this study suggests that mutations in approximately 165 *S. enterica* serovar Typhimurium genes result in attenuation of more than 1,000-fold in mice. In a similar study, a bank of 260 *S. enterica* serovar Typhimurium transposon mutants was screened using intra gastric infection of mice which identified 16 mutants that were defective in their ability to compete with an isogenic wild type for colonization of murine tissues (Tsolis

'Salmonella' Infections: Clinical, Immunological and Molecular Aspects, ed. Pietro Mastroeni and Duncan Maskell. Published by Cambridge University Press.

et al., 1999). Extrapolation of this more sensitive screen suggests that the *S. enterica* serovar Typhimurium genome contains approximately 250 virulence genes that are required for organ colonization in mice. The total number of *S. enterica* serovar Typhimurium virulence genes is likely to be larger than suggested by these estimates because some genes required for transmission to susceptible hosts, genes exhibiting functional redundancy and some genes required for infection of other host species may not be readily identified by screening a mutant bank for the ability to colonize and cause disease in mice. This article will review how *S. enterica* virulence genes have been identified in vivo and provide an overview of what is known about their function and the conditions for their expression.

7.2 IDENTIFICATION OF VIRULENCE GENES IN VIVO

7.2.1 General

Two decades of extensive work has addressed the identification of virulence factors of *Salmonella* both in vitro and in vivo. Over 70 genes involved in virulence have been identified and many of these genes have been studied extensively in vitro. In vitro identification of virulence factors in *S. enterica* has been very successful but is limited by our incomplete understanding of, and inability to recreate, the intricate and variable environments present in vivo. In addition, results of experiments using in vitro models of infection do not always parallel results obtained from in vivo experimentation. The development of in vivo approaches has allowed us to gain better insight into the complicated interaction between pathogen and host. Identification of virulence factors in vivo, however, presents a unique set of challenges compared to in vitro identification. These challenges include the availability of appropriate animal models for human and animal disease syndromes and the ability to screen large numbers of mutants without using large numbers of animals. In vivo identification of virulence factors of *Salmonella* has relied upon scientific innovation and has been based mainly on four approaches: (a) insertional mutagenesis; (b) in vivo expression technology (IVET); (c) signature-tagged transposon mutagenesis (STM); (d) differential fluorescence induction (DFI) (Table 7.1).

7.2.2 Traditional transposon mutagenesis

The identification of virulence genes in in vivo systems begun in the mid 1980s and initially relied heavily on transposon mediated approaches

Table 7.1. *Screens for* in vivo *identification of* Salmonella *virulence genes.*

Reference	Serov.	Method	Model/Route	Genes Identified
Insertional Mutagenesis				
Fields *et al.*, 1986	TM	9516 Tn10 insertions 83 decreased intracellular survival	Peritoneal Macrophages 2 mol Screen in Balb/c mice/IP	See Baumler *et al.* (1994)
Gulig and Curtiss, 1988	TM	Tn5 insertions into virulence plasmid	Balb/C mice/PO Sprauqe Dawley mice/PO	
Miller *et al.*, 1989	TM	150 TnPhoA insertions 15 avirulent	Balb/C mice PO	9/15 LPS defect 6/15 are smooth
Rhen *et al.*, 1989	TM	36 Tn5 insertions on vir. Plasmid 2 with reduced virulence	CBAxC57B6 F1 mice/ IV	
Sizemore *et al.*, 1991	TM	11 Tn5 insertions 9 attenuated	Balb/C mice	
Rubino *et al.*, 1993	AO	95 TnPhoA insertions 23 nonadherent 4 reduced virulence in mice	HeLa, LK cells Balb/c mice PO	
Baumler *et al.*, 1994	TM	Follow up Fields (1986) Tn10		*htrA* *prc* *purD* *fliD nagA* *smpB* *adhE* 14 w/ no homology

(*cont.*)

Table 7.1. (*cont.*)

Reference	Serov.	Method	Model/Route	Genes Identified
Lodge *et al.*, 1995	TM	97 TnPhoA insertions	Hep-2 cells Rabbit ligated ileal loops	*invG* (SPI-1) *invH* (SPI-1) *pagC*
Turner *et al.*, 1998	TM	2800 Tn5-TCI insertions 18 reduced colonization	3 wk. old chicks/ PO	rfaK rfbK rfbB dksA hupA sipC (SPI-1)
Bowe *et al.*, 1998	TM	526 MudJ insertions 12 with 1000x increased LD50	Balb/c mice 330 IP 196 IG	*deoA* LPS biosynth (unnamed) *tolB* *mdoB* *yhdJ* *yabJ/K* *fmt* *pdx* putative malate oxidoredu. 2 w/no homology
In vivo Expression Technology (IVET)				
Mahan *et al.*, 1993	TM	IVET, promoterless *purA* and *lacZY* 10^6 organisms injected 14/273 recovered $purA^+lacZY^-$	Balb/c mice IP	*carAB* *pheSThimA* *rfb* 2 genes w/no homology
Mahan *et al.*, 1995	TM	IVET, promoterles *cat* and *lacZY* 10^5 to 10^6 organisms injected	Balb/c mice IP	*fadB*

Heithoff *et al.*, 1997	TM	IVET, promoterless *purA* and *lacZY* or promoterles *cat* and *lacZY* 10^8 to 10^9 to infect IG 5×10^5 to inject IP 2647 recovered *purA*$^+$*lacZY*$^-$ 476 fusion joints sequenced 100 unique genes	Balb/c mice IG or IP	*phoP* *pmrB* *cadC* *vacB* *vacC* *spvB* *cfa* *otsA* *recD* *hemA* *entF* *fhuA* *cirA* *iviX* put. Heavy metal trans. *ndk* *ivi VI*-A Tia like *iviVI*-B PfemP-1 like 3 with no homology
Conner *et al.*, 1998	TM	Follow up to Heithoff (1997)		No additional genes named
Heithoff *et al.*, 1999	TM	IVET, follow up to Heithoff (1997)		No additional genes named
Janakiram *et al.*, 2000	TM	IVET, promoterless *purA* and *lacZY* 50 *purA*$^+$*lacZY* analyzed 38 independent insertions	Cultured murine hepatocytes	*sitABCD* (*SPI-1 encoded*) *sodA* *pagJ* *ssaE* (SPI-2)
Stanley *et al.*, 2000	TM	IVET, promoterless *purA* and *lacZY* 10^9 organisms to infect orally 90 recovered *purA*$^+$*lacZY*	Mice PO and IP	*gipA*

(*cont.*)

Table 7.1. (*cont.*)

Reference	Serov.	Method	Model/Route	Genes Identified
Signature Tag Mutagenesis (STM)				
Hensel *et al.*, 1995	TM	STM, 1152 unique Tags 40 absent from output pool	Balb/c mice IP	*aroA*
				purD
				purL
				spvA
				spvR
				spvD
				rfbD
				rfbK
				rfbB
				rfbM
				opmR/envZ
				SPI-2- first identification*
Tsolis *et al.*, 1999	TM	STM, 260 unique Tags 17 absent from pool	Balb/c mice PO	*slrP*- in mouse
				O423b-in mouse
		4 absent in only one species	Calves PO	*ipaH* -in mouse
				bcf – in calf
				hilA (SPI-1) in both species
				ptsA both species
				orgA (SPI-1) both species
				spiB (SPI-2) both species
				prgH (SPI-1) both species
				rfaJ both species
				spvB both species
				spvR both species
				yjeP both species
				selB both specices
				3 with no homology

Bispham *et al.*, 2001	D	STM, 5280 unique tags 2 absent from output pool described in detail	Balb/c mice PO Calves/PO	*sseD* (SPI-2)- absent in mouse only *ssaT* (SPI-2)-absent from both species
Lichtensteiger *et al.*, 2003	CS	STM, 45 unique tags 2 absent from output pool	Swine PO and IP	*rfb* *hilA* (SPI-1)
Differential Fluorescence Induction				
Valdivia and Falkow, 1996	TM	DFI, technique development		
Valdivia and Falkow, 1996	TM	DFI, technique development Acid inducible genes in vitro	Raw 264.7 cells	*aas* *dps* *marR* *pagA* *rna* *pbpA* *emrR*
Wendland and Bumann, 2002	TM	DFI, GFP optimization 10^9 organisms to infect recover bacteria from Peyers Patches	Balb/c mice IG	
Bumann, 2002	TM	DFI 10^7 organisms to infect using library	Balb/c mice IV	*pibB* (SPI-V) *sifA* (SPI-2 secreted) *aroQ*
Other Methods				
Gulig and Curtiss, 1987	TM	Cure of virulence plasmid	Balb/c mice PO	*spv*

TM = Typhimurium; AO = Abortusovis; D = Dublin; PO = *per os*
IP = intraperitoneal; IG = intragastric; IV = intravenous; CS = choleraesuis

for the generation of mutants. The subsequent testing of individual mutants in animal models was in some cases preceded by initial screening in cell culture (Bowe *et al.*, 1998; Fields *et al.*, 1986; Lodge *et al.*, 1995; Miller *et al.*, 1989; Rhen *et al.*, 1989; Rubino *et al.*, 1993; Sizemore *et al.*, 1991). These early transposon mediated approaches have the drawback that screening for loss of function in vivo involves large numbers of animals, is highly labor intensive and virulence genes encoding proteins with redundant function cannot be identified.

Early work identified 83 Tn*10* insertion mutants of *S. enterica* serovar Typhimurium, from a pool of 9516, by initial screening in peritoneal macrophages followed by a second screen in mice (Fields *et al.*, 1986). The Tn*10* flanking nucleotide sequence of 23 of these mutants was determined and identified *htrA, prc, purD, fliD, nagA smpB,* and 14 genes with no sequence homology to known genes (Bäumler *et al.*, 1994; Fields *et al.*, 1989; Groisman *et al.*, 1989). Subsequent approaches used Tn*phoA*, Mud*J* and Tn*5* transposons, and insertion mutants were screened for virulence in mice (Bowe *et al.*, 1998; Miller *et al.*, 1989; Rhen *et al.*, 1989; Rubino *et al.*, 1993; Sizemore *et al.*, 1991). Three studies used Tn*5* based mutagenesis to study the in vivo role of different regions of the virulence plasmid of *S. enterica* serovar Typhimurium in infections in mice (Rhen *et al.*, 1989; Sizemore *et al.*, 1991; Turner *et al.*, 1998). These studies followed early work on *S. enterica* serovar Typhimurium that involved curing the bacteria of the virulence plasmid and testing these isolates for virulence in mice (Gulig *et al.*, 1992; Gulig, Curtiss, 1988). Complementation of these mutants with sections of the virulence plasmid, and insertional inactivation of genes on the virulence plasmid ultimately resulted in the identification of the *spv* locus (Gulig *et al.*, 1992; Gulig and Curtiss, 1988; Sizemore *et al.*, 1991). Thirty-six Tn*5* insertion mutants in the *S. enterica* serovar Typhimurium virulence plasmid were generated, and two mutants were found to have reduced virulence in intravenous challenge of mice (Rhen *et al.*, 1989). In a second study, 11 Tn*5* hops were generated on the virulence plasmid and it was found that 9 of these showed an attenuated phenotype in mice (Sizemore *et al.*, 1991). In a more recent study, a Tn*5* based strategy was used to make 2,800 mutants of *S. enterica* serovar Typhimurium that were later screened in 3 week old chicks (Turner *et al.*, 1998). Eighteen mutants were recovered that had reduced colonization in this model and the genes responsible for 6 of these mutations were identified as genes involved in lipopolysaccharide (LPS) biosynthesis and as genes belonging to the *Salmonella* pathogenicity island –1 (SPI-1) type three secretion system (TTSS) (Table 7.1) (Turner *et al.*, 1998).

A Tn*phoA* based approach was used to generate insertion mutants in *S. enterica* serovar Typhimurium that were tested for virulence using oral infections in mice (Miller *et al.*, 1989). Insertion of Tn*phoA* generates translational fusions with the *phoA* gene, and a prescreen for PhoA activity can be used to restrict further in vivo analysis to those mutants that carry an insertion in a gene encoding a protein that is exported from the bacterial cytosol. This approach yielded 15 avirulent mutants (of 150 tested), and 9/15 mutants had defective lipopolysaccharide (Miller *et al.*, 1989). A Tn*phoA* based approach was also used in a study in which 97 insertion mutants in *S. enterica* serovar Typhimurium were generated, subsequently pre-screened in HEP-2 cells, and screened in rabbit ligated ileal loops (Lodge *et al.*, 1995). This approach identified two genes within SPI-1 that are known as *invG* and *invH* and that have previously been shown to be involved in virulence (Lodge *et al.*, 1995). Tn*phoA* has also been used to generate insertion mutations in *S. enterica* serovar Abortusovis and these mutants have been prescreened for epithelial cell adherence to HeLa and LK (lamb kidney) cells, and subsequently screened in oral infection studies in mice (Rubino *et al.*, 1993). From a total of 95 mutants, this study identified 23 with reduced ability to adhere to epithelial cells in vitro, and of these, 4 mutants had reduced virulence in mice after oral challenge (Rubino *et al.*, 1993).

7.2.3 In vivo expression technology (IVET)

IVET is a widely used approach developed for the in vivo identification of virulence genes in *S. enterica.* IVET is a promoter trap strategy where a library of random genomic fragments is ligated to a promoterless reporter gene, and this construct is used to determine transcriptional activity of the reporter that is required for in vivo growth/survival. This strategy was developed for identification of in vivo induced genes (*ivi*) in *S. enterica* serovar Typhimurium (Mahan *et al.*, 1993; Osbourn *et al.*, 1987). Advantages of IVET include the ability to screen positively a large number of promoters using few animals and the ability to identify virulence genes encoding proteins with redundant function. However, there is no particular reason why genes induced in vivo should necessarily be required for causing disease in an animal. The first IVET strategies used a genomic library cloned upstream of a promoterless *purA-lacZ* operon fusion, in a *purA* mutant background strain of *S. enterica* serovar Typhimurium (Mahan, 1993). The bacterial clones in which constitutive promoters had been cloned upstream of *purA-lacZ* would survive in vivo due to transcription of the *purA* gene, and could be identified as Lac^+ in vitro on indicator plates due to the expression of the *lacZ* gene. Conversely, promoters activated only in vivo would induce the transcription of the *purA* gene and allow bacterial survival within the infected animal, but would be Lac$^-$ on indicator plates due to lack of expression of the *lacZ* gene in vitro. This strategy had several limitations including the fact that transiently active promoters could not be identified, and virulence genes expressed both in vitro and in vivo would be screened out using this system. In the IVET study carried out by Mahan *et al.*, a pool of 10^6 clones of *S. enterica* serovar Typhimurium *purA* with integrated *purA-lacZ* fusions was injected intraperitoneally into mice. Five per cent of the recovered strains were both PurA$^+$ and Lac$^-$, and thus the genes regulated by the promoters contained within the *purA-lacZ* fusions of the PurA$^+$ Lac$^-$ recovered strains were termed *ivi* for in vivo induced. These *ivi* genes included *carAB*, *pheSThimA*, the *rfb* locus, and two genes with no homology to known genes (Mahan *et al.*, 1993).

Additional IVET experiments have been completed to identify *ivi* genes of *S. enterica* serovar Typhimurium (Heithoff *et al.*, 1997; Heithoff *et al.*, 1999; Julio *et al.*, 1998; Stanley *et al.*, 2000). A study by Heithoff *et al.* used an IVET approach in mice and cultured murine macrophages to identify *ivi* genes of *S. enterica* serovar Typhimurium (Heithoff *et al.*, 1997). In this study, 2,647 fusions were recovered that were Lac$^-$, 476 fusion joints were sequenced, and 100 unique *ivi* genes were identified. More than 50% of these were genes that had been identified previously, and included several genes implicated in *Salmonella* virulence, including major regulators of SPI-1, *phoP*, *pmrB*, and *spvB* (Table 7.1) (Heithoff *et al.*, 1997). This work was followed up by additional studies (Conner *et al.*, 1998; Heithoff *et al.*, 1999). In the first of these two studies, a collection of *ivi* genes was screened for genes in regions of atypical base composition, for the absence of sequence homology to known genes, and for their presence in *Salmonella* but absence in other selected pathogens (Conner *et al.*, 1998). Seven *ivi* genes were identified that fulfilled the first two criteria and were *Salmonella* specific. In the second study, previously identified *ivi* genes were grouped according to response to defined in vitro conditions, such as low magnesium and pH, and defectiveness of entry into cultured human epithelial cells (Heithoff *et al.*, 1999).

Modifications of the original IVET strategy have been used. One of these includes the use of antibiotic resistance cassettes and recombinase based systems (RIVET) in order to address these limitations (Mahan *et al.*, 1995; Merrell and Camilli, 2000). In a later study, the original IVET strategy was modified to utilize a *cat* gene (chloramphenicol resistance, pIVET8) as the reporter gene, and was used to identify *ivi* genes in *S. enterica* serovar Typhimurium (Mahan *et al.*, 1995). Again, approximately 5% (9/193) of the clones recovered after two days of in vivo growth in Balb/c mice treated with chloramphenicol were chloramphenicol resistant and Lac$^-$ (Mahan *et al.*, 1995). From this study only *fadB* was positively identified, a gene involved in the metabolism of long- and short-chain fatty acids (Mahan *et al.*, 1995).

7.2.4 Signature-tagged transposon mutagenesis (STM)

STM is an innovative and successful approach that has been used to define virulence factors in *Salmonella*. This strategy was developed as an in vivo genetic screen and was originally used for the identification of virulence factors of *S. enterica* serovar Typhimurium upon injection into mice (Hensel *et al.*, 1995). STM uses a comparative hybridization with a collection of uniquely tagged transposons. A pool of tagged mutants is passed through an animal model and the input and output pools are compared. Mutants that are unable to survive in vivo are identified as absent from the output pool. This approach combines the advantages of transposon mutagenesis with the ability to screen a large number of mutants using few animals. The drawbacks of this negative screening approach are that creating the tagged mutants and input pools is laborious, only mutants whose phenotypes cannot be transcomplemented in vivo will be recovered and genes that have redundant function will not be identified (Unsworth and Holden, 2000).

In the original description of STM, 1152 individually tagged mutants of *S. enterica* serovar Typhimurium were generated, and 40 mutants were identified that were absent in the output pool recovered from the spleens of mice. DNA was isolated from 28 of the mutants (Hensel *et al.*, 1995) and 13 transposon insertions were found to be in genes previously known to be involved in virulence (Table 7.1) (Hensel *et al.*, 1995). Five mutants had transposon insertions in 4 unknown genes, that had homology to the *inv-spa* Type III secretion system in *S. enterica* serovar Typhimurium (Hensel *et al.*, 1995). These genes are now known to form a second Type III secretion system in *S. enterica* encoded by *Salmonella* pathogenicity island 2 (SPI-2).

STM has also been used to identify genes necessary for virulence of *S. enterica* serotypes in individual host species (Bispham *et al.*, 2001; Tsolis *et al.*, 1999). A pool of 260 tagged mutants of *S. enterica* serovar Typhimurium was administered orally to calves and mice, and the output pools from the two animal species were compared. In total 17 mutants that were attenuated for virulence were identified and four of these were deficient in growth/survival in one species only (Tsolis *et al.*, 1999). These experiments also confirmed that genes of SPI-2 (*spiB*), SPI-1 (*hilA, orgA, prgH*), LPS biosynthesis (*rfbJ*) and the virulence plasmid (*spvB*, *spvR*) play a role in the virulence of *S. enterica* serovar Typhimurium in both mice and calves. During this work, a new effector gene for type III secretion (*slrP*) was also identified (Tsolis *et al.*, 1999). A similar strategy, using STM in calves and mice was used to identify virulence genes of *Salmonella enterica* serovar Dublin (Bispham *et al.*, 2001). In this study, an effector of the SPI-2 secretion system, *sseD* was absent from the output pool from mice only, while *ssaT* (a structural gene of the SPI-2 TTSS) was absent from the output pool of both species tested. However, mutants in both of these genes were attenuated in calves and mice upon retesting (Bispham *et al.*, 2001).

7.2.5 Differential fluorescence induction (DFI)

Differential fluorescence induction (DFI) is a strategy to identify virulence genes expressed in vivo (Bumann, 2001; Bumann, 2002; Valdivia and Falkow, 1996; Valdivia and Falkow, 1997; Wendland and Bumann, 2002). DFI is a promoter trap strategy where a library of random genomic fragments is ligated to a promoterless *gfp* reporter gene that encodes the green fluorescent protein (GFP), and this construct is used to directly determine transcriptional activity of the reporter in vivo. Advantages and drawbacks of this technique are similar to IVET but DFI has the additional advantage that it allows a direct assessment of in vivo expression levels.

Initial experiments used this technique to assay gene expression under in vitro conditions, and depended on the development of GFP expression vectors for pathogenic bacteria, and the ability to sort GFP expressing bacteria by flow cytometry, (Valdivia and Falkow, 1996; Valdivia and Falkow, 1997). More recently, this technique has been optimized for use in vivo by optimizing GFP detection using two color flow cytometry to distinguish GFP expressing bacteria from host autofluorescence (Bumann, 2002; Wendland and Bumann, 2002). This refined technique has been used to identify novel promoters that are induced in vivo in mice, inoculated intravenously with *S. enterica* serovar Typhimurium containing a library of random *Salmonella* DNA fragments ligated upstream of a promoterless *gfp* gene. Twenty-four hours post infection bacteria recovered from the spleen were sorted for cells abundantly expressing GFP. After sorting recovered clones for low GFP expression during in vitro growth in broth, the candidate pool was used for a second round of screening for high GFP expression in mice. Four in vivo activated promoters were identified, including P_{pibB} (SPI5), P_{sifA}, (SPI2 effector) P_{aroQ}, and the novel promoter P_{3g} (Bumann, 2002). This technique holds promise for identification of further loci that are important for the pathogenesis of *S. enterica*.

7.2.6 Global changes in gene expression assessed by microarray analysis

Microarray technology is a powerful technique that allows researchers to view the bacterial transcriptome as a whole system, rather than strictly focusing on individual parts of various regulatory cascades. Microarrays have been used to map global changes in gene expression occurring between the growth of *S. enterica* serovar Typhimurium in rich medium versus growth in minimal medium (Detweiler *et al.*, 2003) and caused by inactivation of the regulator *csrA* (Lawhon *et al.*, 2003). The technology was recently applied to study host cell- *S. enterica* serovar Typhimurium interaction by comparing the transcriptional profiles of intracellular bacteria recovered from murine macrophage-like J774-A.1 cells with those of bacteria grown in cell culture medium (Eriksson *et al.*, 2003). This study found that 919 of 4451 *S. enterica* serovar Typhimurium coding sequences changed in expression during infection of murine macrophages, a number much larger than the estimated 250 virulence genes present in this pathogen. The transcription of most global regulatory genes was not significantly changed. For example, *rpoS* expression was increased only twofold in intracellular bacteria while PhoP gene expression was unaltered. However, expression of PhoP activated genes was induced (*e.g.* a 50-fold induction of *mgtBC* expression) whilst PhoP repressed genes remained silent. The results further suggest that intracellular *S. enterica* serovar Typhimurium is not starved for amino acids or iron (Fe^{+2}) and that the intravacuolar environment is low in phosphate and magnesium but high in potassium (Eriksson *et al.*, 2003).

7.3 REGULATION OF THE EXPRESSION OF VIRULENCE GENES

7.3.1 General

S. enterica has the ability to sense its surrounding conditions and regulate its gene expression profiles extensively. *S. enterica* must survive a wide range of environmental conditions in its course of infection and during transmission within different animal reservoirs. Extensive research has been conducted both in vivo and in vitro studying these environmental stimuli and the response of *S. enterica* to these signals. *S. enterica* uses a complex regulatory network of two-component regulatory systems, transcriptional regulators and alternate sigma factors to control its virulence gene expression.

7.3.2 PhoPQ as a global regulator of gene expression

The two-component PhoP/PhoQ regulatory system is required for the pathogenesis of *S. enterica* serovar Typhimurium, and controls the transcription of more than 40 genes (Groisman *et al.*, 1989; Miller *et al.*, 1989; Ronson *et al.*, 1987). PhoP/PhoQ is required for virulence in mice, survival in macrophages, resistance to cationic antimicrobial peptides (CAMPs), growth on succinate as a sole carbon source, and growth under limiting amounts of magnesium (Fields *et al.*, 1989; Groisman *et al.*, 1992; Miller *et al.*, 1990). PhoQ is a sensor histidine kinase that phosphorylates PhoP, a response regulator, in response to environmental conditions (Gunn *et al.*, 1996). PhoQ activity is repressed by the divalent cations magnesium and calcium. PhoP activates expression of a set of genes designated *pags* (PhoP activated genes), that promote *Salmonella* survival within host tissues, presumably by increasing the barrier function of the outer membrane against CAMPs, as well as increasing the transport of specific molecules, such as cations.

Proteins encoded by PhoP-activated genes include a nonspecific acid phosphatase, cation transporters, outer membrane proteins, and enzymes important for lipopolysaccharide modification (Guo *et al.*, 1997; Guo *et al.*, 1998; Kier *et al.*, 1979). In addition, PhoP-PO_4 activates expression of the *pmrCAB* operon (polymyxin resistance locus) that is also known as *pagB pmrAB*. This locus encodes a two component regulatory system that is activated in response to an acidic environment (Gunn and Miller, 1996; Soncini and Groisman, 1996). PmrB is a sensor kinase that senses Mg^{2+} indirectly by communicating with the Mg^{2+}-sensitive PhoPQ system whereas PmrA is a transcriptional response regulator. PhoP-PO_4 represses transcription of another set of genes found on SPI-1 (designated *prgs*, for PhoP-repressed genes) that is required for epithelial cell invasion and for the formation of spacious phagosomes. These genes also include *hilA*, which encodes a transcriptional regulator, and the *prgHIJKorgA* operon, which encodes components of a type III secretion system (Alpuche-Aranda *et al.*, 1994; Behlau and Miller, 1993; Hueck *et al.*, 1995; Pegues *et al.*, 1995). In addition to regulating a large number of genes, the *phoPQ* operon is autoregulated, as full expression requires both PhoP and PhoQ (Soncini *et al.*, 1995).

Expression of PhoP-activated genes is maximally induced within nonspacious acidified phagosomes several hours after phagocytosis (Alpuche Aranda *et al.*, 1992; Valdivia and Falkow, 1997). Although the in vivo signals for PhoQ activation are not fully defined, activation can be induced in vitro by low pH, or growth in media containing micromolar concentrations of the divalent cations Mg^{2+} and Ca^{2+} (Alpuche Aranda *et al.*, 1992; Bearson *et al.*, 1998; Garcia Vescovi *et al.*, 1996). PhoQ contains distinct binding sites for Mg^{2+} and Ca^{2+}, and is maximally repressed in the presence of both cations (Vescovi *et al.*, 1997; Waldburger, Sauer, 1996). Therefore, the best-defined signal in vitro is depletion of the divalent cations Mg^{2+} and Ca^{2+}. This finding has led to the hypothesis that PhoQ is also activated in vivo by limiting concentrations of divalent cations. PhoP-activated genes responding to Ca^{2+} and Mg^{2+} levels include three putative loci that encode the magnesium transporters MgtABC. Transcription of *mgtA* and *mgtCB* loci occurs in a PhoP-dependent manner and is repressed by Mg^{2+} and Ca^{2+} (Garcia Vescovi *et al.*, 1996; Soncini *et al.*, 1996).

Expression of *mgtCB*, which is located on a 17-kb pathogenicity island know as SPI-3, is also induced by exposure to acid, even in the presence of high concentrations of Mg^{2+} (Blanc-Potard and Groisman, 1997). This response is also PhoP dependent (Bearson *et al.*, 1998; Tao *et al.*, 1998). Although *mgtA* and *mgtB* are not required for intracellular survival or for virulence, *mgtC* is essential for both functions (Blanc-Potard, Groisman, 1997; Moncrief, Maguire, 1998). Expression of a subset of *pags* in response to low pH is mediated by PmrA and PmrB. Mild acidic growth conditions have been shown to promote transcription of a subset of PhoP-activated genes that are also PmrA-dependent (Soncini *et al.*, 1996). In addition, transcriptional activation of *psiD* (also called *pmrC* or *pagB*) by mild acidification is independent of the PhoQ protein (Garcia Vescovi *et al.*, 1996). A report has indicated that the expression of several proteins induced upon acid shock is dependent on PhoP/PhoQ but not PmrA (Bearson *et al.*, 1998). Although posttranscriptional effects cannot be ruled out, this finding suggests that PhoQ, or some unidentified target of PhoP, can respond to low pH. A recent report shows that exposure to sublethal concentrations of CAMP activates the PhoP/PhoQ and RpoS virulence regulons, while also repressing the transcription of genes required for flagellar synthesis and the invasion associated type III secretion system (Bader *et al.*, 2003). Inducible resistance depends on the presence of PhoP, indicating that the PhoP/PhoQ system also senses sublethal concentrations of CAMPs.

While *phoP* mutants display significant virulence defects, deletions within individual *pag*s (either alone or in combination with other *pag* deletions) fail to attenuate virulence in the mouse model (the exception being the *mgtC* mutant) (Gunn *et al.*, 1998). The complexity of the PhoP regulon suggests that the regulation of cell envelope composition contributes to its role in virulence. These factors may act in concert to promote virulence by resisting the action of host innate immune processes. As the function of many *pags* is still unknown, more detailed knowledge of the genes and functions regulated by PhoP/PhoQ should reveal interesting aspects of host-pathogen interactions.

7.3.3 Expression of the invasion phenotype

Invasion of epithelial cells, which requires the TTSS encoded in SPI-1, is under complex regulatory control. Initial studies demonstrated that invasion is controlled by numerous environmental conditions, including growth phase (Ernst *et al.*, 1990; Lee and Falkow, 1990), DNA structure or supercoiling (Galan and Curtiss, 1990), osmolarity (Ernst *et al.*, 1990) and oxygen availability (Bajaj *et al.*, 1996; Jones and Falkow, 1994; Penheiter *et al.*, 1997). At least one transcriptional regulator from SPI-1, HilA responds to environmental conditions to control expression of the TTSS genes. Mutants lacking HilA, a member of the OmpR/ToxR family of transcriptional regulators, are severely defective in epithelial cell invasion (Bajaj *et al.*, 1995), whereas overexpression of HilA results in a hyperinvasive phenotype (Lee *et al.*, 1992). HilA is coordinately regulated by oxygen, osmolarity, pH, and PhoP (which regulates it) (Bajaj *et al.*, 1996). Maximal transcription from HilA-regulated genes, and thus maximal invasion of epithelial cells, occurs under conditions of high osmolarity and low oxygen, and when the bacteria have an intact PhoP phenotype. HilA regulates a number of genes in SPI-1, including those that encode the TTSS structural components PrgHIJK and OrgA, the secreted proteins Sip/SspBCDA and the transcriptional regulator InvF. InvF is homologous to transcriptional regulators from the AraC and PulD families (Kaniga *et al.*, 1994), and regulates expression of the *sipBCDA* (*sspBCDA*) operon (Eichelberg and Galan, 1999).

The signal transduction pathways leading to activation of *hilA* expression and transcription of SPI-1 genes are still being elucidated. Transcription of *hilA* is repressed in the presence of PhoP-PO_4 but appears to be induced by several other transcriptional regulators. The SirA protein, which belongs to the response regulator family of proteins, positively regulates *hilA* in response to unknown signals and sensor molecules (Johnston *et al.*, 1996). Several unlinked loci (also termed *sir*) that suppress a *sirA* mutation and activate SPI-1 gene expression have also been identified (Johnston *et al.*, 1996). One hypothesis is that each Sir transduces information about a single environmental condition. In this model HilA would function as one central "receiving" center, where various environmental signals are processed. This model does not rule out the possibility that other regulators can directly activate transcription of SPI-1 genes independent of HilA (Johnston *et al.*, 1996).

7.3.4 Acid tolerance response

The alternate sigma factor RpoS (also called KatF or σ^S) is required for expression of over 30 genes during stationary growth phase in *S. enterica* serovar Typhimurium. RpoS also plays an important role in *S. enterica* serovar Typhimurium pathogenesis, as genes regulated by RpoS protect the bacteria against a variety of stressful conditions that might be encountered within the host, including anaerobiosis, nitrogen and phosphate starvation, acid shock, osmotic shock and oxidative stress (Loewen and Hengge-Aronis, 1994). In addition, RpoS regulates expression of several genes on the virulence plasmid that contribute to efficient systemic infections (Guiney *et al.*, 1995). Mutations within *rpoS* have been associated with attenuation of virulence of *S. enterica* serovar Typhimurium in the mouse model (Nickerson and Curtiss, 1997; Wilmes-Riesenberg *et al.*, 1997) and in humans (Robbe-Saule *et al.*, 1995).

A well characterized bacterial response to environmental stress is the induction of the acid tolerance response (ATR) (Bearson *et al.*, 1997). Exposure to low pH (4.4–5.8) for a short period of time initiates a bacterial response that increases resistance to even more acidic conditions (pH 3.3–3.0). Activation of the ATR also induces cross-protection against other environmental conditions, including heat, and osmotic and oxidative stress. Bacterial response to acid shock differs depending on the growth state of the bacteria, and requires at least three different transcriptional regulators including RpoS, PhoP and Fur.

Activation of RpoS during the ATR is regulated by the *mviA* gene product, which decreases the turnover rate of RpoS when the intracellular pH decreases, presumably by regulating expression of a specific protease (Bearson *et al.*, 1996). Although RpoS is not required for initiation of the ATR during logarithmic growth phase, it is required for maintenance of acid tolerance (Lee *et al.*, 1995). Initiation of the ATR during logarithmic growth requires the ferric uptake regulator Fur, which also regulates genes required for the acquisition of iron (Foster and Spector, 1995). The role of Fur during ATR is not entirely understood and the domains of Fur that are required for ATR are distinct from those needed for iron uptake (Hall and Foster, 1996). The PhoP protein is also induced during acid shock and is responsible for expression of three other acid shock-induced proteins (Bearson *et al.*, 1998).

Although the ATR is thought to contribute to the ability of *Salmonella* to survive acidic environments in vivo, there is currently no evidence for a specific role in virulence. The pleiotropic effects of mutations within the global regulators RpoS and PhoP have hampered investigations of the importance of the ATR in vivo. Furthermore, multiple mechanisms for its induction indicate that there is some functional redundancy in the ATR (Garcia-del Portillo *et al.*, 1993; Riesenberg-Wilmes *et al.*, 1996). Mutations within a number of individual acid-induced genes that disrupt the ATR in the avirulent *S. enterica* serovar Typhimurium laboratory strain LT2 have little effect on either acid resistance or virulence in the mouse model when transduced into wild-type virulent strains of *S. enterica* serovar Typhimurium (Riesenberg-Wilmes *et al.*, 1996). However, strains carrying multiple mutations in acid-induced genes are no longer acid tolerant and are attenuated in vivo, implicating acid tolerance as an important virulence mechanism.

7.3.5 Genes regulated by OmpR/EnvZ

Responses to external osmolarity are controlled by the two-component regulators OmpR (response regulator) and EnvZ (sensor kinase). OmpR reciprocally regulates the expression of the outer membrane porins OmpC and OmpF (Dorman *et al.*, 1989). When *S. enterica* serovar Typhimurium is exposed to conditions of high osmolarity, the level of OmpC in the membrane is increased whilst the level of OmpF is decreased. As OmpC forms smaller pores than OmpF, the preferential expression of OmpC in the membrane when bacteria are exposed to high osmolarity may decrease the permeability of the membrane to potentially harmful substances. Single mutations in *ompC* or *ompF* do not attenuate the virulence of *S. enterica* serovar Typhimurium (Dorman *et al.*, 1989); however, a double *ompC ompF* mutant is severely attenuated when administered orally to mice (Chatfield *et al.*, 1991). *OmpR* mutants are also severely attenuated on both oral and intravenous inoculation in mice (Dorman *et al.*, 1989). Interestingly, OmpR/EnvZ is required for late-onset cell death induced by *S. enterica* serovar Typhimurium in macrophages (Lindgren *et al.*, 1996). OmpR directly binds to the *ssrA* promoter, thereby controlling the activity of the SsrAB two component regulatory system (Lee *et al.*, 2000). In turn, SsrAB controls expression of the type III secretion system encoded on SPI-2, which is required for inducing late-onset cell death in macrophages (van der Velden *et al.*, 2000). OmpR and EnvZ are also required for the formation of filamentous tubular lysosomes (Mills *et al.*,

1998), raising the hypothesis that OmpR/EnvZ-regulated genes may affect trafficking of *Salmonella*-containing phagosomes within infected eukaryotic cells.

7.3.6 Gene regulation in response to temperature changes

The *tlpA* gene on the large virulence plasmid of *S. enterica* serovar Typhimurium, encodes an autoregulated repressor protein that responds to changes in temperature by shifting from an inactive unfolded monomer to an active folded coiled-coil dimer (Hurme *et al.*, 1997). The TlpA protein binds to the *tlpA* gene promoter and represses it. At 22 °C, the promoter is repressed, and derepression occurs as the temperature increases. The shift in the structure of the tlpA protein is reversible and therefore TlpA does not become denatured permanently by increases in temperature. No accessory proteins are required for TlpA action at the binding site of the *tlpA* promoter. It has been pointed out that a DNA sequence motif characteristic of the *tlpA* promoter is found in several virulence gene sequences, including *spvA* on the *S. enterica* virulence plasmid, and on the PhoP/PhoQ-regulated genes *prgH* and *pagC* on the chromosome (Hurme *et al.*, 1997).

7.4 FUNCTIONS OF VIRULENCE GENES INVOLVED IN GASTROENTERITIS AND SYSTEMIC DISEASE

7.4.1 General

The majority of infections caused by *S. enterica* in mammals fall into two major categories of disease. The first category, exemplified by *S. enterica* serovar Typhimurium infection in humans or cattle, consists of a localized intestinal infection manifesting as severe diarrhoea with infiltration of polymorphonuclear granulocytes in infected intestinal mucosa. The second category consists of systemic infections that are exemplified by typhoid fever caused by *S. enterica* serovar Typhi in humans and *S. enterica* serovar Typhimurium in mice. The functions of key genes and virulence factors important for the induction of systemic and gastroenteritic *Salmonella* diseases are briefly summarized below.

7.4.2 Interaction of *S. enterica* with the intestine

An important property that distinguishes *S. enterica* serovar Typhimurium from bacteria present in the normal intestinal flora is its ability to invade the intestinal mucosa. The genes required for invasion are encoded within the SPI-1 TTSS locus (Galán and Curtiss, 1989).

The main mechanism for SPI-1-mediated invasion of tissue culture cells is the induction of actin cytoskeleton rearrangements triggered by SopB, SopE and SopE2 (Mirold *et al.*, 2001; Zhou and Galan, 2001). Inactivation of the genes encoding these three effector proteins is necessary to reduce the invasiveness of *S. enterica* serovar Typhimurium SL1344 to the level of a mutant with a defective TTSS (*i.e. invA* or *invG* mutant) (Mirold *et al.*, 2001; Zhou, Galan, 2001). SipA may also play a minor role in bacterial uptake in vitro since entry is delayed at early time points in a *S.* Typhimurium *sipA* mutant (Zhou *et al.*, 1999). The molecular mechanisms used by SPI-1 TTSS effectors to manipulate the host cell have been extensively studied in vitro. SipA is a SPI-1 TTSS effector protein, that binds and stabilizes actin filaments and modulates the actin-bundling activity of T-plastin, resulting in a more pronounced outward extension of *S. enterica* serovar Typhimurium-induced membrane ruffles in epithelial cells in vitro (Zhou *et al.*, 1999; Zhou *et al.*, 1999). SopE induces membrane ruffling and nuclear responses in human cell lines, presumably by acting as a nucleotide exchange factor in two Rho GTPases, Rac-1 and CDC42 (Hardt *et al.*, 1998). Inactivation of *sopE* does not alter the virulence of *S. enterica* serovar Typhimurium for mice and only modestly reduces its invasiveness for tissue culture cells (Hardt *et al.*, 1998; Wood *et al.*, 1996). The modest invasion defect of a *S. enterica* serovar Typhimurium *sopE* mutant may be explained by the presence of a homologous gene, *sopE2* (Bakshi *et al.*, 2000; Stender *et al.*, 2000). However, SopE2 and SopE do not activate identical sets of Rho GTPase signaling cascades, since SopE2 acts as a nucleotide exchange factor only for CDC42 but not for Rac-1 (Friebel *et al.*, 2001). SopB, encoded within SPI-5, is an inositol-phosphate phosphatase which hydrolyzes phosphatidylinositol 3,4,5-trisphosphate. This enzymatic activity promotes bacterial invasion and, since phosphatidylinositol 3,4,5-trisphosphate is an inhibitor of chloride secretion, SopB may contribute to fluid secretion in vivo (Norris *et al.*, 1998).

The role of SPI-1 mediated invasion in causing *S. enterica* serovar Typhimurium-induced diarrhea has been studied using a calf model (Zhang *et al.*, 2003). This work has revealed that the TTSS encoded by SPI-1 constitutes a major virulence factor contributing to the development of inflammatory diarrhea (Ahmer *et al.*, 1999; Tsolis *et al.*, 1999; Tsolis *et al.*, 2000; Tsolis *et al.*, 1999; Wallis *et al.*, 1999; Watson *et al.*, 1998). A role in eliciting intestinal inflammation and fluid accumulation in the calf model has been demonstrated for SPI-1 TTSS effector proteins, including SipA, SopD, SopA, SopB, SopE and SopE2 (Zhang *et al.*, 2002; Zhang *et al.*, 2002). These effector proteins seem to have a redundant effect in eliciting inflammation and fluid secretion (Zhang *et al.*, 2002).

Upon SPI-1 TTSS-mediated invasion of the intestinal epithelium in calves, *S. enterica* serovar Typhimurium triggers the release of CXC chemokines by epithelial cells (Zhang *et al.*, 2003). In response to the chemotactic gradient created by the release of CXC chemokines, neutrophils infiltrate the lamina propria and are the hallmark of the intestinal pathology elicited by *S. enterica* serovar Typhimurium (Santos *et al.*, 2002; Tsolis *et al.*, 1999). Experiments in cell culture suggest that this inflammatory response is likely to be a consequence of innate recognition of pathogen associated molecular patterns (PAMPs), like flagella, by Toll-like receptors (TLRs) (Gewirtz *et al.*, 2001; Gewirtz *et al.*, 2001; Zeng *et al.*, 2003). In bovine ligated ileal loops infected with a *S. enterica* serovar Typhimurium strain carrying mutations in both flagellin genes (a *fljB fliC* mutant), the neutrophil influx is reduced, supporting a role for flagella in triggering inflammation in vivo (Schmitt *et al.*, 2001). Another PAMP that may contribute to inflammation in vivo is the lipid A domain of the *S. enterica* serovar Typhimurium LPS because a mutation in *msbB* (which results in reduced acylation of lipid A, thereby abrogating its PAMP activity) reduces fluid accumulation in bovine ligated ileal loops (Watson *et al.*, 2000).

7.4.3 Key genes needed for systemic infection

Mechanisms involved in the development of systemic infections have been elucidated using tissue culture and mouse models. The ability of *S. enterica* serovar Typhimurium to survive and multiply in macrophages is essential for causing lethal systemic infections in mice (Fields *et al.*, 1986). Virulence genes important for the systemic phase of infection caused by *S. enterica* serovar. Typhimurium in mice confer the ability to alter vesicular trafficking in macrophages (TTSS encoded by SPI-2), to interfere with actin polymerization (*spvB*), to resist macrophage killing mechanisms (*sodC1, sapABCDF*) and to obtain nutrients in the macrophage intracellular environment (*mgtBC*) (Blanc-Potard and Groisman, 1997; DeGroote *et al.*, 1997; Gulig *et al.*, 1998; Ochman and Groisman, 1996; Roudier *et al.*, 1990).

The SPI-2 encoded TTSS has been shown to be a major determinant involved in the replication of bacteria and their systemic dissemination in mice (Hensel *et al.*, 1995; Shea *et al.*, 1996). SPI-2 contributes to the avoidance of monocyte-macrophage killing mechanisms (Hensel *et al.*, 1998; Ochman and Groisman, 1996; Shea *et al.*, 1999). Although the precise mechanism used by the SPI-2 encoded TTSS is unknown, current evidence suggests that interference with intracellular trafficking plays an important role in this process. The SPI-2 encoded TTSS mediates *Salmonella* containing vacuole (SCV) formation and inhibits its fusion with lysosomes (Beuzon *et al.*, 2000; Uchiya *et al.*, 1999; Yu *et al.*, 2002). Several SPI-2 TTSS effectors seem to be involved in the process of actin rearrangement and vacuolar trafficking (Uchiya *et al.*, 1999; Vazquez-Torres *et al.*, 2000; Yu *et al.*, 2002). SPI-2 gene products prevent the trafficking of NADPH oxidase-containing vesicles to the *Salmonella* containing phagosome and to interfere with the killing mediated by nitrogen derived radicals (iNOS) (Chakravortty *et al.*, 2002; Vazquez-Torres *et al.*, 2000).

The virulence plasmid of *S. enterica* serovar Typhimurium promotes systemic dissemination and multiplication of the bacteria in the reticuloendothelial system of the mouse in a SPI-2 independent mechanism. Although virulence plasmids vary in size between different *S. enterica* serotypes (50–90 kb), an 8 kb region of the virulence plasmid, containing the *spvABCD* operon and its positive regulator *spvR* are highly conserved (Popoff *et al.*, 1984; Poppe *et al.*, 1989; Roudier *et al.*, 1990; Woodward, 1989). SpvB is a mono(ADP-ribosyl)transferase which ADP-ribosylates actin, thereby interfering with actin polymerization and contributing to full mouse virulence (Lesnick *et al.*, 2001; Otto *et al.*, 2000; Tezcan-Merdol *et al.*, 2001).

Additional virulence genes of *S. enterica* have also been implicated in resistance to macrophage killing mechanisms and acquisition of nutrients in the SCV. For example, the *sapABCDF* operon encodes a peptide transporter that mediates resistance to CAMPs and promotes protamine resistance in vitro, survival within cultured macrophages, and virulence in the mouse model (Groisman *et al.*, 1992; Parra-Lopez *et al.*, 1993). Expression of two additional genes, *sapG* and *sapJ*, are thought to be required for protamine resistance (Parra-Lopez *et al.*, 1994). Another example is the *sodCI* gene that is located on a bacteriophage and encodes a periplasmic superoxide dismutase that is required for full mouse virulence, macrophage survival and protection of *S. enterica* serovar Typhimurium from phagocyte-derived reactive oxygen and reactive nitrogen species (DeGroote *et al.*, 1997; Fang *et al.*, 1999).

The functions of *mgtA*, *mgtB*, and *mgtC* are currently unclear. MgtA and MgtB are homologous to P-type ATPases that use the energy of ATP hydrolysis to move molecules across cell membranes. These proteins may act as magnesium transporters for intracellular *S. enterica* serovar Typhimurium as suggested by the finding that both MgtA and MgtB can transport magnesium (and other divalent cations) and are regulated by magnesium and PhoP. However, this hypothesis has been challenged for several reasons (Smith, Maguire, 1998). Firstly, a constitutively active magnesium transporter (CorA) has a much higher affinity for magnesium and should be able to provide ample amounts of magnesium for intracellular bacterial growth. Secondly, it seems unusual that MgtA and MgtB should have to hydrolyze ATP in order to move magnesium across the membrane down its own concentration gradient. Thirdly, although MtgC has also been hypothesized to be a magnesium transporter, there is no direct evidence that this protein can actually function as a magnesium transporter (Moncrief, Maguire, 1998). An alternative hypothesis is that MgtA, MgtB and MgtC may function as a counter- or cotransporter for another molecule, such as an antimicrobial peptide, using magnesium as a signal. Although there is no direct evidence for this hypothesis, it is an intriguing model that is consistent with the role of PhoP in defending *S. enterica* serovar Typhimurium against the host immune response.

7.5 CONCLUSIONS

Salmonella enterica serovars are associated with two main human disease syndromes, gastroenteritis and typhoid fever. A large number of virulence factors important for the pathogenesis of these diseases have been identified using animal models and a variety of elegant genetic methods, including signature-tagged transposon mutagenesis, in vivo expression technology and differential fluorescence induction. In vivo expression of virulence genes is tightly controlled by an intricate network of regulatory elements that are beginning to be elucidated. Finally, recent work has identified eukaryotic targets of *S. enterica* virulence factors, thereby providing new insights into the pathogenesis of gastroenteritis and typhoid fever.

7.6 ACKNOWLEDGEMENTS

Work in A. J. Baumler's laboratory is supported by USDA/NRICGP grant #2002-35204-12247 and Public Health Service grants AI40124 and AI44170. H. Andrews-Polymenis is currently supported by Public Health Service grant AI052250.

REFERENCES

Ahmer, B. M., van Reeuwijk, J., Watson, P. R., Wallis, T. S. and Heffron, F. (1999). *Salmonella* SirA is a global regulator of genes mediating enteropathogenesis. *Mol Microbiol*, **31**, 971–82.

Alpuche-Aranda, C. M., Racoosin, E. L., Swanson, J. A. and Miller, S. I. (1994). *Salmonella* stimulate macrophage macropinocytosis and persist within spacious phagosomes. *J Exp Med*, **179**, 601–8.

Alpuche Aranda, C. M., Swanson, J. A., Loomis, W. P. and Miller, S. I. (1992). *Salmonella typhimurium* activates virulence gene transcription within acidified macrophage phagosomes. *Proc Natl Acad Sci USA*, **89**, 10079–83.

Bader, M. W., Navarre, W. W., Shiau, W. *et al.* (2003). Regulation of *Salmonella typhimurium* virulence gene expression by cationic antimicrobial peptides. *Mol Microbiol*, **50**, 219–30.

Bajaj, V., Hwang, C. and Lee, C. A. (1995). *hilA* is a novel *ompR/toxR* family member that activates the expression of *Salmonella typhimurium* invasion genes. *Mol Microbiol*, **18**, 715–27.

Bajaj, V., Lucas, R. L., Hwang, C. and Lee, C. A. (1996). Co-ordinate regulation of *Salmonella typhimurium* invasion genes by environmental and regulatory factors is mediated by control of *hilA* expression. *Mol Microbiol*, **22**, 703–14.

Bakshi, C. S., Singh, V. P., Wood, M. W. *et al.* (2000). Identification of SopE2, a *Salmonella* secreted protein which is highly homologous to SopE and involved in bacterial invasion of epithelial cells. *J Bacteriol*, **182**, 2341–4.

Baümler, A. J., Kusters, J. G., Stojiljkovic, I. and Heffron, F. (1994). *Salmonella typhimurium* loci involved in survival within macrophages. *Infect Immun*, **62**, 1623–30.

Bearson, B. L., Wilson, L. and Foster, J. W. (1998). A low pH-inducible, PhoPQ-dependent acid tolerance response protects *Salmonella typhimurium* against inorganic acid stress. *J Bacteriol*, **180**, 2409–17.

Bearson, S., Bearson, B. and Foster, J. W. (1997). Acid stress responses in enterobacteria. *FEMS Microbiol Lett*, **147**, 173–80.

Bearson, S. M., Benjamin, W. H., Jr., Swords, W. E. and Foster, J. W. (1996). Acid shock induction of RpoS is mediated by the mouse virulence gene *mviA* of *Salmonella typhimurium*. *J Bacteriol*, **178**, 2572–9.

Behlau, I. and Miller, S. I. (1993). A PhoP-repressed gene promotes *Salmonella typhimurium* invasion of epithelial cells. *J Bacteriol*, **175**, 4475–84.

Beuzon, C. R., Meresse, S., Unsworth, K. E. *et al.* (2000). *Salmonella* maintains the integrity of its intracellular vacuole through the action of SifA. *Embo J*, **19**, 3235–49.

Bispham, J., Tripathi, B. N., Watson, P. R. and Wallis, T. S. (2001). *Salmonella* pathogenicity island 2 influences both systemic salmonellosis and *Salmonella*-induced enteritis in calves. *Infect Immun*, **69**, 367–77.

Blanc-Potard, A. B. and Groisman, E. A. (1997). The *Salmonella selC* locus contains a pathogenicity island mediating intramacrophage survival. *Embo J*, **16**, 5376–85.

Bowe, F., Lipps, C. J., Tsolis, R. M. *et al.* (1998). At least four percent of the *Salmonella typhimurium* genome is required for fatal infection of mice. *Infect Immun*, **66**, 3372–7.

Bumann, D. (2001). In vivo visualization of bacterial colonization, antigen expression, and specific T-cell induction following oral administration of live recombinant *Salmonella enterica* serovar Typhimurium. *Infect Immun*, **69**, 4618–26.

(2002). Examination of *Salmonella* gene expression in an infected mammalian host using the green fluorescent protein and two-colour flow cytometry. *Mol Microbiol*, **43**, 1269–83.

Chakravortty, D., Hansen-Wester, I. and Hensel, M. (2002). *Salmonella* pathogenicity island 2 mediates protection of intracellular *Salmonella* from reactive nitrogen intermediates. *J Exp Med*, **195**, 1155–66.

Chatfield, S. N., Dorman, C. J., Hayward, C. and Dougan, G. (1991). Role of *ompR*-dependent genes in *Salmonella typhimurium* virulence: mutants deficient in both *ompC* and *ompF* are attenuated in vivo. *Infect Immun*, **59**, 449–52.

Conner, C. P., Heithoff, D. M. and Mahan, M. J. (1998). In vivo gene expression: contributions to infection, virulence, and pathogenesis. *Curr Top Microbiol Immunol*, **225**, 1–12.

DeGroote, M. A., Ochsner, U. A., Shiloh, M. U. *et al.* (1997). Periplasmic superoxide dismutase protects *Salmonella* from products of phagocyte NADPH-oxidase and nitric oxide synthase. *Proc Natl Acad Sci USA*, **94**, 13997–4001.

Detweiler, C. S., Monack, D. M., Brodsky, I. E., Mathew, H. and Falkow, S. (2003). *virK*, *somA* and *rcsC* are important for systemic *Salmonella enterica* serovar Typhimurium infection and cationic peptide resistance. *Mol Microbiol*, **48**, 385–400.

Dorman, C. J., Chatfield, S., Higgins, C. F., Hayward, C. and Dougan, G. (1989). Characterization of porin and *ompR* mutants of a virulent strain of *Salmonella typhimurium*: *ompR* mutants are attenuated in vivo. *Infect Immun*, **57**, 2136–40.

Eichelberg, K. and Galan, J. E. (1999). Differential regulation of *Salmonella typhimurium* type III secreted proteins by pathogenicity island 1 (SPI-1)-encoded transcriptional activators InvF and hilA. *Infect Immun*, **67**, 4099–105.

Eriksson, S., Lucchini, S., Thompson, A., Rhen, M. and Hinton, J. C. (2003). Unravelling the biology of macrophage infection by gene expression profiling of intracellular *Salmonella enterica*. *Mol Microbiol*, **47**, 103–18.

Ernst, R. K., Dombroski, D. M. and Merrick, J. M. (1990). Anaerobiosis, type 1 fimbriae, and growth phase are factors that affect invasion of HEp-2 cells by *Salmonella typhimurium*. *Infect Immun*, **58**, 2014–16.

Fang, F. C., DeGroote, M. A., Foster, J. W. *et al.* (1999). Virulent *Salmonella typhimurium* has two periplasmic cu, Zn-superoxide dismutases. *Proc Natl Acad Sci USA*, **96**, 7502–7.

Fields, P. I., Groisman, E. A. and Heffron, F. (1989). A *Salmonella* locus that controls resistance to microbicidal proteins from phagocytic cells. *Science*, **243**, 1059–62.

Fields, P. I., Swanson, R. V., Haidaris, C. G. and Heffron, F. (1986). Mutants of *Salmonella typhimurium* that cannot survive within the macrophage are avirulent. *Proc. Natl. Acad. Sci. USA*, **83**, 5189–93.

Foster, J. W. and Spector, M. P. (1995). How *Salmonella* survive against the odds. *Annu Rev Microbiol*, **49**, 145–74.

Friebel, A., Ilchmann, H., Aepfelbacher, M. *et al.* (2001). SopE and SopE2 from *Salmonella typhimurium* activate different sets of RhoGTPases of the host cell. *J Biol Chem*, **276**, 34035–40.

Galán, J. E. and Curtiss, R., III (1989). Cloning and molecular characterization of genes whose products allow *Salmonella typhimurium* to penetrate tissue culture cells. *Proc Natl Acad Sci USA*, **86**, 6383–7.

(1990). Expression of *Salmonella typhimurium* genes required for invasion is regulated by changes in DNA supercoiling. *Infect Immun*, **58**, 1879–85.

Garcia Vescovi, E., Soncini, F. C. and Groisman, E. A. (1996). $Mg2^{+}$ as an extracellular signal: environmental regulation of *Salmonella* virulence. *Cell*, **84**, 165–74.

Garcia-del Portillo, F., Foster, J. W. and Finlay, B. B. (1993). Role of acid tolerance response genes in *Salmonella typhimurium* virulence. *Infect Immun*, **61**, 4489–92.

Gewirtz, A. T., Navas, T. A., Lyons, S., Godowski, P. J. and Madara, J. L. (2001). Cutting edge: bacterial flagellin activates basolaterally expressed TLR5 to induce epithelial proinflammatory gene expression. *J Immunol*, **167**, 1882–5.

Groisman, E. A., Chiao, E., Lipps, C. J. and Heffron, F. (1989). *Salmonella typhimurium phoP* virulence gene is a transcriptional regulator. *Proc Natl Acad Sci USA*, **86**, 7077–81.

Groisman, E. A., Parra-Lopez, C., Salcedo, M., Lipps, C. J. and Heffron, F. (1992). Resistance to host antimicrobial peptides is necessary for *Salmonella* virulence. *Proc Natl Acad Sci USA*, **89**, 11939–43.

Guiney, D. G., Fang, F. C., Krause, M. *et al.* (1995). Biology and clinical significance of virulence plasmids in *Salmonella* serovars. *Clin Infect Dis*, **21** (suppl. 2), S146–S151.

Gulig, P. A. and Curtiss, R., III (1987). Plasmid-associated virulence of *Salmonella* typhimurium. *Infect Immun*, **55**, 2891–901.

(1988). Cloning and transposon insertion mutagenesis of virulence genes of the 100-kilobase plasmid of *Salmonella typhimurium*. *Infect Immun*, **56**, 3262–71.

Gulig, P. A., Caldwell, A. L. and Chiodo, V. A. (1992). Identification, genetic analysis and DNA sequence of a 7.8-kb virulence region of the *Salmonella typhimurium* virulence plasmid. *Mol Microbiol*, **6**, 1395–411.

Gulig, P. A., Doyle, T. J., Hughes, J. A. and Matsui, H. (1998). Analysis of host cells associated with the Spv-mediated increased intracellular growth rate of *Salmonella typhimurium* in mice. *Infect Immun*, **66**, 2471–85.

Gunn, J. S. and Miller, S. I. (1996). PhoP-PhoQ activates transcription of *pmrAB*, encoding a two-component regulatory system involved in *Salmonella typhimurium* antimicrobial peptide resistance. *J Bacteriol*, **178**, 6857–64.

Gunn, J. S., Belden, W. J. and Miller, S. I. (1998). Identification of PhoP-PhoQ activated genes within a duplicated region of the *Salmonella typhimurium* chromosome. *Microb Pathog*, **25**, 77–90.

Gunn, J. S., Hohmann, E. L. and Miller, S. I. (1996). Transcriptional regulation of *Salmonella* virulence: a PhoQ periplasmic domain mutation results in increased net phosphotransfer to PhoP. *J Bacteriol*, **178**, 6369–73.

Guo, L., Lim, K. B., Gunn, J. S. *et al.* (1997). Regulation of lipid A modifications by *Salmonella typhimurium* virulence genes *phoP-phoQ*. *Science*, **276**, 250–3.

Guo, L., Lim, K. B., Poduje, C. M. *et al.* (1998). Lipid A acylation and bacterial resistance against vertebrate antimicrobial peptides. *Cell*, **95**, 189–98.

Hall, H. K. and Foster, J. W. (1996). The role of fur in the acid tolerance response of *Salmonella typhimurium* is physiologically and genetically separable from its role in iron acquisition. *J Bacteriol*, **178**, 5683–91.

Hardt, W. D., Chen, L. M., Schuebel, K. E., Bustelo, X. R. and Galan, J. E. (1998). *S. typhimurium* encodes an activator of Rho GTPases that induces membrane ruffling and nuclear responses in host cells. *Cell*, **93**, 815–26.

Heithoff, D. M., Conner, C. P., Hanna, P. C. *et al.* (1997). Bacterial infection as assessed by in vivo gene expression. *Proc Natl Acad Sci USA*, **94**, 934–9.

Heithoff, D. M., Conner, C. P., Hentschel, U. *et al.* (1999). Coordinate intracellular expression of *Salmonella* genes induced during infection. *J Bacteriol*, **181**, 799–807.

Hensel, M., Shea, J. E., Gleeson, C. *et al.* (1995). Simultaneous identification of bacterial virulence genes by negative selection. *Science*, **269**, 400–3.

Hensel, M., Shea, J. E., Waterman, S. R. *et al.* (1998). Genes encoding putative effector proteins of the type III secretion system of *Salmonella* pathogenicity island 2 are required for bacterial virulence and proliferation in macrophages. *Mol Microbiol*, **30**, 163–74.

Hueck, C. J., Hantman, M. J., Bajaj, V. *et al.* (1995). *Salmonella typhimurium* secreted invasion determinants are homologous to *Shigella* Ipa proteins. *Mol Microbiol*, **18**, 479–90.

Hurme, R., Berndt, K. D., Normark, S. J. and Rhen, M. (1997). A proteinaceous gene regulatory thermometer in *Salmonella*. *Cell*, **90**, 55–64.

Johnston, C., Pegues, D. A., Hueck, C. J., Lee, A. and Miller, S. I. (1996). Transcriptional activation of *Salmonella typhimurium* invasion genes by a member of the phosphorylated response-regulator superfamily. *Mol Microbiol*, **22**, 715–27.

Jones, B. D. and Falkow, S. (1994). Identification and characterization of a *Salmonella typhimurium* oxygen-regulated gene required for bacterial internalization. *Infect Immun*, **62**, 3745–52.

Julio, S. M., Conner, C. P., Heithoff, D. M. and Mahan, M. J. (1998). Directed formation of chromosomal deletions in *Salmonella typhimurium*: targeting of specific genes induced during infection. *Mol Gen Genet*, **258**, 178–81.

Kaniga, K., Bossio, J. C. and Galan, J. E. (1994). The *Salmonella typhimurium* invasion genes *invF* and *invG* encode homologues of the AraC and PulD family of proteins. *Mol Microbiol*, **13**, 555–68.

Kier, L. D., Weppelman, R. M. and Ames, B. N. (1979). Regulation of nonspecific acid phosphatase in *Salmonella*: *phoN* and *phoP* genes. *J Bacteriol*, **138**, 155–61.

Knuth, K., Niesella, H., Hueck, C. J. and Fuchs, T. M. (2004). Large-scale identification of essential *Salmonella* by trapping lethal insertions. *Mol Microbiol*, **51**, 1729–44.

Lawhon, S. D., Frye, J. G., Suyemoto, M. *et al.* (2003). Global regulation by CsrA in *Salmonella typhimurium*. *Mol Microbiol*, **48**, 1633–45.

Lee, C. A. and Falkow, S. (1990). The ability of *Salmonella* to enter mammalian cells is affected by bacterial growth state. *Proc Natl Acad Sci USA*, **87**, 4304–8.

Lee, A. K., Detweiler, C. S. and Falkow, S. (2000). OmpR regulates the two-component system SsrA-ssrB in *Salmonella* pathogenicity island 2. *J Bacteriol*, **182**, 771–81.

Lee, C. A., Jones, B. D. and Falkow, S. (1992). Identification of a *Salmonella typhimurium* invasion locus by selection for hyperinvasive mutants. *Proc Natl Acad Sci USA*, **89**, 1847–51.

Lee, I. S., Lin, J., Hall, H. K., Bearson, B. and Foster, J. W. (1995). The stationary-phase sigma factor sigma S (RpoS) is required for a sustained acid tolerance response in virulent *Salmonella typhimurium*. *Mol Microbiol*, **17**, 155–67.

Lesnick, M. L., Reiner, N. E., Fierer, J. and Guiney, D. G. (2001). The *Salmonella spvB* virulence gene encodes an enzyme that ADP-ribosylates actin and destabilizes the cytoskeleton of eukaryotic cells. *Mol Microbiol*, **39**, 1464–70.

Lichtensteiger, C. A. and Vimr, E. R. (2003). Systemic and enteric colonization of pigs by a hilA signature – tagged mutant of *Salmonella* choleraesuis. *Microb Pathog*, **34**, 149–54.

Lindgren, S. W., Stojiljkovic, I. and Heffron, F. (1996). Macrophage killing is an essential virulence mechanism of *Salmonella typhimurium*. *Proc Natl Acad Sci USA*, **93**, 4197–201.

Lodge, J., Douce, G. R., Amin, I. I. *et al.* (1995). Biological and genetic characterization of TnphoA mutants of *Salmonella typhimurium* TML in the context of gastroenteritis. *Infect Immun*, **63**, 762–9.

Loewen, P. C. and Hengge-Aronis, R. (1994). The role of the sigma factor sigma S (KatF) in bacterial global regulation. *Annu Rev Microbiol*, **48**, 53–80.

Mahan, M. J., Slauch, J. M. and Mekalanos, J. J. (1993). Selection of bacterial virulence genes that are specifically induced in host tissues. *Science*, **259**, 686–8.

Mahan, M. J., Tobias, J. W., Slauch, J. M. *et al.* (1995). Antibiotic-based selection for bacterial genes that are specifically induced during infection of a host. *Proc Natl Acad Sci USA*, **92**, 669–73.

McClelland, M., Sanderson, K. E., Spieth, J. *et al.* (2001). Complete genome sequence of *Salmonella enterica* serovar Typhimurium LT2. *Nature*, **413**, 852–6.

Merrell, D. S. and Camilli, A. (2000). Detection and analysis of gene expression during infection by *in vivo* expression technology. *Philos Trans R Soc Lond B Biol Sci*, **355**, 587–99.

Miller, I., Maskell, D., Hormaeche, C., Johnson, K., Pickard, D. and Dougan, G. (1989). Isolation of orally attenuated *Salmonella typhimurium* following TnphoA mutagenesis. *Infect Immun*, **57**, 2758–63.

Miller, S. I., Pulkkinen, W. S., Selsted, M. E. and Mekalanos, J. J. (1990). Characterization of defensin resistance phenotypes associated with mutations in the *phoP* virulence regulon of *Salmonella typhimurium*. *Infect Immun*, **58**, 3706–10.

Mills, S. D., Ruschkowski, S. R., Stein, M. A. and Finlay, B. B. (1998). Trafficking of porin-deficient *Salmonella typhimurium* mutants inside HeLa cells: *ompR* and *envZ* mutants are defective for the formation of *Salmonella*-induced filaments. *Infect Immun*, **66**, 1806–11.

Mirold, S., Ehrbar, K., Weissmuller, A. *et al.* (2001). *Salmonella* host cell invasion emerged by acquisition of a mosaic of separate genetic elements, including

Salmonella pathogenicity island 1 (SPI1), SPI5, and sopE2. *J Bacteriol*, **183**, 2348–58.

Moncrief, M. B. and Maguire, M. E. (1998). Magnesium and the role of MgtC in growth of *Salmonella typhimurium. Infect Immun*, **66**, 3802–9.

Nickerson, C. A. and Curtiss, R., III (1997). Role of sigma factor RpoS in initial stages of *Salmonella typhimurium* infection. *Infect Immun*, **65**, 1814–23.

Norris, F. A., Wilson, M. P., Wallis, T. S., Galyov, E. E. and Majerus, P. W. (1998). SopB, a protein required for virulence of *Salmonella dublin*, is an inositol phosphate phosphatase. *Proc Natl Acad Sci USA*, **95**, 14057–9.

Ochman, H. and Groisman, E. A. (1996). Distribution of pathogenicity islands in *Salmonella* spp. *Infect Immun*, **64**, 5410–12.

Osbourn, A. E., Barber, C. E. and Daniels, M. J. (1987). Identification of plant-induced genes of the bacterial pathogen *Xanthomonas campestris pathovar campestris* using a promoter probe plasmid. *EMBO J*, **6**, 23–8.

Otto, H., Tezcan-Merdol, D., Girisch, R. *et al.* (2000). The spvB gene-product of the *Salmonella enterica* virulence plasmid is a mono(ADP-ribosyl)transferase. *Mol Microbiol*, **37**, 1106–15.

Parra-Lopez, C., Baer, M. T. and Groisman, E. A. (1993). Molecular genetic analysis of a locus required for resistance to antimicrobial peptides in *Salmonella typhimurium. Embo J*, **12**, 4053–62.

Parra-Lopez, C., Lin, R., Aspedon, A. and Groisman, E. A. (1994). A *Salmonella* protein that is required for resistance to antimicrobial peptides and transport of potassium. *Embo J*, **13**, 3964–72.

Pegues, D. A., Hantman, M. J., Behlau, I. and Miller, S. I. (1995). PhoP/PhoQ transcriptional repression of *Salmonella typhimurium* invasion genes: evidence for a role in protein secretion. *Mol Microbiol*, **17**, 169–81.

Penheiter, K. L., Mathur, N., Giles, D., Fahlen, T. and Jones, B. D. (1997). Non-invasive *Salmonella typhimurium* mutants are avirulent because of an inability to enter and destroy M cells of ileal Peyer's patches. *Mol Microbiol*, **24**, 697–709.

Popoff, M. Y., Miras, I., Coynault, C., Lasselin, C. and Pardon, P. (1984). Molecular relationships between virulence plasmids of *Salmonella* serotypes Typhimurium and Dublin and large plasmids of other *Salmonella* serotypes. *Annal Microbiol*, **135A**, 389–98.

Poppe, C., Curtiss, R., III, Gulig, P. A. and Gyles, C. L. (1989). Hybridization with a DNA probe derived from the virulence region of the 60 Mdal plasmid of *Salmonella typhimurium. Can J Vet Res*, **53**, 378–84.

Rhen, M., Virtanen, M. and Makela, P. H. (1989). Localization by insertion mutagenesis of a virulence-associated region on the *Salmonella typhimurium* 96 kilobase pair plasmid. *Microb Pathog*, **6**, 153–8.

Riesenberg-Wilmes, M. R., Bearson, B., Foster, J. W. and Curtis, R., III. (1996). Role of the acid tolerance response in virulence of *Salmonella typhimurium*. *Infect Immun*, **64**, 1085–92.

Robbe-Saule, V., Coynault, C. and Norel, F. (1995). The live oral typhoid vaccine Ty21a is a *rpoS* mutant and is susceptible to various environmental stresses. *FEMS Microbiol Lett*, **126**, 171–6.

Ronson, C. W., Nixon, B. T. and Ausubel, F. M. (1987). Conserved domains in bacterial regulatory proteins that respond to environmental stimuli. *Cell*, **49**, 579–81.

Roudier, C., Krause, M., Fierer, J. and Guiney, D. G. (1990). Correlation between the presence of sequences homologous to the *vir* region of *Salmonella dublin* plasmid pSDL2 and the virulence of twenty-two *Salmonella* serotypes in mice. *Infect Immun*, **58**, 1180–5.

Rubino, S., Leori, G., Rizzu, P. *et al.* (1993). TnphoA *Salmonella abortusovis* mutants unable to adhere to epithelial cells and with reduced virulence in mice. *Infect Immun*, **61**, 1786–92.

Santos, R. L., Zhang, S., Tsolis, R. M., Bäumler, A. J. and Adams, L. G. (2002). Morphologic and molecular characterization of *Salmonella typhimurium* infection in neonatal calves. *Vet Pathol*, **39**, 200–15.

Schmitt, C. K., Ikeda, J. S., Darnell, S. C. *et al.* (2001). Absence of all components of the flagellar export and synthesis machinery differentially alters virulence of *Salmonella enterica* serovar Typhimurium in models of typhoid fever, survival in macrophages, tissue culture invasiveness, and calf enterocolitis. *Infect Immun*, **69**, 5619–25.

Shea, J. E., Beuzon, C. R., Gleeson, C., Mundy, R. and Holden, D. W. (1999). Influence of the *Salmonella typhimurium* pathogenicity island 2 type III secretion system on bacterial growth in the mouse. *Infect Immun*, **67**, 213–19.

Shea, J. E., Hensel, M., Gleeson, C. and Holden, D. W. (1996). Identification of a virulence locus encoding a second type III secretion system in *Salmonella typhimurium*. *Proc Natl Acad Sci USA*, **93**, 2593–7.

Sizemore, D. R., Fink, P. S., Ou, J. T. *et al.* (1991). Tn5 mutagenesis of the *Salmonella typhimurium* 100 kb plasmid: definition of new virulence regions. *Microb Pathog*, **10**, 493–9.

Smith, R. L. and Maguire, M. E. (1998). Microbial magnesium transport: unusual transporters searching for identity. *Mol Microbiol*, **28**, 217–26.

Soncini, F. C. and Groisman, E. A. (1996). Two-component regulatory systems can interact to process multiple environmental signals. *J Bacteriol*, **178**, 6796–801.

Soncini, F. C., Garcia Vescovi, E. and Groisman, E. A. (1995). Transcriptional autoregulation of the *Salmonella typhimurium phoPQ* operon. *J Bacteriol*, **177**, 4364–71.

Soncini, F. C., Garcia Vescovi, E., Solomon, F. and Groisman, E. A. (1996). Molecular basis of the magnesium deprivation response in *Salmonella typhimurium*: identification of PhoP-regulated genes. *J Bacteriol*, **178**, 5092–9.

Stanley, T. L., Ellermeier, C. D. and Slauch, J. M. (2000). Tissue-specific gene expression identifies a gene in the lysogenic phage Gifsy-1 that affects *Salmonella enterica* serovar Typhimurium survival in Peyer's patches. *J Bacteriol*, **182**, 4406–13.

Stender, S., Friebel, A., Linder, S. *et al.* (2000). Identification of SopE2 from *Salmonella typhimurium*, a conserved guanine nucleotide exchange factor for Cdc42 of the host cell. *Mol Microbiol*, **36**, 1206–21.

Tao, T., Grulich, P. F., Kucharski, L. M., Smith, R. L. and Maguire, M. E. (1998). Magnesium transport in *Salmonella typhimurium*: biphasic magnesium and time dependence of the transcription of the *mgtA* and *mgtCB* loci. *Microbiology*, **144**, 655–64.

Tezcan-Merdol, D., Nyman, T., Lindberg, U. *et al.* (2001). Actin is ADP-ribosylated by the *Salmonella enterica* virulence-associated protein SpvB. *Mol Microbiol*, **39**, 606–19.

Tsolis, R. M., Adams, L. G., Ficht, T. A. and Baumler, A. J. (1999). Contribution of *Salmonella typhimurium* virulence factors to diarrheal disease in calves. *Infect Immun*, **67**, 4879–85.

Tsolis, R. M., Adams, L. G., Hantman, M. J. *et al.* (2000). SspA is required for lethal *Salmonella typhimurium* infections in calves but is not essential for diarrhea. *Infect Immun*, **68**, 3158–63.

Turner, A. K., Lovell, A., Hulme, S. D., Zhang-Barber, L. and Barrow, P. A. (1998). Identification of *Salmonella typhimurium* genes required for colonization of the chicken alimentary tract and for virulence in newly hatched chicks. *Infect Immun*, **66**, 2099–106.

Uchiya, K., Barbieri, M. A., Funato, K. *et al.* (1999). A *Salmonella* virulence protein that inhibits cellular trafficking. *Embo J*, **18**, 3924–33.

Unsworth, K. E. and Holden, D. W. (2000). Identification and analysis of bacterial virulence genes *in vivo*. *Philos Trans R Soc Lond B Biol Sci*, **355**, 613–22.

Valdivia, R. H. and Falkow, S. (1996). Bacterial genetics by flow cytometry: rapid isolation of *Salmonella typhimurium* acid-inducible promoters by differential fluorescence induction. *Mol Microbiol*, **22**, 367–78.

(1997). Fluorescence-based isolation of bacterial genes expressed within host cells. *Science*, **277**, 2007–11.

van der Velden, A. W., Lindgren, S. W., Worley, M. J. and Heffron, F. (2000). *Salmonella* pathogenicity island 1-independent induction of apoptosis in infected macrophages by *Salmonella enterica* serotype Typhimurium. *Infect Immun*, **68**, 5702–9.

Vazquez-Torres, A., Xu, Y., Jones-Carson, J. *et al.* (2000). *Salmonella* Pathogenicity Island 2-dependent evasion of the phagocyte NADPH oxidase. *Science*, **287**, 1655–8.

Vescovi, E. G., Ayala, Y. M., Di Cera, E. and Groisman, E. A. (1997). Characterization of the bacterial sensor protein PhoQ. Evidence for distinct binding sites for $Mg2^{+}$ and $Ca2^{+}$. *J Biol Chem*, **272**, 1440–3.

Waldburger, C. D. and Sauer, R. T. (1996). Signal detection by the PhoQ sensor-transmitter. Characterization of the sensor domain and a response-impaired mutant that identifies ligand-binding determinants. *J Biol Chem*, **271**, 26630–6.

Wallis, T. S., Wood, M., Watson, P. *et al.* (1999). Sips, Sops, and SPIs but not *stn* influence *Salmonella* enteropathogenesis. *Adv Exp Med Biol*, **473**, 275–80.

Watson, P. R., Benmore, A., Khan, S. A. *et al.* (2000). Mutation of *waaN* reduces *Salmonella enterica* serovar Typhimurium-induced enteritis and net secretion of type III secretion system 1-dependent proteins. *Infect Immun*, **68**, 3768–71.

Watson, P. R., Galyov, E. E., Paulin, S. M., Jones, P. W. and Wallis, T. S. (1998). Mutation of *invH*, but not *stn*, reduces *Salmonella*-induced enteritis in cattle. *Infect Immun*, **66**, 1432–8.

Wendland, M. and Bumann, D. (2002). Optimization of GFP levels for analyzing *Salmonella* gene expression during an infection. *FEBS Lett*, **521**, 105–8.

Wilmes-Riesenberg, M. R., Foster, J. W. and Curtiss, R., III (1997). An altered *rpoS* allele contributes to the avirulence of *Salmonella typhimurium* LT2. *Infect Immun*, **65**, 203–10.

Wood, M. W., Rosqvist, R., Mullan, P. B., Edwards, M. H. and Galyov, E. E. (1996). SopE, a secreted protein of *Salmonella dublin*, is translocated into the target eukaryotic cell via a *sip*-dependent mechanism and promotes bacterial entry. *Mol Microbiol*, **22**, 327–38.

Woodward, M. J., McLaren, I. and Wray, C. (1989). Distribution of virulence plasmids within salmonellae. *J Gen Microbiol*, **135**, 503–11.

Yu, X. J., Ruiz-Albert, J., Unsworth, K. E. *et al.* (2002). SpiC is required for secretion of *Salmonella* Pathogenicity Island 2 type III secretion system proteins. *Cell Microbiol*, **4**, 531–40.

Zeng, H., Carlson, A. Q., Guo, Y. *et al.* (2003). Flagellin is the major proinflammatory determinant of enteropathogenic *Salmonella*. *J Immunol*, **171**, 3668–74.

Zhang, S., Adams, L. G., Nunes, J. *et al.* (2003). Secreted effector proteins of *Salmonella enterica* serotype Typhimurium elicit host-specific chemokine profiles in animal models of typhoid fever and enterocolitis. *Infect Immun*, **71**, 4795–803.

Zhang, S., Santos, R. L., Tsolis, R. M. *et al.* (2002). Phage mediated horizontal transfer of the *sopE1* gene increases enteropathogenicity of *Salmonella enterica* serotype Typhimurium for calves. *FEMS Microbiol Lett*, **217**, 243–7.

Zhou, D. and Galan, J. (2001). *Salmonella* entry into host cells: the work in concert of type III secreted effector proteins. *Microbes Infect*, **3**, 1293–8.

Zhou, D., Mooseker, M. S. and Galan, J. E. (1999a). An invasion-associated *Salmonella* protein modulates the actin-bundling activity of plastin. *Proc Natl Acad Sci USA*, **96**, 10176–81.

(1999b). Role of the *S. typhimurium* actin-binding protein SipA in bacterial internalization. *Science*, **283**, 2092–5.

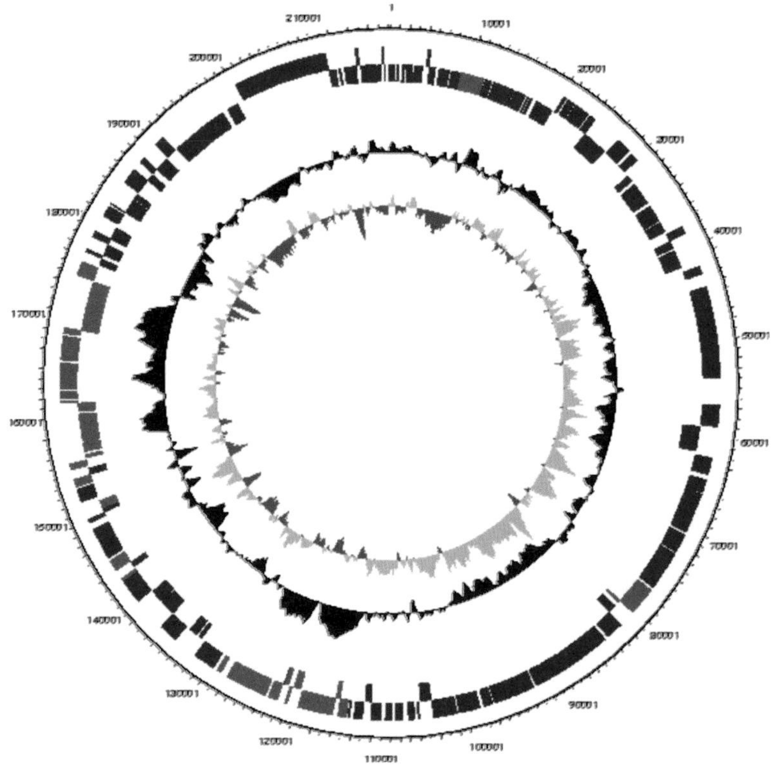

Plate 2.1. A map of plasmid pHCM1 from the sequenced strain of *S. enterica* serovar Typhi showing regions of similarity with plasmid R27.

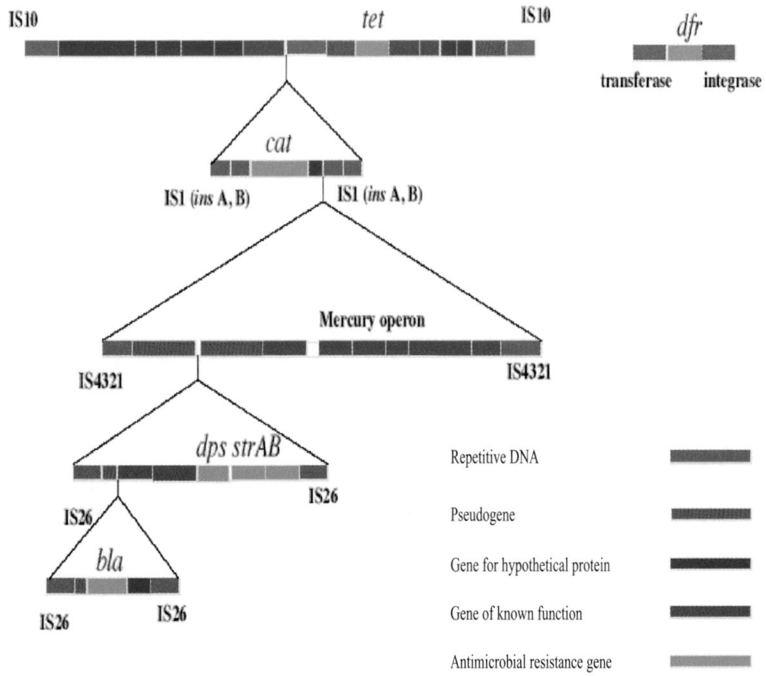

Plate 2.2. A diagram of the large insertion region of plasmid pHCM1.

Plate 6.1. (*cont.*) invasion and enteropathogenesis, the SPI-2-encoded TTSS is required for the intracellular proliferation of *S. enterica* and systemic pathogenesis. SPI-1 and SPI-2 both contain portions that are not related to TTSS function but have functions in iron uptake (*sit* gene in SPI-1) or anaerobic respiration (*ttr* genes in SPI-2). Maps of SPI-1 and SPI2 were adapted from Hansen-Wester and Hensel (Hansen-Wester and Hensel, 2001). The SPI-5 map was adapted from (Wood *et al.*, 1998). B) SPI-3, SPI-7 and SGI-1 harbor DNA sequences and genes involved in DNA mobility. SPI-3 encodes the high affinity Mg^{2+} uptake system MgtCB, required for intracellular proliferation. A central portion of SPI-3 (*rmbA* to *marT*) is flanked by IS element-like DNA sequences. SGI-1 encodes five antibiotic resistances. Various genes with similarity to integrases, transposases and conjugational transfer systems are present, as well as cryptic prophage. SGI-1 is flanked by direct repeats, and further direct repeats flank the *floR* gene. SPI-7 is composed of various regions with different functions. The *pil* gene cluster encodes a type IVB pilus assembly system and the *viaB* region is required for biosynthesis of the Vi antigen capsule. Furthermore, the *sopE* phage harboring the *sopE* gene for an effector protein of the SPI-1-encoded TTSS is present within SPI-7. Maps of SPI-3 (Blanc-Potard *et al.*, 1999), SGI-1 (Boyd *et al.*, 2001) and SPI-7 (Pickard *et al.*, 2003) were adapted from the original publications.

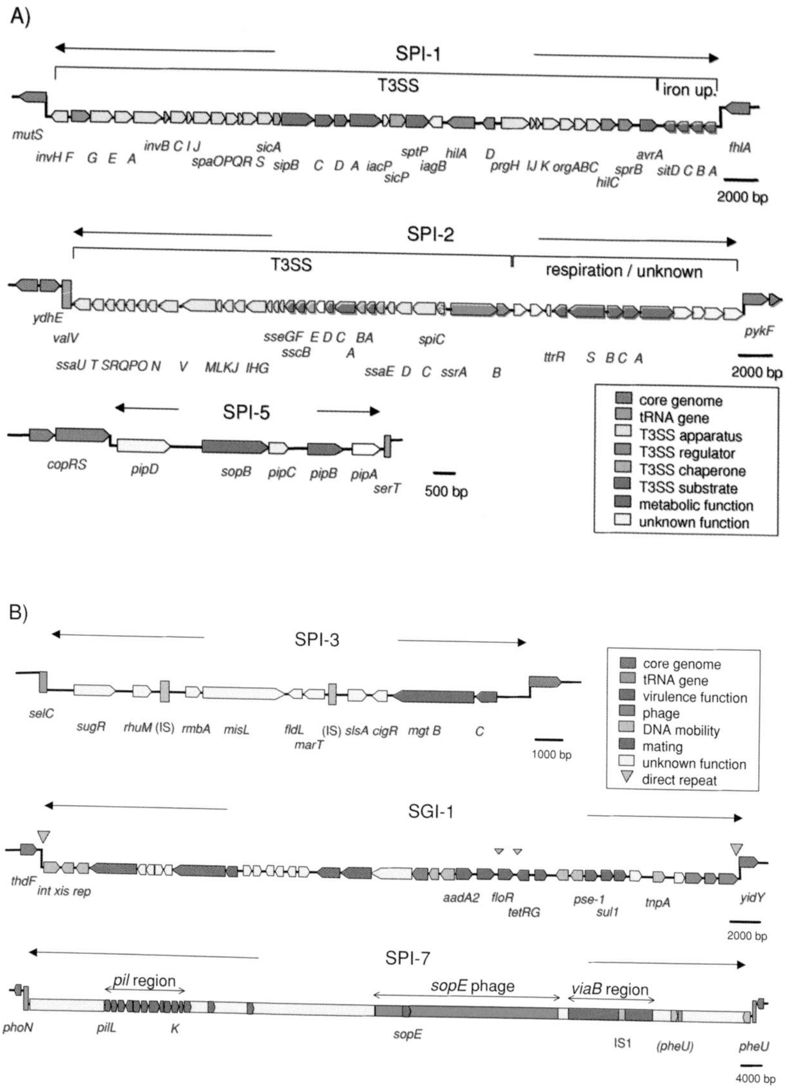

Plate 6.1. Topologies of representative *Salmonella* pathogenicity islands.
Genetic organization, gene functions and the role in virulence are shown. Gene designations of relevant loci are indicated. Incomplete elements (IS, tRNA) are indicated by parenthesis.
A) SPI encoding type III secretion systems (TTSS) and cognate effector proteins. SPI-1 and SPI-2 encode TTSS for the secretion and translocation of virulence proteins into eukaryotic target cells. Both loci encode the TTSS apparatus, regulatory systems and a subset of substrate proteins, *i.e.* secreted proteins involved in translocation and translocated effector proteins. While the SPI-1-encoded TTSS is involved in host cell

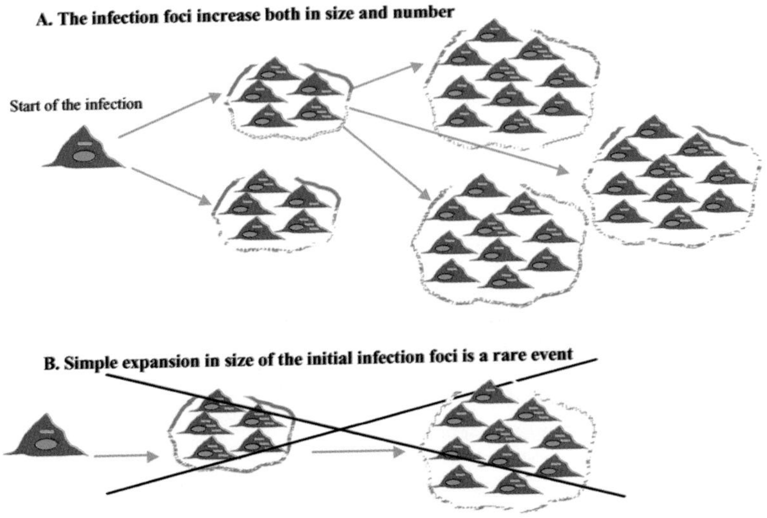

Plate 8.3. Variation in the size and numbers of pathological lesions during the growth of *S. enterica* in the tissues.
Schematic representation of bacterial distribution at the level of the infected foci (pathological lesions) in mice infected with a *S. enterica* strain that would increase in numbers about tenfold per day. Bacterial growth results in an increase in the numbers of pathological lesions and in the number of infected phagocytes per lesion. B. A scenario where the pathological lesions increase solely in size trapping the bacteria in the initial foci of infection is a rare event.

Plate 9.1. (*cont.*) NADPH oxidase by expressing an array of superoxide dismutases (SOD), catalases and scavengers such as reduced glutathione (GSH). In addition, this intracellular pathogen stores iron in an inactive form inside iron storage proteins (ISP) to prevent formation of hydroxyl radical (OH) from hydrogen peroxide, and repairs damage incurred to the DNA by nucleotide excision repair enzymes. (C) Recognition of LPS on the surface of *S. enterica* by TLR4 induces the expression of iNOS. Analogous to functional NADPH oxidase complexes, iNOS-containing vesicles are excluded from the *S. enterica* phagosome by SPI-2 effectors. NO however diffuses freely across membranes. Reactive nitrogen species (RNS) such as NO, dinitrogen trioxide, nitrogen dioxide, nitrosothiols and dinitrosyl iron complexes exert their toxicity by reacting with metal cofactors, DNA, lipids and thiol groups. *S. enterica* detoxifies NO via the enzymatic activity of the Hm*p* flavohaemoprotein, and uses a battery of scavengers such as GSH for the detoxification of a variety of RNS. The SPI-2 type system prevents fusion of the *S. enterica* phagosome with lysosomes along the degradative pathway. The *S. enterica* phagosome nevertheless maintains a dynamic interaction with late endosomes and the trans Golgi network (TGN), acquiring immature cathepsin D and lysosomal-associated membrane glycoproteins (LAMP). (D) Acquisition of the Nramp1 divalent metal transporter reduces the intraphagosomal concentration of iron and manganese. *S. enterica* expresses a battery of metal acquisition systems that compete with the host for divalent metals such as iron and manganese. The Nramp1-associated diminution in intraphagosomal iron concentration hastens the maturation of the *S. enterica* phagosome, promoting its interaction with early endosomal markers, mature hydrolases and the mannose 6-phosphate receptor (M6PR).

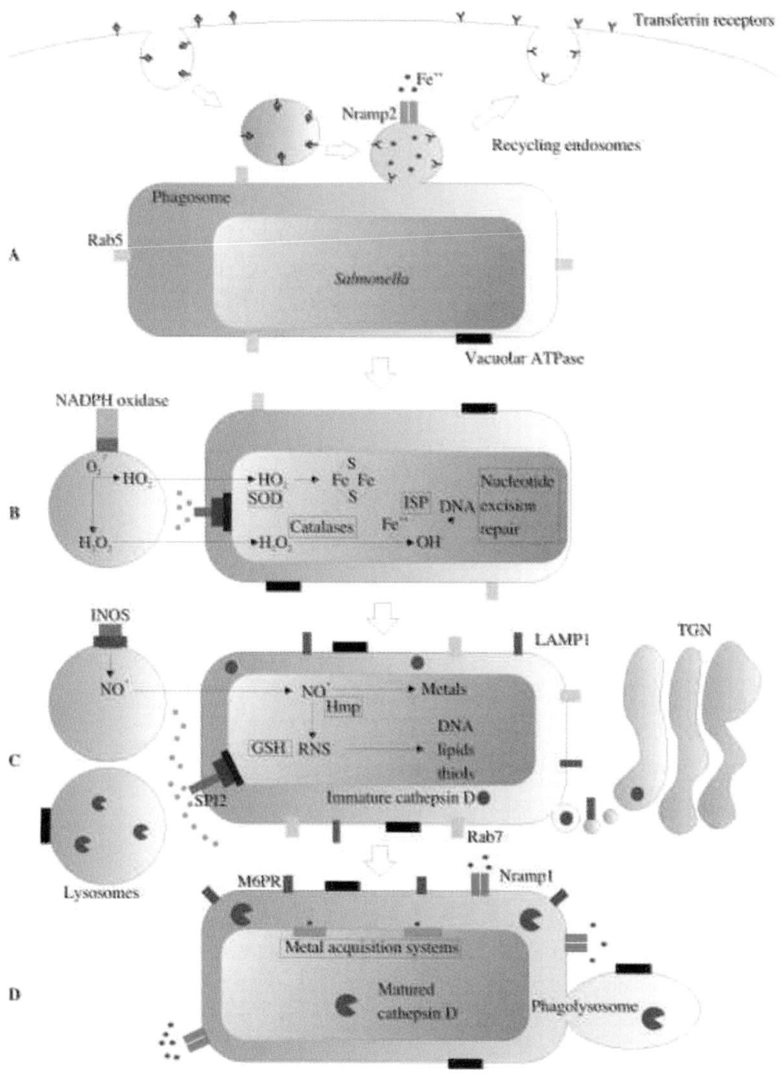

Plate 9.1. A model of the bidirectional interactions between *S. enterica* virulence factors and phagocytic host defenses.

(A) *S. enterica* is internalized by professional phagocytes into a membrane-bound vacuole called the phagosome. The young phagosome expresses the small GTPase Rab5 that allows this vesicle to maintain a dynamic relationship with early endosomes containing transferrin (pink small spheres). In the acidic environment of recycling endosomes, transferrin is released from its receptor and iron is pumped out to the cytosol by the Nramp2 transporter. Shortly after its formation, the *S. enterica* phagosome associates with vacuolar ATPases (black rectangles) that acidify the lumen. (B) Uptake of *S. enterica* activates the assembly of membrane-bound and cytosolic components of the NADPH oxidase in small vesicles. Effectors (green squares) of the type III secretion system encoded within the *S. enterica* pathogenicity island 2 (SPI-2) reduce the oxidative stress that *S. enterica* has to endure by blocking fusion of NADPH oxidase-containing vesicles with the phagosome. Some membrane soluble ROS such as hydroperoxyl (HO_2) and hydrogen peroxide (H_2O_2) however gain access to *S. enterica*, oxidizing iron sulfur clusters, proteins and DNA. *S. enterica* protects vital targets from the cytotoxicity associated with the

CHAPTER 8

Mechanisms of immunity to *Salmonella* infection

Pietro Mastroeni

8.1 INTRODUCTION

Salmonella enterica affects humans and animals worldwide. It can be found in sewage-, sea-, and river- water and can contaminate food. Asymptomatic carriage in domestic animals can result in the introduction of the bacteria into the food chain.

Interest in understanding the mechanisms of pathogenesis and immunity that operate in *S. enterica* infections is twofold. Firstly, development of vaccines against salmonellosis has been too empirical due to insufficient understanding of how the host controls these infections, and how the bacteria evade immune surveillance. The fact that *S. enterica*-based vaccines are also being evaluated as systems to deliver recombinant antigens or DNA vaccines to the immune system and as new tools for the therapy of cancer has further increased the need to study how these vaccines work (Chabalgoity *et al.*, 2002; Mastroeni *et al.*, 2001; Reisfeld *et al.*, 2004).

Secondly, *S. enterica* provides a model to understand how bacterial pathogens interact with the immune system. *S. enterica* is an intriguing bacterium in the way it interacts with the immune system and the immunological requirements for host resistance to this bacterium are affected by a very large number of variables.

8.2 MODELS FOR THE STUDY OF IMMUNITY TO *S. ENTERICA*

The study of the immunobiology of *S. enterica* infections has been facilitated by the availability of reliable models and by improved genetic tools

'Salmonella' Infections: Clinical, Immunological and Molecular Aspects, ed. Pietro Mastroeni and Duncan Maskell. Published by Cambridge University Press.

that allow identification of polymorphic differences or mutations in genes involved in immune functions.

In vitro models have been used for studying direct interactions between bacteria and individual cell populations such as epithelial cells, phagocytes and antigen presenting cells (dendritic cells and B-cells). The mouse model has been extensively used for the study of the mechanisms of host resistance to *S. enterica* in a whole host organism. Infections in domestic animals such as chickens, pigs and cattle, as well as clinical and genetic studies in patients with salmonellosis have extended the fundamental work performed in infection models. The following sections will give a brief overview of the knowledge generated about immunity to *S. enterica* in the mouse model of infection. Later in this chapter, the correlates and mechanisms of immunity to salmonellosis in humans will be discussed.

The severity and outcome of *S. enterica* infections depend on the combination of the virulence of the infecting bacterial strain, the infectious dose, the route of infection and the genetic makeup and immunological status of the host. All of these variables can profoundly influence the assessment of the relative importance of different immunological mechanisms in host resistance to the disease. Therefore, the experimental set up must be carefully considered when interpreting results.

The main immunological factors that control the course of a *S. enterica* infection in mice are illustrated in Figure 8.1.

8.3 EARLY EVENTS IN THE INTERACTION BETWEEN *S. ENTERICA* AND THE IMMUNE SYSTEM

8.3.1 Invasion of the gut and early inflammatory responses

S. enterica usually enters the host by the oral route. A proportion of the organisms resists stomach acid and reaches the distal ileum and the caecum. The resident microbial flora protects the host in these early stages of the infection as shown by the increased susceptibility to oral infection in mice pre-treated with streptomycin (Bohnoff and Miller, 1962). *S. enterica* invades epithelial and M cells in the gut using the type III secretion system encoded by genes within the *Salmonella* pathogenicity island 1 (SPI-1) (Lee and Schneewind, 1999). During invasion, the SPI-1- encoded sipB protein activates Caspase-1 within resident macrophages. Caspase-1 is a cysteine protease that induces apoptosis in macrophages resulting in escape of *S. enterica* from these cells. Caspase-1 also cleaves the precursors of IL1β and IL18 to produce bioactive cytokines. IL1β and IL18 enhance local inflammation

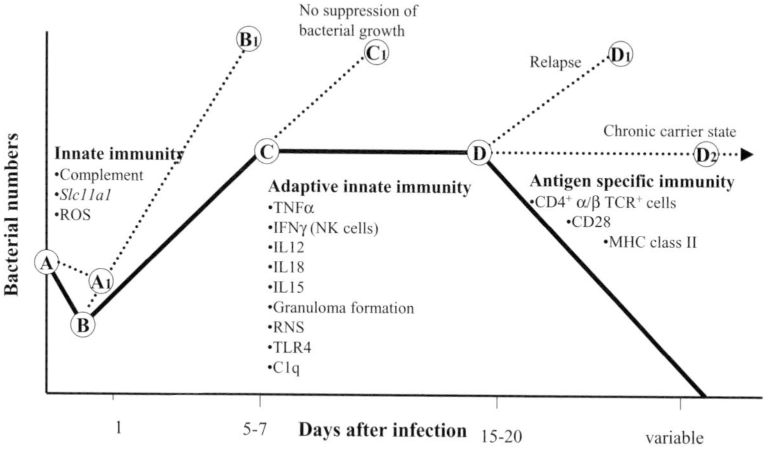

Figure 8.1. Immunological mechanisms that control the progression of a sublethal *S. enterica* infection in the mouse model.
The solid line in the graph shows the course of a sublethal infection in wild type immunocompetent mice. The dotted lines show the course of the infection when immunological mechanisms required at points A-D are absent. Lack of complement and/or ROS affects the early blood clearance and killing of the bacteria with a shift of the growth curve from A-B to A-A1. The *Slc11a1* gene and ROS influence the net growth rate and their absence determines a shift of the curve from B-C to B-B1. Point C coincides with the onset of the adaptive innate immune response. The lack of any of the immunological factors that form part of adaptive innate immunity determines failure to suppress bacterial growth in the tissues (C-D) and the unrestrained progression of the infection process (C-C1). Point D coincides with the intervention of antigen-specific immunity that is required to clear the infection and prevent relapse (D-D1) or the establishment of a chronic carrier state (D-D2).

and promote infiltration of polymorphonuclear phagocytes (PMN) into the gut mucosa resulting in internalization of the bacteria by these cells. SipB-mediated activation of Caspase-1 ultimately results in enhanced bacterial colonization of the Peyer's patches and mesenteric lymph nodes indicating that escape from macrophages and localization in the extracellular compartment and/or within PMNs is beneficial for *S. enterica* at this stage of the infection (Monack *et al.*, 2000). Products of the *S. enterica sipA* and *sopABDE2* genes can also trigger inflammation in the gut by eliciting the production of CC and CXC chemokines such as GROα, GROγ, IL8 and GCP-2 (Zhang *et al.*, 2003). In vitro experiments in intestinal cells also suggest that the *S. enterica* SspH1 and SptP proteins can inhibit NF-κB-dependent gene expression resulting in the down regulation of IL8 production (Haraga and Miller, 2003).

An alternative SPI-1-independent mechanism of invasion has been described where *S. enterica* does not interact with M cells but is engulfed by dendritic cells (DC) that open the tight junctions between epithelial cells and sample bacteria at the mucosal surface (Rescigno *et al.*, 2001).

8.3.2 The journey from the gut to the systemic compartment

Once the invasion process has been completed the bacteria are transported from the gastrointestinal tract to the bloodstream presumably passing *via* the mesenteric lymph nodes. In the blood, *S. enterica* can be found either as extracellular bacteria or associated with $CD18^{+}$ cells (Vazquez-Torres *et al.*, 1999). In the circulation, extracellular bacteria are opsonized but not lysed by complement factors. The complement cascade is activated by *S. enterica* LPS (resulting in opsonization), but the C5-9 lytic complex fails to insert into the bacterial cell membranes due to the presence of the long LPS-O antigen side chains that protrude from the bacterial surface. The opsonized bacteria are rapidly cleared from the blood (Biozzi *et al.*, 1960). Differences in the structure of the LPS O-side chains between *S. enterica* serovars determine the rate of blood clearance by affecting the efficiency of complement activation at the bacterial surface (Liang-Takasaki *et al.*, 1983; Saxen *et al.*, 1987). Complement activation presumably follows the alternate pathway. However, studies in $C1qa^{-/-}$ gene targeted mice have revealed that C1q (a component of the classical pathway) also contributes to the clearance of *S. enterica* from the blood (Warren *et al.*, 2002).

8.4 *S. ENTERICA* REACHES THE PHAGOCYTIC CELLS IN THE INFECTED TISSUES

8.4.1 *S. enterica* resides intracellularly

After being cleared from the blood, *S. enterica* is found mainly in the spleen, liver and bone marrow where it resides intracellularly in macrophages, PMNs and DC (Dunlap *et al.*, 1992; Richter-Dahlfors *et al.*, 1997; Salcedo *et al.*, 2001; Sheppard *et al.*, 2003; Yrlid *et al.*, 2001). In the spleen, red pulp macrophages ($F4/80^{+}$, MSR-A^{low}) and marginal zone macrophages (MSR-A^{+}) appear to contain the majority of the bacteria (Salcedo *et al.*, 2001) and a minor proportion of the bacteria are found in $B220^{+}$ B-cells (Yrlid *et al.*, 2001). In the liver, *S. enterica* localizes preferentially in the resident Kupffer cells. Some of the salmonellae are likely to be, at least transiently, present in the extracellular space throughout the infection. In fact, administration of gentamicin, an antibiotic that poorly penetrates inside eukaryotic cells, can reduce the bacterial load in the tissues of mice infected with *S. enterica* (Bonina *et al.*, 1998). Bacteria can be seen in the extracellular space, as well as in non phagocytic cells, when high bacterial numbers are reached in the tissues, such as after injection of large inocula or in the terminal stages of lethal infections (Hsu, 1989). A proportion of the bacteria can also be cultured from the blood (Collins, 1969).

8.5 DYNAMICS OF *S. ENTERICA* SPREAD AND DISTRIBUTION AT THE SINGLE CELL LEVEL

8.5.1 General

In many infections, microbes grow within the tissues to increase their initial numbers before this process is aborted either by the host immune system or by pharmacological intervention. The amplification of the infection can lead to: (a) an increase in the number of micro-organisms in the initial foci of infection established soon after invasion; (b) the spread and distribution of bacteria to new infection foci and/or new locations in the body; (c) a combination of the expansion of bacterial numbers in the initial infection foci and their spread to new foci. The patterns and mechanisms of distribution of a given micro-organism in the body have important implications for understanding how the immune system controls the infection and for improving therapeutic and preventive measures to fight infectious diseases.

In mice, in the initial stages of the infection, *S. enterica* is seen in individual cells scattered in the infected tissues. Later in the course of the disease, the infected macrophages and PMNs are found in well-defined pathological lesions that are separated by normal tissue (Richter-Dahlfors *et al.*, 1997; Sheppard *et al.*, 2003). It is within these cells and lesions that the interplay between host and pathogen takes place.

8.5.2 Individual bacterial populations segregate to different infected phagocytes within the infected tissues

Fluorescence microscopy techniques have recently allowed the visualization of individual bacteria in vivo in infected cells (Sheppard *et al.*, 2003). This has provided new approaches for studying how individual *S. enterica* bacteria, and bacterial populations derived from the growth of these organisms, distribute in the same host. It has been established that during the growth of *S. enterica* in the tissues, different bacterial populations always segregate to individual infected cells and pathological lesions. For example, when mice are infected simultaneously with two nearly isogenic *S. enterica* serovar Typhimurium strains (representing two bacterial populations within the host) that express different LPS-O antigens and can therefore be differentially immuno-stained, infection foci always contain only one of the two strains but never both. Spread of each *S. enterica* strain (bacterial population) from one infected focus to a different established focus that already contains the other strain does not occur. At any time during the infection, individual phagocytes contain clusters of *S. enterica* all of the same strain that are very likely to have resulted from clonal expansion of individual bacteria. With the progression of the disease, pathological lesions form at foci of infection. Each lesion contains only one bacterial population consisting of either one or the other infecting strain and residing mostly within $CD18^+$ phagocytes. Spread of bacteria from one lesion to another established lesion, and mixing of different bacterial populations, are thus very rare events (Sheppard *et al.*, 2003). These observations indicate that in *S. enterica* infections each infectious focus is a separate unit that results from the clonal growth of an individual bacterium and evolves as a spatially independent entity throughout the infection process. The spatial segregation of *S. enterica* populations to different infection foci within the same organ supports the view that the mechanisms of host resistance within each infection focus, and not just the systemic activation of generalized immune responses, play a major role in controlling the infection. In fact, in support of this view it has been shown that the growth curves of different *S. enterica* strains, simultaneously inoculated into the same animal, progress independently from one another (Maskell *et al.*, 1987).

8.5.3 *S. enterica* growth results in an increase in the number of infection foci and low bacterial numbers are seen within infected phagocytes

In the initial stages of a *S. enterica* infection, each infection focus consists of an individual phagocyte containing only one bacterium. *S. enterica* growth in the tissues results in the distribution of the bacteria to uninfected cells and in their spread to new infection foci. Bacterial growth in the tissues results in an increase in the number of infected cells and does not

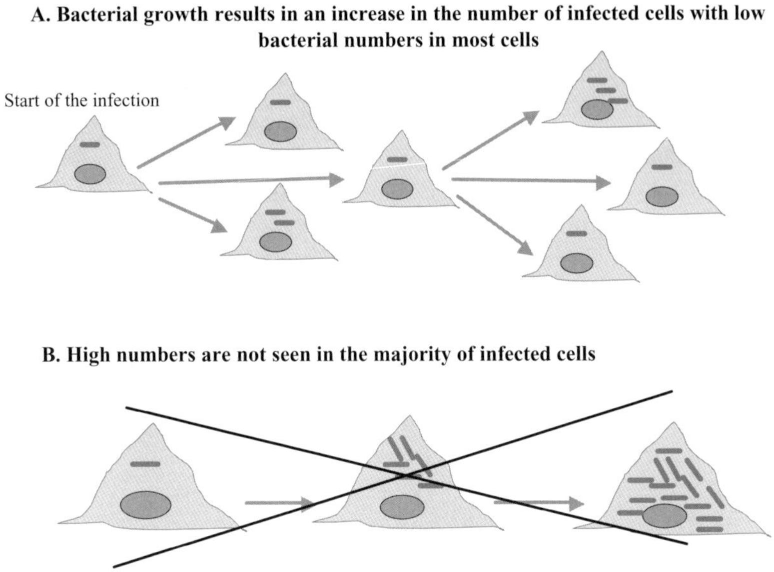
A. Bacterial growth results in an increase in the number of infected cells with low bacterial numbers in most cells
Start of the infection
B. High numbers are not seen in the majority of infected cells

Figure 8.2. *S. enterica* spreads to uninfected cells with low bacterial numbers in the majority of phagocytes.
A. Schematic representation of bacterial distribution at the level of individual cells in mice infected with a *S. enterica* strain that would increase in numbers about tenfold per day. As the net bacterial numbers increase in the tissues the number of infected cells increases and the numbers of bacteria remains low in the majority of cells. B. *S. enterica* growth to high numbers in infected phagocytes is not a frequent event.

involve growth to high bacterial numbers in the majority of infected phagocytes. Under experimental conditions that do not allow bacterial numbers to increase (no net bacterial growth) the number of infected phagocytes remains unchanged. Regardless of the net growth of the bacteria in the tissues, the average number of salmonellae per phagocyte throughout the infection is very low with the majority of phagocytes containing one bacterium (Figure 8.2) (Sheppard *et al.*, 2003).

A similar picture is observed when analyzing the size and numbers of pathological lesions in the infected tissues (Figure 8.3). In fact, the number of infected lesions increases in parallel with the net bacterial growth rate with a small increase in the size of the lesion and in the numbers of infected cells per lesion. Thus, a scenario where the pathological lesions increase solely in size trapping the bacteria in the initial foci of infection is very unlikely (Sheppard *et al.*, 2003).

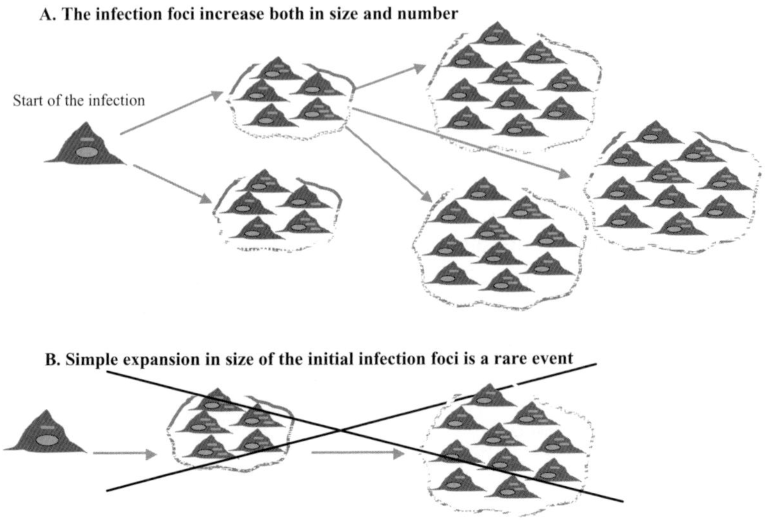

Figure 8.3. Variation in the size and numbers of pathological lesions during the growth of *S. enterica* in the tissues.
Schematic representation of bacterial distribution at the level of the infected foci (pathological lesions) in mice infected with a *S. enterica* strain that would increase in numbers about tenfold per day. Bacterial growth results in an increase in the numbers of pathological lesions and in the number of infected phagocytes per lesion. B. A scenario where the pathological lesions increase solely in size trapping the bacteria in the initial foci of infection is a rare event. (See colour plate section.)

8.5.4 Escape from infected cells and distribution in the tissues as a virulence trait?

Escape from infected cells and from already-formed pathological lesions is a key feature of *S. enterica* growth in the tissues and is likely to represent one of the mechanisms used by the bacteria to counteract and evade the host immune response. A plausible scenario would one whereby the host strives to restrain *S. enterica* to discrete foci in the tissues. The bacteria would therefore need to escape from these foci and disseminate throughout the organ to be one step ahead of the local immune response that is activated at the level of each infection focus. This would indicate that, contrary to what is suggested by in vitro studies (Fields *et al.*, 1986; Harrington and Hormaeche, 1986; Lissner *et al.*, 1983), the ability to grow inside the majority of infected phagocytes and to avoid intracellular killing are not the only prerequisites for *S. enterica* virulence. Escape from infected cells, followed by bacterial spread and distribution in the tissues is required for the expansion of the initial infecting dose.

8.6 INNATE IMMUNITY AND CONTROL OF THE EARLY GROWTH OF *S. ENTERICA* IN THE TISSUES

8.6.1 General

A proportion of bacteria that reach the intracellular compartment are rapidly killed by resident macrophages within 6–12 hours from the start of the infection. From this point of the infection onwards, the fate of the surviving bacteria is variable and the bacterial load in the tissues can either increase or remain constant or decrease. The early net growth rate of the bacteria in the tissues of the infected animal is not affected by the initial infectious dose and depends on the virulence of the bacteria and on the level of innate resistance of the host (Mastroeni, 2002).

8.6.2 Recognition of *S. enterica* by phagocytic cells

S. enterica is likely to interact with receptors on the surface of phagocytes from the early stages of the infection.

As mentioned before, the bacteria are opsonized by complement factors present in the blood and in other tissues. The absence of a functional complement cascade (*e.g.* in C3 deficient animals) reduces host resistance early in infection (P. Mastroeni, unpublished observations).

Toll-like receptors (TLR) are a group of PAMPs found in plants, insects, birds and mammals and they recognize bacterial components including lipopolysaccharide (LPS), peptidoglycan, lipoproteins, flagella proteins and DNA. Ligand activation of TLRs induces pro-inflammatory cytokines such as IL1, IL6, IL8 and TNFα (Medzhitov *et al.*, 1997).

TLR 4 in conjunction with CD14, LPS binding protein (LBP) and MD2 is involved in the recognition of the Lipid A portion of bacterial LPS (Takeuchi *et al.*, 1999) and mediates the inflammatory response of macrophages to *S. enterica* (Rosenberger *et al.*, 2000; Royle *et al.*, 2003). C3H/HeJ mice have a Pro712His dominant-negative mutation in TLR4 that renders them hyporesponsive to LPS (Poltorak *et al.*, 1998; Qureshi *et al.*, 1999; Vogel *et al.*, 1999). TLR4 deficient mice are impaired in their ability to mount innate immune responses to *S. enterica* and are very susceptible to *S. enterica* infection (see also section 8.8.7) (Eisenstein *et al.*, 1982; Hormaeche, 1990). Cells from TLR4 deficient mice show reduced production of TNFα and other inflammatory cytokines due to reduced transcription of the TNFα gene and complete inhibition of TNFα mRNA translation (Beutler *et al.*, 1986). Macrophages from endotoxin non-responder C3H/HeJ mice and TLR4$^{-/-}$ mice can produce TNFα and IL6 in response to whole *S. enterica* micro-organisms suggesting the involvement of receptors other than TLR4 in the induction of cytokines in response to *S. enterica* (Freudenberg and Galanos, 1991; Lembo *et al.*, 2003). Interestingly, TLR2 is up-regulated during *S. enterica* infection in vivo (Totemeyer *et al.*, 2003) and is involved in cytokine production by cultured macrophages and mice exposed to *S. enterica*. In fact, TNFα and IL6 production is reduced in TLR $4^{-/-}$ $TLR2^{-/-}$ mice as compared to mice lacking only TLR4 (Lembo *et al.*, 2003). $TLR4^{-/-}$ $TLR2^{-/-}$ mice and macrophages harvested from these animals can still produce small amounts of TNFα and IL6 in response to *S. enterica*, indicating a possible role for other receptors in cytokine production in response to whole *S. enterica* micro-organisms (Lembo *et al.*, 2003). TLR4 and TLR2 also play a role in phagocytosis of *S. enterica* by macrophages and enhance the fusion of lysosomes with phagosomes containing *S. enterica* (Blander and Medzhitov, 2004).

The role of the other TLRs in infection, such as TLR9 and TLR5, is unclear. TLR9 mediates the host inflammatory response to bacterial DNA (Bauer *et al.*, 2001; Hemmi *et al.*, 2000; Takeshita *et al.*, 2001). There is no information yet available to suggest that animals or patients with a functionally defective TLR9 have altered susceptibility to *S. enterica* infection. TLR5 is involved in the host response to *S. enterica* infections by responding to bacterial flagellin protein (Hayashi *et al.*, 2001; Moors *et al.*, 2001; Sebastiani *et al.*, 2000). TLR5 is expressed on the surface of epithelial cells (Adamo *et al.*, 2003) and is likely to be one of the detection mechanisms for *S. enterica* in the gut during the initial stages of enteric infections. Interestingly, MOLF/Ei mice have a deficiency in TLR5 that might be the basis of their increased susceptibility to *S. enterica* infections (Sebastiani *et al.*, 2000). However, a polymorphism in the TLR5 gene that introduces a premature stop codon does not appear to increase susceptibility to typhoid fever in humans (Dunstan *et al.*, 2005).

8.6.3 Role of the *Slc11a1* gene in innate resistance to *S. enterica*

The *Slc11a1* gene (formerly known as *Nramp1*) is located on mouse chromosome 1 and controls the exponential growth rate of *S. enterica* in the tissues (Vidal *et al.*, 1993). *Slc11a1* encodes a phosphoglycoprotein of 90–100 kD that is preferentially localized in the membranes of phagosomes containing bacteria. In mice, *Slc11a1* is expressed mainly in macrophages and in cells of the granulocyte lineage where it functions as a divalent metal (Fe^{2+}, Zn^{2+}, Mn^{2+}) ion pump at the membrane of the late endosome. The gene has two allelic forms, $Slc11a1^{resistant}$ and $Slc11a1^{susceptible}$, and the resistance allele is dominant. Susceptibility is associated with a substitution at position 169 where glycine is replaced by aspartic acid (Blackwell *et al.*, 2001; Forbes and Gros, 2001). The principal effect of the *Slc11a1* gene is on the control of the rate of division of intracellular *S. enterica* with little effect on bacterial killing by the host (Benjamin *et al.*, 1990; Hormaeche, 1980).

8.6.4 Reactive oxygen intermediates

Phagocytic cells control the growth of *S. enterica* in the first few days of the infection using reactive oxygen species [ROS (generated *via* the phagocyte NADPH-oxidase)]. The growth of *S. enterica* is faster in the tissues of gene-targeted mice lacking the gp91 subunit of the NADPH-oxidase (gp91*phox*$^{-/-}$ mice) than in tissues of immunocompetent mice (Mastroeni *et al.*, 2000b). In vitro studies further indicate that ROS are needed for the rapid killing of ingested *S. enterica* by macrophages (Vazquez-Torres *et al.*, 2000a). Reactive nitrogen species (RNS) play a role in host resistance in the later stages of *S. enterica* infection (see below).

The production of the cytokine tumor necrosis factor alpha (TNFα) and its binding to the receptor TNFR55 is needed for the localization of vesicles containing NADPH oxidase to vacuoles containing *S. enterica*. This process is inefficient in macrophages lacking TNFR55, leading to impairment in bacterial killing (Vazquez-Torres *et al.*, 2001).

Lysosomal enzymes and defensins have also been implicated in macrophage dependent killing of *S. enterica* (Groisman *et al.*, 1992).

8.6.5 Which cells control bacterial growth early in the infection?

Host resistance in the early phases of *S. enterica* infections is mainly controlled by innate immune functions of those phagocytes that are present initially at the site of infection (resident macrophages) and those cells that infiltrate the infection foci in the first few days of the disease (PMNs). Resident macrophages are the first phagocytes encountered by *S. enterica* in the spleen and liver. Thereafter the majority of the bacteria are found within PMNs (Richter-Dahlfors *et al.*, 1997; Salcedo *et al.*, 2001; Sheppard *et al.*, 2003).

It is still unclear which cell populations control the growth of the bacteria early in infection. Studies in radiation chimera mice suggest that the *Slc11a1* genotype of resident macrophages determines the phenotype of the animal. In these studies, X-irradiated *Slc11a1*$^{susceptible/resistant}$ F1 mice, when grafted with bone marrow cells from *Slc11a1*^{susceptible} mice acquired the innate susceptibility of the bone marrow donors by 3 months, but not by one month after grafting. The data therefore suggest that radiation-resistant cells with a slow turnover (*i.e.* macrophages) determine the Slc11a1 phenotype of the host (Hormaeche, 1979). Furthermore, administration of substances such as silica, known to impair macrophage functions, greatly exacerbates early growth of *S. enterica* in the early *Slc11a1*-dependent phase of the infection (O'Brien *et al.*, 1979). The fact that the in vivo Slc11a1 resistant phenotype of a mouse is paralleled by the level of resistance expressed by its resident macrophages (splenic, peritoneal and hepatic) in vitro further suggests that these cells might be key players in the control of the early stages of the infection (Harrington and Hormaeche, 1986; Lissner *et al.*, 1983). However, evidence for a role played by PMNs in early host resistance to *S. enterica* has also been provided. PMNs can express *Slc11a1* and early in the infection, *S. enterica* is found inside PMNs in the liver and spleen (Forbes and Gros, 2001; Richter-Dahlfors *et al.*, 1997). Mice rendered neutropenic by treatment with the granulocyte-depleting monoclonal antibody RB6–8C5 are more susceptible to *S. enterica* infections than immunocompetent mice (Conlan, 1997). These observations indicate that PMNs might be important for host resistance to *Salmonella*. On the contrary, whole body X-irradiation, which severely impairs the circulating numbers of PMN, but does not affect resident macrophages, has no effect on the early net growth rate of the organisms in the RES (Hormaeche *et al.*, 1990).

8.6.6 *S. enterica* can evade killing by phagocytes

The growth of *S. enterica* in the tissues requires multiplication in at least a proportion of the infected phagocytes and the avoidance of killing within these cells. Several *S. enterica* genes are upregulated upon entry inside phagocytes and some of these genes are needed to adapt to the intracellular environment and to counteract the killing mechanisms of the cells (Cirillo *et al.*, 1998; Eriksson *et al.*, 2003). *S. enterica* counteracts killing by cultured macrophages by inhibiting the localization of ROS and RNS to the phagosome. This involves genes contained within *Salmonella* pathogenicity island 2 (SPI-2) (Vazquez-Torres *et al.*, 2000b). *S. enterica* can also inhibit the production of NO within cells in vitro (Eriksson *et al.*, 2000). Inhibition of phagolysosome fusion and resistance to defensins have also been described as potential evasion mechanisms (Groisman *et al.*, 1992).

Escape from infected macrophages is an additional potential mechanism of immunoevasion. For example, the *S. enterica* sipB protein induces death of macrophages in the Peyer's patches, thus enabling the bacteria to localize in PMNs that may be less efficient than macrophages in controlling bacterial growth (Monack *et al.*, 2000). Apoptosis of infected cells is seen in the liver of mice inoculated with *S. enterica* (Richter-Dahlfors *et al.*, 1997). *S. enterica* produces cytotoxins that induce necrosis of phagocytic cells (Ashkenazi *et al.*, 1988; Koo *et al.*, 1984; Reitmeyer *et al.*, 1986).

8.7 PROGRESSIVE BACTERIAL GROWTH IN THE TISSUES RESULTS IN LETHAL INFECTIONS

8.7.1 General

Innate immunity regulates, but it is often unable to stop, the exponential growth of *S. enterica* in the early stages of the infection. A progressive increase in bacterial numbers leads to the death of the animal if the microbial load in the spleen and liver reaches about 10^8–10^9 live bacteria per organ. Lethal systemic infections can result from a combination of innately susceptible animals and virulent salmonellae, from high initial doses of bacteria or from the suppression the immune system of the animal by genetic or immunological manipulations (Collins, 1974).

8.7.2 Involvement of LPS and endotoxic shock as the cause of death in lethal infections

It is not clear whether the death of animals infected with *S. enterica* is due to the classic endotoxic septic shock syndrome characterized by bacteraemia, clinical signs of infection, hypoxia, hypoperfusion and disseminated intravascular coagulation and multi-organ failure. Cytokines such as TNFα, IL1β, IL6, IL12, IFNγ, platelet activating factor, chemokines and eicosanoids are all produced by host cells in response to LPS and are involved in the pathogenesis of endotoxic shock (Van Amersfoort *et al.*, 2003). The levels of these mediators are also elevated in the late stages of lethal *S. enterica* infections, indicating that the toxicity of *S. enterica* LPS may be contributing to the death of the host. The observation that *S. enterica waaN* mutants, that biosynthesize LPS molecules lacking a fatty acyl chain, can grow to unusually high numbers with most animals surviving the infection seems to support these views (Khan *et al.*, 1998). The *waaN* mutants induce lower levels of TNFα and IL1 and reduced production of iNOS as compared to wild type salmonellae. This may further suggest that death in systemic salmonellosis is directly dependent on the production of some cytokines known to be involved in LPS-induced shock.

However, the overall picture suggested by published literature indicates that the lethal effects induced by *S. enterica* LPS are mediated by mechanisms that are more complex, and perhaps in some aspects different, than those involved in the classic endotoxic shock induced by other Gram-negative bacteria such as *E. coli*. Firstly, TNFα and lymphotoxin-α (LT) play a central role in lethality due to *E. coli* LPS, whilst the lethal effects of *S. enterica* LPS also involve other cytokines such as IFNγ, IL12, IL1 and IL18 (Mastroeni *et al.*, 1998; Netea *et al.*, 2001). Secondly, TNFα, one of the principal mediators of endotoxic shock, is undetectable in the late stages of lethal *S. enterica* infections in mice and calves (Mastroeni *et al.*, 1991; Peel *et al.*, 1990). Thirdly, despite mice infected with *Mycobacterium tuberculosis* BCG becoming hypersusceptible to the lethal effects of LPS or TNFα, these sensitized mice die of *S. enterica* infections with the same terminal bacterial numbers in the tissues as unsensitized controls (Senterfitt and Shands, 1968). Fourthly, similar bacterial loads are present in the tissues of LPS non-responders and LPS responders in the late stages of lethal infections, just before death of the animal (P. Mastroeni, unpublished).

8.8 THE ACTIVATION OF THE ADAPTIVE INNATE IMMUNE RESPONSE AND THE SUPPRESSION OF BACTERIAL GROWTH IN SUBLETHAL INFECTIONS

8.8.1 General

In sublethal infections, a complex adaptive response suppresses the exponential growth of *S. enterica* in the tissues resulting in survival of the host. This adaptive response is not antigen specific, leads to the migration of bone marrow derived inflammatory cells in the tissues and requires the concerted action of several immunological mediators including cytokines.

8.8.2 An adaptive response that does not require the contribution of T-cells or B-cells

The adaptive response that aborts the exponential growth of *S. enterica* in the infected tissues does not require the presence of functional T-cells or B-cells. In fact, X-irradiated animals reconstituted with T-cell depleted bone marrow cells and further injected with depleting anti-CD4 and anti-CD8 monoclonal antibodies can still control the growth of *S. enterica* in the early phases of the infection. Similarly, *nu/nu* mice (athymic T-cell deficient), H2I-A$\beta^{-/-}$ mice (lacking $CD4^+$ $TCR\alpha\beta^+$ T-cells), $TCR\beta^{-/-}$ mice (lacking $TCR\alpha\beta^+$ T-cells), *rag* $1^{-/-}$ mice (lacking T-cells and B-cells) and *Igh-6*$^{-/-}$ mice (B-cell deficient) are still able to mount the adaptive immune response that suppresses bacterial growth (Hess *et al.*, 1996; Hormaeche *et al.*, 1990; Mastroeni *et al.*, 2000a).

8.8.3 Cell migration in the tissues and granuloma formation

As mentioned earlier, PMNs infiltrate the foci of infection in the tissues during the first few days of the infection (exponential growth phase) to form pathological lesions. The suppression of bacterial growth coincides with the development of splenomegaly and with the migration of mononuclear cells into the organs. These mononuclear cells gradually replace the PMNs in the pathological lesions and are organized in such a way that they confine the bacteria to localized foci of infection. Lesion formation is a dynamic process that requires an influx of cells from the bone marrow (Hormaeche *et al.*, 1990), the presence of adhesion molecules such as ICAM 1 (Clare *et al.*, 2003) and the balanced action of cytokines (TNFα, IFNγ, IL12, IL18, IL4 and IL15) (Mastroeni, 2002). Failure to form pathological lesions results in abnormal growth and dissemination of the bacteria in the infected tissues (Everest *et al.*, 1998; Mastroeni *et al.*, 1998; Mastroeni *et al.*, 1995).

8.8.4 Cytokines in the recruitment and activation of phagocytic cells at the site of infection

The recruitment of mononuclear cells at the site of infection and the formation of macrophage-rich pathological lesions is a crucial event in the onset of adaptive innate immunity and in the containment and suppression of growth of *S. enterica* in the tissues. TNFα is essential for the formation and persistence of the pathological lesions. Mice treated with anti-TNFα antibodies or gene-targeted mice lacking TNFα receptor 55 (TNFR55$^{-/-}$ mice) become extremely susceptible to salmonellosis and bacterial growth in their tissues proceeds unrestrained. In the absence of biologically active TNFα or functional TNFR55, poorly organized pathological lesions and reduced numbers of inflammatory mononuclear cells are seen in the infected tissues (Everest *et al.*, 1998; Mastroeni *et al.*, 1991; Mastroeni *et al.*, 1995) resulting in necrosis and widespread dissemination of salmonellae extracellularly and to non-phagocytic cells (Mastroeni *et al.*, 1995). TNFα is constantly required for the persistence of the pathological lesions once these have formed in the tissues (Mastroeni *et al.*, 1993c).

Cell recruitment in the tissues is not sufficient for the expression of adaptive innate immunity. The recruited cells need to be exposed to signals that enhance their antibacterial functions. IFNγ is one of the mediators able to activate mouse macrophages to kill intracellular *S. enterica*, probably by enhancing phagosome-lysosome fusion and increasing the production of nitric oxide derivatives (Kagaya *et al.*, 1989). IFNγ production in this phase of early adaptive innate immunity is triggered by the presence of IL12, IL18 and IL15 that are produced mainly by macrophages and dendritic cells exposed to *S. enterica*. In this phase of the infection, natural killer (NK) cells, macrophages and neutrophils are the main sources of IFNγ. In fact, IFNγ can be produced in response to IL12 and IL18 by *rag1-/-* mice and *scid* mice that lack T-cells (Hirose *et al.*, 1999; Mastroeni *et al.*, 1999; Mastroeni *et al.*, 1996; Mastroeni *et al.*, 1998; Nishimura *et al.*, 1996; Ramarathinam *et al.*, 1993).

Experiments in IFN$\gamma^{-/-}$ mice, IFNγ receptor$^{-/-}$ (IFNγR$^{-/-}$) mice or in mice injected with neutralizing anti-IFNγ antibodies have shown the absolute requirement for this cytokine in the expression of adaptive innate immunity to *S. enterica* (Hess *et al.*, 1996; Muotiala and Makela, 1990; Nauciel and Espinasse-Maes, 1992). The unavailability of biologically active IFNγ results in dramatic effects on the infection due to lack of macrophage activation resulting in overwhelming bacterial growth leading to death of the host (Mastroeni *et al.*, 1998). Under these experimental conditions the formation of pathological lesions is disrupted and there is widespread infiltration of macrophages into the tissues that nevertheless are unable to restrain bacterial growth (Mastroeni *et al.*, 1998).

8.8.5 The balance between activation and suppression signals in adaptive innate immunity

A number of cytokines capable of down regulating macrophage antibacterial functions are produced during the adaptive innate immunity phase of the host response to *S. enterica* and are likely to counterbalance the activation pathways.

IL4 is produced by NK1.1$^+$ TCR$\alpha\beta^+$ cells following CD1-mediated cell interactions with phagocytic or dendritic cells (Enomoto *et al.*, 1997; Naiki *et al.*, 1999; Pie *et al.*, 1996; Pie *et al.*, 1997). IL4 can suppress macrophage functions in salmonellosis. For example, expression of the murine IL4 gene in recombinant *S. enterica* exacerbates the course of the infection and impairs macrophage killing of the bacteria (Denich *et al.*, 1993). Conversely, the absence of IL4 increases resistance to *S. enterica* infection as indicated by the delayed time to death seen in IL4$^{-/-}$ mice compared to IL4$^{+/+}$ mice (Everest *et al.*, 1997). Abscesses containing large numbers of bacteria can be seen in mice infected with *Salmonella*, and IL4$^{-/-}$ mice show reduced formation of these abscesses (Everest *et al.*, 1997). Furthermore, macrophage functions and IL12 production are enhanced following IL4 neutralization by specific antibodies and are suppressed by exposure to rIL4 (John *et al.*, 2002;

Naiki *et al.*, 1999). IL10 is another cytokine that downregulates the functions of phagocytes in *S. enterica* infections and in vivo neutralization of IL10 enhances host resistance to *S. enterica* (Arai *et al.*, 1995; Pie *et al.*, 1996).

It appears that during *S. enterica* infection there is a fine balance between the phagocyte activation systems involving IL12, IL18, IL15 and IFNγ, and the inhibitory signals delivered by IL10 and IL4. Whenever IL12 is deficient, the levels of IL10 and IL4 rise resulting in inhibition of the antibacterial functions of phagocytes (Mastroeni *et al.*, 1998). Conversely, absence of IL4 and IL10 results in increased production of IL12, TNFα, IFNγ and enhancement of macrophage antibacterial functions (Arai *et al.*, 1995; Naiki *et al.*, 1999).

8.8.6 Adaptive innate immunity and antibacterial functions of phagocytes

The antibacterial functions of phagocytic cells in the early phases of the infection are mainly based on the production of ROS generated *via* the phagocyte NADPH-oxidase (see section 8.6.4). Conversely, during the phase of adaptive innate immunity, the bactericidal activity of ROS is replaced by the bacteriostatic activity of reactive nitrogen substances (RNS) that are produced following the activation of the inducible nitric oxide synthase (iNOS) encoded by the *NOS2* gene. *NOS2* expression can be seen in macrophages within the pathological lesions and is upregulated by IFNγ, IL12 and IL18 (Khan *et al.*, 2001; Mastroeni *et al.*, 1998). RNS are needed for the expression of adaptive innate immunity to *S. enterica*. In fact, gene targeted $NOS2^{-/-}$ mice cannot control the growth of *S. enterica* despite efficient granuloma formation, macrophage activation and cytokine production (Mastroeni *et al.*, 2000). Similarly, cultured macrophages from $NOS2^{-/-}$ mice have reduced bacteriostatic activity (Vazquez-Torres *et al.*, 2000).

8.8.7 Responsiveness to LPS and adaptive innate immunity

Recognition of *S. enterica* LPS by host cells triggers activation signals that lead to the production of cytokines (*e.g.* TNFα, IL1β, IL6, IL12 and IFNγ), chemokines, eicosanoids and RNS (Van Amersfoort *et al.*, 2003). The ability to respond to LPS appears to be important for expression of adaptive resistance to *S. enterica*. C3H/HeJ mice (TLR4 deficient) cannot efficiently suppress the growth of virulent *S. enterica* in the spleen and liver due to defects in macrophage effector functions (O'Brien *et al.*, 1982).

8.8.8 Activation of adaptive innate immunity mechanisms induces transient suppression of T-cell and B-cell dependent antigen specific immunity

Animals that are infected with *S. enterica*, transiently show reduced ability to mount effective antibody responses to several antigens (*e.g.* tetanus toxoid, sheep red blood cells) and have impaired B- and T-cell responses to mitogens. This transient immunosuppression requires IFNγ and IL12, is mediated by macrophages through the production of nitric oxide, and can be reversed by administration of rIL4 (al-Ramadi *et al.*, 1992; MacFarlane *et al.*, 1999; Schwacha and Eisenstein, 1997).

8.9 THE CLEARANCE OF A PRIMARY INFECTION REQUIRES THE PRESENCE OF T-CELLS

8.9.1 General

Suppression of the growth of *S. enterica* is followed by the elimination of the bacteria from the tissues. If the bacteria are not cleared, a late resurgence of bacterial growth can occur (relapse) or a chronic carrier state can develop. Chronic carriage of *S. enterica* is undesirable and is a serious problem in farm animals and in typhoid patients where it constitutes a reservoir of infection.

8.9.2 Clearance of the infection requires $CD4^+$ $TCR\alpha\beta^+$ T-cells

Clearance of *S. enterica* from infection foci and/or the prevention of the resurgence of bacterial growth requires CD28-dependent activation of T-cells. In fact, athymic *nu/nu* mice and $CD28^{-/-}$ gene-targeted mice do not clear *S. enterica* infections and are killed by slow progressive growth of the bacteria in their organs (Mittrucker *et al.*, 1999; O'Brien and Metcalf, 1982; Sinha *et al.*, 1997). $CD4^+$ $TCR\alpha\beta^+$ T-cells appear to be the principal mediators of late resistance and bacterial clearance. In fact H2I-$A\beta^{-/-}$ mice (lacking mature $CD4^+$ $TCR\alpha\beta^+$ T-cells) and $TCR\beta^{-/-}$ mice (lacking $TCR\alpha\beta^+$ T-cells) can suppress the early growth of *S. enterica* in the tissues but die in the later stages of the infection (Hess *et al.*, 1996). Similarly, in vivo $CD4^+$ T-cell depletion by administration of T-cell depleting antibodies impairs bacterial clearance (Nauciel, 1990). The mechanisms used by $CD4^+$ T-cells to control bacterial growth in the late stages of the infection are still unclear. Although *S. enterica*-specific T-cells produce IFNγ and TNFα, neutralization of these cytokines with antibodies has little effect on the clearance of the bacteria (Mastroeni *et al.*, 1993; Muotiala and Makela, 1993). Expression of the IL2 gene by *S. enterica*, results in enhanced bacterial clearance from the host tissues, indicating a possible role for this cytokine in the elimination of the bacteria from the organs (al-Ramadi *et al.*, 2001).

8.9.3 TCR$\gamma\delta^+$ T-cells in *S. enterica* infection

The role of TCR$\gamma\delta^+$ T-cells in salmonellosis is controversial. TCR$\gamma\delta^+$ T-cells and NK1.1$^+$ TCR$\gamma\delta^+$ cells increase in numbers and produce IFNγ in response to IL15 in the peritoneal cavity of mice infected with *S. enterica* (Emoto *et al.*, 1992; Nishimura *et al.*, 1996; Nishimura *et al.*, 1999; Skeen and Ziegler, 1993). A role for TCR$\gamma\delta^+$ T-cells in host resistance has been reported in a limited number of host-pathogen combinations such as in *Slc11a1*^{susceptible} mice infected orally with virulent salmonellae (Mixter *et al.*, 1994).

8.9.4 Host resistance in the late stages of *S. enterica* infections and the rate of clearance of the bacteria from the infected tissues is under genetic control

Genes within and outside the MHC complex (H-2) control host resistance in the late, T-cell dependent stages of *S. enterica* infections. For example, congenic B10 mouse strains with different H-2 haplotypes vary in their ability to control the numbers of *S. enterica* in their tissues from the fourth week of infection onwards (Hormaeche *et al.*, 1985). These H-2 congenic mice also differ in their resistance to the relapse of infection that may occur after interruption of antibiotic treatment (Maskell and Hormaeche, 1986). Resistance to *S. enterica* in the late stages of the infection has been mainly mapped to the I-Eα subregion of the mouse H-2 system, with also the contribution of genes outside H-2 (Hormaeche *et al.*, 1985; Nauciel *et al.*, 1988, 1990).

The ability of mice to clear attenuated strains of *S. enterica* from the infection foci is also controlled by MHC genes and involves the D and Aα regions in the mouse H-2 system (Nauciel *et al.*, 1990). *S. enterica* can downregulate the expression of MHC class II molecules on immune cells in vitro and this effect is mediated by the bacterial protein sifA (Mitchell *et al.*, 2004). It is not known whether this effect is relevant to the in vivo dowregulation of immune responses to the bacterium.

8.10 THE INITIATION AND DEVELOPMENT OF ANTIGEN-SPECIFIC IMMUNITY

8.10.1 General

Antigen-specific immune mechanisms are activated from the early stages of a primary *S. enterica* infection eventually leading to the clearance of the bacteria from the tissues and to the establishment of immunological memory. Natural infection or immunization with live attenuated vaccines leads to the development of both humoral and cell mediated immunity against the bacterium.

8.10.2 Antibody responses in *S. enterica* infections

S. enterica infections induce early IgM antibody responses followed by IgG and IgA production (Mastroeni, 2002). The antibody is directed against a multitude of antigens including LPS determinants, the Vi surface polysaccharide, porins, outer membrane proteins, lipoproteins, heat shock proteins, flagella and fimbriae (Mastroeni, 2002).

8.10.3 T-cell responses in S. enterica infections

S. enterica infections induce the proliferation of $CD4^+$ T-cells that have an activated phenotype ($CD44^{high}$, $CD62L^{low}$) and are specific for the bacterium. These T-cells produce IFNγ, TNFα and IL2 and therefore display a T-helper 1 (Th1) phenotype (Mittrucker *et al.*, 2002; Srinivasan *et al.*, 2004; Ugrinovic *et al.*, 2003). Delayed type hypersensitivity (DTH) T-cell responses to *S. enterica* antigens can also be detected in vivo after exposure to live bacteria (Doucet and Bernard, 1997; Gohin *et al.*, 1997; Harrison *et al.*, 1997; McSorley *et al.*, 2000; Murphy *et al.*, 1987; Murphy *et al.*, 1989; Robertsson *et al.*, 1982; Sztein *et al.*, 1994; Thatte *et al.*, 1993; Villarreal *et al.*, 1992; Villarreal-Ramos *et al.*, 1998). Exposure to live *S. enterica* is needed for the induction of Th1 responses. In fact, administration of killed *S. enterica* or bacterial components (*e.g.* porins) gives rise to IL4-dominated T-helper 2 (Th2) type responses with low levels of DTH and high levels of specific antibodies of the IgG1 isotype (Harrison *et al.*, 1997; Thatte *et al.*, 1993). *S. enterica*-specific memory Th1 type T-cells and DTH reactivity persist long after the clearance of the bacteria from the infection foci (Mastroeni *et al.*, 1993a; Villarreal *et al.*, 1992).

The antigen specificity of the T-cell response appears to be broad. *S. enterica*-specific T-cell responses are directed towards a multitude of protein antigens amongst which porins, flagellar epitopes and pilin have been identified (Mastroeni, 2002).

$CD8^+$ T-cells with cytolytic activity (CTLs) against *S. enterica*-infected target cells are present in animals and humans exposed to live salmonellae (Lo *et al.*, 1999; Lo *et al.*, 2000; Pope *et al.*, 1994; Salerno-Goncalves *et al.*, 2002; Salerno-Goncalves *et al.*, 2003; Sztein *et al.*, 1995). In mice, the *S. enterica*-specific CTL responses are class Ib restricted with the involvement of the Qa-1 molecule (Lo *et al.*, 1999). CTL responses to heterologous antigens/epitopes expressed in live recombinant *S. enterica* vaccines have been documented (Mastroeni *et al.*, 2001), and an epitope derived from the *S. enterica* serovar Typhimurium GroEL molecule is known to be recognised by *S. enterica*-specific CTLs (Lo *et al.*, 2000).

8.10.4 Interactions between T-cells and B-cells in the activation and development of immune responses to *S. enterica*

The development of humoral and cell mediated immunity to *S. enterica* relies on the cross-talk between T-cells and B-cells.

In *S. enterica* infections, T-cells modulate antibody production, contribute to the establishment of long lasting humoral responses to the bacterium and determine the isotype profile of the specific antibodies. The presence of T-cells is also needed for the development of antibody responses to *S. enterica* protein antigens. In fact, athymic *nu/nu* (T-cell deficient) and $CD28^{-/-}$ mice (with impaired T-cell activation and reduced T-B-cell co-operation) produce lower levels of IgM and IgG3, and little or no IgG1, IgG2a and IgG2b antibodies against *S. enterica* LPS or proteins (Mittrucker *et al.*, 1999; Sinha *et al.*, 1997).

Until recently, the role of B-cells in salmonellosis had been thought to be largely confined to the production of antibodies (Casadevall, 1998; Collins, 1974; Hochadel, Keller, 1977). Recent work has unveiled a new role for B-cells in the modulation of T-cell immunity to *S. enterica*. B-cells can be infected by *S. enterica* in vitro and in vivo (Sztein *et al.*, 1995; Yrlid *et al.*, 2001), can upregulate the expression of CD86 co-stimulatory molecules and can present bacterial antigens to *S. enterica*-specific $CD4^+$ T-cells. Furthermore, B-cells are essential for the generation of Th1 type T-cell responses to *S. enterica* in vivo. Studies in *Igh-6*$^{-/-}$ B-cell deficient gene-targeted mutant mice have shown that these animals fail to develop protective Th1 type T-cell immunity to the bacterium due to reduced numbers of $CD4^+$ T-cells able to produce IFNγ and to an increase in IL4-producing T-cells (Mastroeni *et al.*, 2000; McSorley and Jenkins, 2000; Mittrucker *et al.*, 2000).

8.10.5 DC and the activation of T-cell immunity

S. enterica can infect DC in vitro and in vivo and can induce activation of these cells and subsequent cytokine production (Hopkins and Kraehenbuhl, 1997; Svensson *et al.*, 2000). DC exposed to *S. enterica* in vitro can prime bacterium-specific $CD4^+$ and $CD8^+$ T-cells following adoptive transfer into mice, suggesting a role for these cells in the initiation of an immune response to *S. enterica* (Yrlid *et al.*, 2001). *S. enterica* can also induce apoptosis of DC via activation of Caspase-1 and this may represent a possible mechanism of immune evasion (van der Velden *et al.*, 2003). The biology of the interaction between *S. enterica* and DC is described in detail in Chapter 10.

8.11 MECHANISMS OF HOST RESISTANCE IN SECONDARY INFECTIONS

8.11.1 General

In sections 8.10.2 and 8.10.3 it has been outlined that primary infections with *S. enterica* generate $CD4^+$ and $CD8^+$ T-cell responses as well as antibody responses directed towards a broad range of *S. enterica* antigens. The relative importance of each of these immune responses in protection against reinfection has been a matter of long-standing debate mainly due to the use of very different experimental models, routes of infection, administration doses and *S. enterica* strains. In general, although anti-*S. enterica* antibody alone or T-cells alone can confer a low level of protection, a combination of cell-mediated and humoral immunity is needed for the expression of high level resistance against virulent *S. enterica*.

8.11.2 Antibody and host resistance in secondary infections

Antibody alone can confer protection in secondary systemic salmonellosis. For example, adoptive transfer of *S. enterica*-immune B-cells or immune serum can further increase host resistance in animals that already possess a high level of genetically determined innate resistance to *S. enterica* infections (e.g. $Slc11a1^{resistant}$ mice). Antibodies specific for the LPS O-antigen, porins and flagellar epitopes have been implicated in serum-mediated protection, with IgM being more protective than IgG (Mastroeni, 2002). Humoral immunity alone can also protect animals that are genetically susceptible to *S. enterica* if these are infected with *S. enterica* strains of low virulence or with very low doses of virulent micro-organisms (Collins, 1974; Eisenstein *et al.*, 1984; Xu *et al.*, 1993). The modest, but albeit significant, protective ability of antibody in salmonellosis is probably the basis of the efficacy of *S. enterica* killed vaccines and some subunit vaccines, such as the currently licensed human Vi polysaccharide typhoid vaccine, that induce humoral responses but are unable to induce good Th1 type T-cell immunity (Klugman *et al.*, 1987).

8.11.3 T-cells and host resistance in secondary infections

S. enterica are facultative intracellular bacteria that localize and grow within phagocytes. It is therefore unsurprising that acquired resistance against this bacterium relies heavily on cell-mediated immunity. For example, mice immunized with a live *S. enterica* vaccine and later subjected to in vivo depletion of $CD4^+$ and/or $CD8^+$ T-cells show impaired recall of immunity to reinfection with virulent *S. enterica*, indicating that $CD4^+$ and $CD8^+$ T-cells are essential for protection. One of the hallmarks of cell-mediated immunity is its transferability from immune donors to naïve recipients using antigen-specific immune T-cells. Indeed, acquired resistance to other intracellular pathogens, such as *Listeria monocytogenes* and *Mycobacterium tuberculosis* can be readily transferred into recipient animals using immune T-cells. Most surprisingly, adoptive transfer of immunity to *S. enterica* using T-cells from immunized animals or *S. enterica*-specific T-cell lines has been very difficult and has been achieved only in some experimental conditions, where the recipient animals were challenged with very low doses of virulent bacteria or with bacteria of low virulence (Guilloteau *et al.*, 1993; Paul *et al.*, 1985; Paul *et al.*, 1988). This indicates that the immunological requirements for host resistance to *S. enterica* in immunized animals are different from what normally is seen for other intracellular pathogens.

8.11.4 Both antibody and T-cells are needed for host resistance in severe infections

T-cells or antibody alone can confer only a low level of immunity against *S. enterica*. The expression of a high level of host resistance requires the concerted action of both anti-*S. enterica* antibody and T-cells. In fact, in adoptive transfer experiments, it has been shown that simultaneous transfer of both immune serum and immune T-cells is able to confer full protection against mortality after oral challenge with virulent *S. enterica* in mice (Mastroeni *et al.*, 1993b).

These finding have important applications for the rational development of new vaccines against *S. enterica*. In fact, it is reasonable to predict that those vaccines able to elicit anti-*S. enterica* antibodies in addition to *S. enterica*-specific $CD4^+$ and $CD8^+$ T-cells will be able to confer the highest level of protection against the disease.

8.11.5 Cytokines in acquired immunity to *S. enterica*

Similar to what is seen in primary infections (section 8.8.4), cytokine networks are essential for the expression of acquired resistance.

In secondary infections, both TNFα and IFNγ are needed for host resistance to infection and for the formation of discrete granulomas. IFNγ and IL12 are required for macrophage activation (Mastroeni *et al.*, 1998; Mastroeni *et al.*, 1992).

8.11.6 Protective antigens in *S. enterica* infections

Infection with *S. enterica* triggers T-cell and B-cell responses to a broad range of antigens (see sections 8.10.2 and 8.10.3). In animal models, low levels of protection against *S. enterica* infection can be induced by administration of flagella, porins, or polysaccharide fractions (*e.g.* LPS, O-antigen) (Matsui and Arai, 1989; McSorley *et al.*, 2000; Svenson *et al.*, 1979). In humans, vaccination with the Vi polysaccharide induces moderate protection against typhoid fever (Klugman *et al.*, 1987). These observations suggest that both protein and polysaccharide antigens play a role in protection against salmonellosis. In vivo studies in the mouse typhoid model and in calves have investigated the role of LPS O-antigens in protection induced by live attenuated *S. enterica* vaccines. These studies involved vaccination of animals with *S. enterica* serotypes expressing different LPS O-antigen specificities followed by challenge with strains expressing homologous or heterologous LPS-O antigens. The work showed that both immuno-dominant LPS O-antigens and non-LPS-O determinants are required to achieve high levels of protection against fully virulent *S. enterica* (Hormaeche *et al.*, 1996; Segall and Lindberg, 1993).

8.12 IMMUNITY TO *S. ENTERICA* INFECTION IN HUMANS

8.12.1 General

Current knowledge about the mechanisms of immunity to *S. enterica* infections in humans derives mainly from: a) analysis of immune responses generated in vaccinated individuals during vaccine trials; b) genetic studies on the association between *S. enterica* infections and polymorphic alleles of genes or gene promoter regions involved in immune functions; c) identification and analysis of immunological defects that predispose humans to *S. enterica* infections.

A comparative analysis in mice and man illustrates many similarities in the immunological determinants of resistance/susceptibility to *S. enterica* infections in the two species. Work in mice has guided investigations in to immunity to human *S. enterica* infections providing a lead on which

Table 8.1. *Comparison of the known mechanisms of immunity needed for resistance to salmonellosis in the mouse typhoid model and in humans*

Mechanism	Mouse	Humans
Low stomach pH	No function	protects
Defensins	Protect transgenic mice	Not known
Slc11a1 gene	Confers resistance	Apparently no function
TNFα/TNFR55	Formation of macrophage-rich pathological lesions, localization of NADPH oxidase (ROS) to phagosome	Genetic association with resistance/susceptibility
RNS	Bacteriostatic in macrophages	Not known
ROS	Bactericidal in macrophages	CGD patients are more susceptible to salmonellosis
IL12, IL12R, IFNγ, IFNR, STAT-1	Impaired activation of phagocytes	Genetic defects in these factors predispose humans to salmonellosis
Plasmodium infections	Decreased phagocyte functions	Decreased phagocyte functions
$CD4^+$ T-cells	Needed for bacterial clearance and resistance in secondary infection	Increased susceptibility to salmonellosis in patients with HIV infections, SCID, MHC class II defiency and Good's syndrome
$CD8^+$ T-cells	Needed for resistance in secondary infections	*S. enterica* specific $CD8^+$ T-cells are present in humans infected with serovar *typhi*. Function unknown
B-cells and antibody	Needed for resistance in secondary infections	Increased susceptibility to salmonellosis in patients with XLA, CVID and Good's syndrome
Interaction between T-cells and B-cells	Isotype switching in B-cells. Initiation of T-cell immunity	Increased susceptibility to salmonellosis in patients with CD154 deficiency

immunological paramenters to study and indicating criteria for the testing of new vaccines and antimicrobial treatments.

8.12.2 Early defence mechanisms in the gut

The normal function of the physiological factors operating in the gastrointestinal system protects against *S. enterica* infection probably by reducing the numbers of viable bacteria that reach the sites of the enteric mucosa suitable for invasion. Reduced gastric acid secretion by treatment with proton-pump inhibitors or rapid gastric emptying time results in increased susceptibility to *S.enterica* (Gianella *et al.*, 1971). Gastric acidity may directly kill the bacteria or it may operate by activating the proteolytic activity of pepsin which is required for the cleavage of Histone 2A (derived from gastric epithelial cells) into the antibacterial peptide Buforin 1 (Kim *et al.*, 2000). Human defensin 5 (HD5), an antimicrobial peptide produced by Paneth cells in the crypts of Lieberkuhn may also play an important role in human salmonellosis (Salzman *et al.*, 2003; Zasloff, 2002).

8.12.3 Phagocytes and innate immunity

Defects in phagocyte numbers or functions are predisposing factors for *S. enterica* infections in humans.

Neutropenia has been described as a predisposing factor for bacteraemia caused by *S. enterica* (Weinberger and Pizzo, 1992).

Chronic granulomatous disease (CGD) comprises a group of disorders due to mutations in genes encoding for membrane-bound (gp91*phox* and p21*phox*) or cytosolic (p67*phox* and p47*phox*) subunits of the NADPH oxidase. These defects render neutrophils and macrophages unable to produce ROS and impair their ability to kill microorganisms. CGD patients have an increased risk of developing septicaemia due to *S. enterica* and other Gram-negative enteric bacteria (Lazarus and Neu, 1975). In a study involving 48 individuals with CGD, over 50% of the patients had peritoneal and enteric infections with *S. enterica* (Mouy *et al.*, 1989).

Plasmodium falciparum infection causes macrophage dysfunction by the production of haemozoin from the breakdown of haemoglobin. Haemozoin is taken up by phagocytic cells leading to defective phagocytosis and decreased NADPH oxidase-dependent production of ROS. An association between *Plasmodium falciparum* malaria and non-typhoidal *S. enterica* (NTS) bacteraemia was first recognized in the 1920s in British Guiana and has more recently been confirmed in the Gambia (Graham *et al.*, 2000).

Sickle cell disease (SCD) is another condition that leads to defects in macrophage function and is associated with *S. enterica* septicaemia, meningitis and osteomyelitis (Anand, Glatt, 1994; Landesman *et al.*, 1982).

SLC11A1 is the human homologue of *Slc11a1* that in mice controls early innate immunity to *S. enterica* infection. *SLC11A1* is expressed by phagocytes in the spleen, liver and lungs, as well as by circulating monocytes and PMNs (Forbes and Gros, 2001). Although *SLC11A1* has been associated with, or linked to, resistance or susceptibility to infectious diseases such as tuberculosis, leprosy, meningococcal meningitis and leishmaniasis (Blackwell *et al.*, 2001), a role for *SLC11A1* in resistance/susceptibility to human typhoid fever has not been established. In Southern Vietnam, patients with typhoid fever and healthy controls were tested for single base pair polymorphisms within and near *SLC11A1*, for a (GT)n repeat in the promoter region of the gene and for a microsatellite marker downstream of *SLC11A1*. No association was found between these polymorphic alleles and susceptibility to typhoid fever (Dunstan *et al.*, 2001a).

8.12.4 Cytokines and resistance to *S. enterica* in humans

Human peripheral blood mononuclear cells produce several cytokines including IFNγ, TNFα, IL1β, IL6 and IL10 in response to *S. enterica* (House *et al.*, 2002; Wyant *et al.*, 1999)

Some cytokines, known to be essential for resistance to infection in animals models, appear to play an important role in resistance to *S. enterica* in humans.

A genetic study performed in Southern Vietnam suggests a role for TNFα in resistance to human typhoid fever. In this study, the polymorphic allele variant of the human TNFα gene, TNFA*2 (-308) was associated with susceptibility to typhoid fever, whereas TNFA*1 (-308) was associated with resistance to the disease (Dunstan *et al.*, 2001b).

IL12 and IFNγ are essential in resistance to *S. enterica* infections in humans. Genetic mutations that result in partial or total loss of function of IL12, IFNγ or their respective receptors and signalling molecules lead to recurrent *S. enterica* infections that are difficult to treat with antibiotic therapy. Complete IFNγR and complete STAT-1 deficiencies result in severe infections with a poor outcome due to total loss of IFNγ-mediated responses. In contrast, IL12R, IL12p40, partial IFNγR and partial STAT-1 deficiencies all have a milder clinical phenotype probably due to residual IFNγ function (Doffinger *et al.*, 2002). Identification of IFNγ deficiencies as a result of defects in IL12 production or signalling has led to successful targeted treatments using combinations of IFNγ therapy and anti-microbials (Altare *et al.*, 1998; de Jong *et al.*, 1998; Picard *et al.*, 2001).

8.12.5 T-cell responses to *S. enterica* in humans

Lymphoproliferative responses can be detected in patients with a history of clinically proven typhoid fever or in individuals immunized with live attenuated *S. enterica* serovar Typhi vaccines. Oral administration of the new generation of live attenuated serovar Typhi vaccine strains, with deletions in the *htrA, aroC* and *aroD* genes has been shown to generate Th1 type T-cell immunity, as indicated by IFNγ production from leukocytes in response to flagellin antigens (Sztein *et al.*, 1994). The *htrA, aroC, aroD* vaccines, as well as the currently licensed *S. enterica* serovar Typhi Ty21a vaccine, also induce $CD8^+$ T-cells capable of both producing IFNγ and lysing targets cells infected with *S. enterica* (Salerno-Goncalves *et al.*, 2002; Salerno-Goncalves *et al.*, 2003; Sztein *et al.*, 1994).

8.12.6 Antibody responses to *S. enterica* in humans

Antibody responses to LPS O-antigen, proteins and the Vi surface polysaccharide antigen can be detected in patients with typhoid fever or in individuals immunized with live *S. enterica* vaccines, as well as whole cell killed or subunit (Vi) *S. enterica* vaccines. Production of IgM is usually followed by IgG and IgA (Forrest *et al.*, 1991; Levine *et al.*, 1989; Murphy *et al.*, 1987). In humans, *S. enterica*-specific antibody-secreting cells, in the absence of circulating antibodies, can be detected in response to vaccination with live attenuated *S. enterica* (Kantele *et al.*, 1986). A rapid rise in serum anti-O and anti-H (flagella) antibodies is considered as an indication of active immunity against *S. enterica* serovar Typhi. Antibody titres against the Vi antigen are low in patients with acute typhoid fever and high in chronic carriers (Robbins, Robbins, 1984).

A good correlation between the presence of antibodies alone and resistance to typhoid fever has been difficult to establish. For example, whole cell killed vaccines induce good antibody responses, but do not always confer satisfactory protection (Levine *et al.*, 1989). In some field trials (*e.g.* Vi vaccine trials) (Klugman *et al.*, 1987; Klugman *et al.*, 1996), protection was paralleled by an increase in antibody titres, whilst in other trials (e.g. *S. enterica* serovar Typhi Ty21a vaccine in Chile) seroconversion paralleled the percentage of protection with some vaccine preparations, but not with others (Levine *et al.*, 1989).

8.12.7 T-cells, B-cells and antibody in resistance to *S. enterica* in humans

T-cell and B-cell immunity have an important role in protection against *S. enterica* in humans, as illustrated by the increased incidence of *S. enterica* infections in individuals with deficiency in humoral and cell-mediated immunity.

8.12.7.1 Common Variable Immunodeficiency (CVID)

CVID is an immunodeficiency syndrome that is characterized by hypogammaglobulinaemia, low B-cell numbers, inadequate capacity to mount specific antibody responses and an increased tendency of affected individuals to develop autoimmune disease and malignancy. Recurrent infections, predominantly with extracellular bacteria are often associated with CVID. CVID patients are also at increased risk of infection with *S. enterica* (Hermaszewski and Webster, 1993; Leen *et al.*, 1986).

8.12.7.2 X-linked agammaglobulinaemia (XLA)

XLA is a syndrome characterized by an arrest in B-cell development caused by mutations of the tyrosine kinase Btk, resulting in B-cell deficiency and hypogammaglobulinaemia. Although these patients are mainly susceptible to encapsulated bacterial sepsis, persistent diarrhoea due to *S. enterica* was reported in 3 out of 44 patients with XLA (Hermaszewski and Webster, 1993).

8.12.7.3 Major histocompatibility complex Class II (MHC Class II) deficiency

MHC class II genes are associated with resistance and susceptibility to typhoid fever in humans. A study in Southern Vietnam showed that the polymorphic allele variants HLA-DRB1*0301/6/8 and HLA-DQB 1*0201–3 are associated with susceptibility to typhoid fever, whereas HLADRB1*04 and HLADQB1*0401/2 are associated with resistance to the disease (Dunstan *et al.*, 2001).

MHC class II deficiency is a rare disease that arises from defects in transacting factors (CIITA, RFXANK, RFX5 and RFXAP) that are essential for normal MHC class II gene expression (Klein *et al.*, 1993). Lack of MHC-II expression results in reduced antimicrobial antibody responses and severe $CD4^+$ T-cell lymphopenia. The majority of patients die before puberty from respiratory and/or gastrointestinal infections, unless they receive bone marrow transplantation. *S. enterica* is one of the most common pathogens isolated from these patients.

The genetic studies described above indicate that MHC Class II dependent antigen presentation and normal $CD4^+$ T-cell function are essential for anti-*S. enterica* immunity.

8.12.7.4 Good's syndrome (GS)

GS refers to a range of adult-onset immunodeficiencies characterized by B-cell lymphopenia, hypogammaglobulinaemia, $CD4^+$ T-cell lymphopenia and an inverted $CD4^+/CD8^+$ T-cell ratio (Kelleher and Misbah, 2003; Tarr *et al.*, 2001). GS patients suffer mainly from respiratory tract infections caused by extracellular bacteria but also have an increased risk of infections usually associated with reduced CMI. Diarrhoea due to non-typhoidal *S. enterica* (NTS) has been described in GS patients (Kelleher and Misbah, 2003; Tarr *et al.*, 2001).

8.12.7.5 X-Linked Hyper IgM syndrome

X-Linked Hyper IgM syndrome is a rare genetic disorder of T-cells characterized by decreased serum levels of IgG, IgA and IgE but normal or elevated levels of IgM. The disease is due to mutations in the CD154 (CD40 ligand) gene expressed on activated T-cells. This results in a lack of appropriate interactions between T-cells and B-cells leading to failure of B-cell isotype switching and memory generation. Patients suffer with infections usually associated with antibody deficiency, as well as with opportunistic infections that are associated with impaired CMI. In a series of 56 patients with X-linked hyper IgM syndrome, 2 patients were infected with *S. enterica* (Levy *et al.*, 1997). The increased susceptibility to infection in patients with X-Linked Hyper IgM syndrome confirms the importance of the cross-talk between T-cells and B-cells in immunity to *S. enterica*, as already indicated by work in the mouse model (see section 8.10.4).

8.12.7.6 HIV infection

Due to profound suppressive effects on T-cell and macrophage-mediated immunity, HIV infection results in markedly increased susceptibility to NTS infection (Celum *et al.*, 1987; Hart *et al.*, 2000). *S. enterica* gastroenteritis in HIV-infected individuals is protracted and relapsing, often with associated septicaemia and focal extraintestinal infections including osteomyelitis, visceral abcesses, meningitis, endocarditis, pyomyositis and pneumonia. The tendency for NTS infections to relapse and disseminate systemically in these patients indicates an important role for cell mediated immunity in bacterial clearance during *S. enterica* gastroenteritis and for preventing the dissemination of the bacteria to extraintestinal sites.

8.12.7.7 Ectodermal dysplasia with immunodeficiency

X-linked ectodermal dysplasia with associated immunodeficiency (EDA-ID) is caused by mutations in the gene encoding NFκB essential modulator (NEMO), a protein essential for activation of the transcription factor NF-κB and the subsequent transcription of genes involved in a variety of developmental and immunological pathways (Doffinger *et al.*, 2001). EDA-ID patients have poor antibody responses to bacterial polysaccharide antigens, with low levels of IgG or IgG2 due to defective immunoglobulin class switching, and are suceptible to bacterial sepsis. One patient with *S. enterica* serovar Enteritidis infection had impaired cellular responses to IL18 and IL1β presumably leading to impaired IFNγ production (Doffinger *et al.*, 2001).

8.12.8 Immunodeficiencies and clinical syndromes

It is interesting to observe that different immunodeficiencies predispose humans to different clinical syndromes. For example, physiological factors operating in the gastro-intestinal tract and antibodies appear to prevent intestinal colonization and enteric infection. Killing mechanisms operating within neutrophils and macrophages are important in overcoming the septicaemic phase of *S. enterica* infection and prevent metastatic disease. Optimal operation of IL12- and IFNγ-dependent immunity and T-cell function appear to be essential for the eradication of *S. enterica* from the tissues preventing spread and recrudescence of the infection.

8.13 CONCLUSIONS

Several immunological aspects of *S. enterica* infections are unique and puzzling. *S. enterica* causes a very broad range of diseases in humans and animals and the relative importance of each mechanism of immunity to this bacterium can be different depending on individual host-pathogen combinations.

During infections, there is the usual progression from innate immunity to adaptive innate immunity leading finally to the development and expression of antigen-specific T-cell dependent acquired resistance. At least in mice, it is puzzling to observe that T-lymphocytes are dispensable for survival of the host in the first three weeks of the infection process making the cytokine-driven adaptive response (section 8.8) sufficient for the suppression of early bacterial growth in the tissues. T-cell immunity is initiated early in the disease but remains dispensable for the control of the infection until later when T-cells are involved in clearance of the pathogens from the tissues.

The interaction of *S. enterica* with phagocytes is rather interesting. The bacteria are likely to grow within phagocytes and to use immunoevasion mechanisms to counteract phagocyte killing. However, work done in mice indicates that the majority of infected cells in vivo contain low numbers of salmonellae. This can partially be explained by taking into account the growth dynamics at the single cell level where heavily infected cells would release the micro-organisms that would then distribute to other phagocytes. It is interesting to note that the number of infected phagocytes varies in relation to bacterial growth, with an increase in infected cells seen in those phases of the infections where the overall numbers of bacteria in the tissues are increasing. A similar increase is seen in the number of pathological lesions at the foci of infection. All these observations indicate that escape from infected cells and from pathological lesions to reach new sites in the tissues is likely to be an additional mechanism of immunoevasion and virulence for *S. enterica*.

The requirement for antibodies in protection against salmonellosis is an unusual feature given the fact that *S. enterica* is generally believed to be an intracellular organism. One likely explanation is that virulent bacteria are transiently present in the extracellular compartment. This view is supported by the observation that a low-grade bacteriaemia can be detected during salmonellosis in mice and humans (Collins 1969; Tacket *et al.*, 1992), and by the fact that some antibiotics unable to penetrate inside cells (e.g. gentamicin) can reduce the bacterial load in the spleen and liver of animals infected with *S. enterica* (Bonina *et al.*, 1998). Furthermore, virulent bacteria rapidly spread and distribute in the tissues continuously forming new foci of infection. The spread from infected to uninfected cells is likely to occur *via* the release of bacteria into the extracellular space followed by internalization by other phagocytes (Sheppard *et al.*, 2003). It remains unclear why antibodies contribute to host resistance only in secondary *S. enterica* infections, and are dispensable for the control and clearance of primary infections.

A comparison of the experimental work done in the mouse model with the observations in humans illustrates numerous similarities in the immunological determinants of resistance/susceptibility to *S. enterica* infections in the two species. This confirms the validity of mouse typhoid as a model to produce advanced fundamental knowledge to be used as a crucial guide for the understanding of the pathogenesis and immunity to these infections in other animal species, including humans, and for the testing of new vaccines and antimicrobial treatments. It is surprising that a large proportion of immunodeficiencies that exacerbate the course of systemic *S. enterica* infection in mice, are mainly associated with non-typhoidal *S. enterica* infections in humans. This indicates that different types of *S. enterica* disease in different animal species are controlled by common immunological mechanisms. Consequently, laboratory animal models of *S. enterica* infection that do not strictly reproduce all the clinical features of salmonellosis seen in human disease or in diseases of domestic animals, can still provide a great deal of useful general information for the understanding of the immunobiology of *S. enterica* infections.

8.14 ACKNOWLEDGEMENTS

Our work is supported by the Wellcome Trust, BBSRC, MRC, The Royal Society and Diabetes UK.

REFERENCES

Adamo, R., Sokol, S., Soong, G., Gomez, M. and Prince, A. (2003). *P. aeruginosa* flagella activate airway epithelial cells through asialoGM1 and TLR2 as well as TLR5. *Am J Respir Cell Mol Biol*, **30**, 627–34.

al-Ramadi, B. K., Al-Dhaheri, M. H., Mustafa, N. *et al.* (2001). Influence of vector-encoded cytokines on anti-*Salmonella* immunity: divergent effects of interleukin-2 and tumor necrosis factor alpha. *Infect Immun*, **69**, 3980–8.

al-Ramadi, B. K., Meissler, J. J., Jr., Huang, D. and Eisenstein, T. K. (1992). Immunosuppression induced by nitric oxide and its inhibition by interleukin-4. *Eur J Immunol*, **22**, 2249–54.

Altare, F., Lammas, D., Revy, P. *et al.* (1998). Inherited interleukin 12 deficiency in a child with bacille Calmette-Guerin and *Salmonella enteritidis* disseminated infection. *J Clin Invest*, **102**, 2035–40.

Anand, A. J. and Glatt, A. E. (1994). *Salmonella* osteomyelitis and arthritis in sickle cell disease. *Semin Arthritis Rheum*, **24**, 211–21.

Arai, T., Hiromatsu, K., Nishimura, H. *et al.* (1995). Effects of in vivo administration of anti-IL10 monoclonal antibody on the host defence mechanism against murine *Salmonella* infection. *Immunology*, **85**, 381–8.

Ashkenazi, S., Cleary, T. G., Murray, B. E., Wanger, A. and Pickering, L. K. (1988). Quantitative analysis and partial characterization of cytotoxin production by *Salmonella* strains. *Infect Immun*, **56**, 3089–94.

Bauer, S., Kirschning, C. J., Hacker, H. *et al.* (2001). Human TLR9 confers responsiveness to bacterial DNA via species-specific CpG motif recognition. *Proc Natl Acad Sci USA*, **98**, 9237–42.

Benjamin, W. H., Jr., Hall, P., Roberts, S. J. and Briles, D. E. (1990). The primary effect of the *Ity* locus is on the rate of growth of *Salmonella typhimurium* that are relatively protected from killing. *J Immunol*, **144**, 3143–51.

Beutler, B., Krochin, N., Milsark, I. W., Luedke, C. and Cerami, A. (1986). Control of cachectin (tumor necrosis factor) synthesis: mechanisms of endotoxin resistance. *Science*, **232**, 977–80.

Biozzi, G., Howard, J. G., Halpern, B. N., Stiffel, C. and Mouton, D. (1960). The kinetics of blood clearance of isotopically labelled *Salmonella enteritidis* by the reticuloendothelial system in mice. *Immunology*, **3**, 74–89.

Blackwell, J. M., Goswami, T., Evans, C. A. *et al.* (2001). SLC11A1 (formerly NRAMP1) and disease resistance. *Cell Microbiol*, **3**, 773–84.

Blander, J. M. and Medzhitov, R. (2004). Regulation of phagosome maturation by signals from Toll-like receptors. *Science*, **304**, 1014–18.

Bohnoff, M. and Miller, P. (1962). Enhanced susceptibility to *Salmonella* infection in streptomycin-treated mice. *J Infec Dis*, **111**, 117–27.

Bonina, L., Costa, G. B. and Mastroeni, P. (1998). Comparative effect of gentamicin and pefloxacin treatment on the late stages of mouse typhoid. *New Microbiol*, **21**, 9–14.

Casadevall, A. (1998). Antibody-mediated protection against intracellular pathogens. *Trends Microbiol*, **6**, 102–7.

Celum, C. L., Chaisson, R. E., Rutherford, G. W., Barnhart, J. L. and Echenberg, D. F. (1987). Incidence of salmonellosis in patients with AIDS. *J Infect Dis*, **156**, 998–1002.

Chabalgoity, J. A., Dougan, G., Mastroeni, P. and Aspinall, R. J. (2002). Live bacteria as the basis for immunotherapies against cancer. *Expert Rev Vaccines*, **1**, 495–505.

Cirillo, D. M., Valdivia, R. H., Monack, D. M. and Falkow, S. (1998). Macrophage-dependent induction of the *Salmonella* pathogenicity island 2 type III secretion system and its role in intracellular survival. *Mol Microbiol*, **30**, 175–88.

Clare, S., Goldin, R., Hale, C. *et al.* (2003). Intracellular adhesion molecule 1 plays a key role in acquired immunity to salmonellosis. *Infect Immun*, **71**, 5881–91.

Collins, F. M. (1969). Effect of specific immune mouse serum on the growth of *Salmonella enteritidis* in non-vaccinated mice challenged by various routes. *J Bacteriol*, **97**, 667–75.

(1974). Vaccines and cell-mediated immunity. *Bacteriol Rev*, **38**, 371–402.

Conlan, J. W. (1997). Critical roles of neutrophils in host defense against experimental systemic infections of mice by *Listeria monocytogenes*, *Salmonella typhimurium*, and *Yersinia enterocolitica*. *Infect Immun*, **65**, 630–5.

de Jong, R., Altare, F., Haagen, I. A. *et al.* (1998). Severe mycobacterial and *Salmonella* infections in interleukin-12 receptor-deficient patients. *Science*, **280**, 1435–8.

Denich, K., Borlin, P., O' Hanley, P. D., Howard, M. and Heath, A. W. (1993). Expression of the murine interleukin-4 gene in an attenuated *aroA* strain of *Salmonella typhimurium*: persistence and immune response in BALB/c mice and susceptibility to macrophage killing. *Infect Immun*, **61**, 4818–27.

Doffinger, R., Dupuis, S., Picard, C. *et al.* (2002). Inherited disorders of IL12- and IFNγ-mediated immunity: a molecular genetics update. *Mol Immunol*, **38**, 903–9.

Doffinger, R., Smahi, A., Bessia, C. *et al.* (2001). X-linked anhidrotic ectodermal dysplasia with immunodeficiency is caused by impaired NF-κB signaling. *Nat Genet*, **27**, 277–85.

Doucet, F. and Bernard, S. (1997). In vitro cellular responses from sheep draining lymph node cells after subcutaneous inoculation with *Salmonella abortusovis*. *Vet Res*, **28**, 165–78.

Dunlap, N. E., Benjamin, W. H., Jr., Berry, A. K., Eldridge, J. H. and Briles, D. E. (1992). A "safe-site" for *Salmonella typhimurium* is within splenic polymorphonuclear cells. *Microb Pathog*, **13**, 181–90.

Dunstan, S. J., Ho, V. A., Duc, C. M. *et al.* (2001a). Typhoid fever and genetic polymorphisms at the natural resistance- associated macrophage protein 1. *J Infect Dis*, **183**, 1156–60.

Dunstan, S. J., Stephens, H. A., Blackwell, J. M. *et al.* (2001b). Genes of the class II and class III major histocompatibility complex are associated with typhoid fever in Vietnam. *J Infect Dis*, **183**, 261–8.

Dunstan, S. J., Hawn, T. R., Hue, N. T. *et al.* (2005). Host susceptibility and clinical outcomes in Toll-like receptor 5-deficient patients with typhoid fever in Vietnam. *J Infect Dis*, **191**, 1068–71.

Eisenstein, T. K., Deakins, L. W., Killar, L., Saluk, P. H. and Sultzer, B. M. (1982). Dissociation of innate susceptibility to *Salmonella* infection and endotoxin responsiveness in C3HeB/FeJ mice and other strains in the C3H lineage. *Infect Immun*, **36**, 696–703.

Eisenstein, T. K., Killar, L. M. and Sultzer, B. M. (1984). Immunity to infection with *Salmonella typhimurium*: mouse-strain differences in vaccine- and serum-mediated protection. *J Infect Dis*, **150**, 425–35.

Emoto, M., Danbara, H. and Yoshikai, Y. (1992). Induction of gamma/delta T-cells in murine salmonellosis by an avirulent but not by a virulent strain of *Salmonella* choleraesuis. *J Exp Med*, **176**, 363–72.

Enomoto, A., Nishimura, H. and Yoshikai, Y. (1997). Predominant appearance of NK1.1$^+$ T-cells producing IL4 may be involved in the increased susceptibility of mice with the beige mutation during *Salmonella* infection. *J Immunol*, **158**, 2268–77.

Eriksson, S., Bjorkman, J., Borg, S. *et al.* (2000). *Salmonella typhimurium* mutants that downregulate phagocyte nitric oxide production. *Cell Microbiol*, **2**, 239–50.

Eriksson, S., Lucchini, S., Thompson, A., Rhen, M. and Hinton, J. C. (2003). Unravelling the biology of macrophage infection by gene expression profiling of intracellular *Salmonella enterica*. *Mol Microbiol*, **47**, 103–18.

Everest, P., Allen, J., Papakonstantinopoulou, A. *et al.* (1997). *Salmonella typhimurium* infections in mice deficient in interleukin-4 production: role of IL4 in infection-associated pathology. *J Immunol*, **159**, 1820–7.

Everest, P., Roberts, M. and Dougan, G. (1998). Susceptibility to *Salmonella typhimurium* infection and effectiveness of vaccination in mice deficient in the tumor necrosis factor alpha p55 receptor. *Infect Immun*, **66**, 3355–64.

Fields, P. I., Swanson, R. V., Haidaris, C. G. and Heffron, F. (1986). Mutants of *Salmonella typhimurium* that cannot survive within the macrophage are avirulent. *Proc Natl Acad Sci USA*, **83**, 5189–93.

Forbes, J. R. and Gros, P. (2001). Divalent-metal transport by NRAMP proteins at the interface of host–pathogen interactions. *Trends Microbiol*, **9**, 397–403.

Forrest, B. D., LaBrooy, J. T., Beyer, L., Dearlove, C. E. and Shearman, D. J. (1991). The human humoral immune response to *Salmonella typhi* Ty21a. *J Infect Dis*, **163**, 336–45.

Freudenberg, M. A. and Galanos, C. (1991). Tumor necrosis factor alpha mediates lethal activity of killed Gram-negative and Gram-positive bacteria in D-galactosamine-treated mice. *Infect Immun*, **59**, 2110–15.

Gianella R. A., Broitman, S. A., Zamcheck, N. (1971). *Salmonella enteritis*. Role of reduced gastric secretion in pathogenesis. *Am J Dig Dis*, **16**, 1000.

Gohin, I., Olivier, M., Lantier, I., Pepin, M. and Lantier, F. (1997). Analysis of the immune response in sheep efferent lymph during *Salmonella abortusovis* infection. *Vet Immunol Immunopathol*, **60**, 111–30.

Graham, S. M., Hart, C. A., Molyneux, E. M., Walsh, A. L. and Molyneux, M. E. (2000). Malaria and *Salmonella* infections: cause or coincidence? *Trans R Soc Trop Med Hyg*, **94**, 227.

Groisman, E. A., Parra-Lopez, C., Salcedo, M., Lipps, C. J. and Heffron, F. (1992). Resistance to host antimicrobial peptides is necessary for *Salmonella* virulence. *Proc Natl Acad Sci USA*, **89**, 11939–43.

Guilloteau, L., Buzoni-Gatel, D., Bernard, F., Lantier, I. and Lantier, F. (1993). *Salmonella abortusovis* infection in susceptible BALB/cby mice: importance of Lyt-2^+ and L3T4^+ T-cells in acquired immunity and granuloma formation. *Microb Pathog*, **14**, 45–55.

Haraga, A. and Miller, S. I. (2003). A *Salmonella enterica* serovar Typhimurium translocated leucine-rich repeat effector protein inhibits NF-κB-dependent gene expression. *Infect Immun*, **71**, 4052–8.

Harrington, K. A. and Hormaeche, C. E. (1986). Expression of the innate resistance gene *Ity* in mouse Kupffer cells infected with *Salmonella typhimurium in vitro*. *Microb Pathog*, **1**, 269–74.

Harrison, J. A., Villarreal-Ramos, B., Mastroeni, P., Demarco de Hormaeche, R. and Hormaeche, C. E. (1997). Correlates of protection induced by live Aro-*Salmonella typhimurium* vaccines in the murine typhoid model. *Immunology*, **90**, 618–25.

Hart, C. A., Beeching, N. J., Duerden, B. I. *et al.* (2000). Infections in AIDS. *J Med Microbiol*, **49**, 947–67.

Hayashi, F., Smith, K. D., Ozinsky, A. *et al.* (2001). The innate immune response to bacterial flagellin is mediated by Toll-like receptor 5. *Nature*, **410**, 1099–103.

Hemmi, H., Takeuchi, O., Kawai, T. *et al.* (2000). A Toll-like receptor recognizes bacterial DNA. *Nature*, **408**, 740–5.

Hermaszewski, R. A. and Webster, A. D. (1993). Primary hypogammaglobulinaemia: a survey of clinical manifestations and complications. *Q J Med*, **86**, 31–42.

Hess, J., Ladel, C., Miko, D. and Kaufmann, S. H. (1996). *Salmonella typhimurium aroA*$^-$ infection in gene-targeted immunodeficient mice: major role of $CD4^+$ TCR-alpha beta cells and IFNγ in bacterial clearance independent of intracellular location. *J Immunol*, **156**, 3321–6.

Hirose, K., Nishimura, H., Matsuguchi, T. and Yoshikai, Y. (1999). Endogenous IL15 might be responsible for early protection by natural killer cells against infection with an avirulent strain of *Salmonella choleraesuis* in mice. *J Leukoc Biol*, **66**, 382–90.

Hochadel, J. F. and Keller, K. F. (1977). Protective effects of passively transferred immune T- or B-lymphocytes in mice infected with *Salmonella typhimurium*. *J Infect Dis*, **135**, 813–23.

Hopkins, S. A. and Kraehenbuhl, J. P. (1997). Dendritic cells of the murine Peyer's patches colocalize with *Salmonella typhimurium* avirulent mutants in the subepithelial dome. *Adv Exp Med Biol*, **417**, 105–9.

Hormaeche, C. E. (1979). The natural resistance of radiation chimeras to *S. typhimurium* C5. *Immunology*, **37**, 329–32.

(1980). The *in vivo* division and death rates of *Salmonella typhimurium* in the spleens of naturally resistant and susceptible mice measured by the superinfecting phage technique of Meynell. *Immunology*, **41**, 973–9.

(1990). Dead salmonellae or their endotoxin accelerate the early course of a *Salmonella* infection in mice. *Microb Pathog*, **9**, 213–18.

Hormaeche, C. E., Harrington, K. A. and Joysey, H. S. (1985). Natural resistance to salmonellae in mice: control by genes within the major histocompatibility complex. *J Infect Dis*, **152**, 1050–6.

Hormaeche, C. E., Mastroeni, P., Arena, A., Uddin, J. and Joysey, H. S. (1990). T-cells do not mediate the initial suppression of a *Salmonella* infection in the RES. *Immunology*, **70**, 247–50.

Hormaeche, C. E., Mastroeni, P., Harrison, J. A. *et al.* (1996). Protection against oral challenge three months after i.v. immunization of BALB/c mice with live Aro *Salmonella typhimurium* and *Salmonella enteritidis* vaccines is serotype (species)-dependent and only partially determined by the main LPS O antigen. *Vaccine*, **14**, 251–9.

House, D., Chinh, N. T., Hien, T. T. *et al.* (2002). Cytokine release by lipopolysaccharide-stimulated whole blood from patients with typhoid fever. *J Infect Dis*, **186**, 240–5.

Hsu, H. S. (1989). Pathogenesis and immunity in murine salmonellosis. *Microbiol Rev*, **53**, 390–409.

John, B., Rajagopal, D., Pashine, A. *et al.* (2002). Role of IL12-independent and IL12-dependent pathways in regulating generation of the IFNγ component of T-cell responses to *Salmonella typhimurium*. *J Immunol*, **169**, 2545–52.

Kagaya, K., Watanabe, K. and Fukazawa, Y. (1989). Capacity of recombinant gamma interferon to activate macrophages for *Salmonella*-killing activity. *Infect Immun*, **57**, 609–15.

Kantele, A., Arvilommi, H. and Jokinen, I. (1986). Specific immunoglobulin-secreting human blood cells after peroral vaccination against *Salmonella typhi*. *J Infect Dis*, **153**, 1126–31.

Kelleher, P. and Misbah, S. A. (2003). What is Good's syndrome? Immunological abnormalities in patients with thymoma. *J Clin Pathol*, **56**, 12–16.

Khan, S. A., Everest, P., Servos, S. *et al.* (1998). A lethal role for lipid A in *Salmonella* infections. *Mol Microbiol*, **29**, 571–9.

Khan, S. A., Strijbos, P. J., Everest, P. *et al.* (2001). Early responses to *Salmonella typhimurium* infection in mice occur at focal lesions in infected organs. *Microb Pathog*, **30**, 29–38.

Kim, H. S., Yoon, H., Minn, I. *et al.* (2000). Pepsin-mediated processing of the cytoplasmic histone H2A to strong antimicrobial peptide buforin I. *J Immunol*, **165**, 3268–74.

Klein, C., Lisowska-Grospierre, B., LeDeist, F., Fischer, A. and Griscelli, C. (1993). Major histocompatibility complex class II deficiency: clinical manifestations, immunologic features, and outcome. *J Pediatr*, **123**, 921–8.

Klugman, K. P., Gilbertson, I. T., Koornhof, H. J. *et al.* (1987). Protective activity of Vi capsular polysaccharide vaccine against typhoid fever. *Lancet*, **2**, 1165–9.

Klugman, K. P., Koornhof, H. J., Robbins, J. B. and Le Cam, N. N. (1996). Immunogenicity, efficacy and serological correlate of protection of *Salmonella typhi* Vi capsular polysaccharide vaccine three years after immunization. *Vaccine*, **14**, 435–8.

Koo, F. C., Peterson, J. W., Houston, C. W. and Molina, N. C. (1984). Pathogenesis of experimental salmonellosis: inhibition of protein synthesis by cytotoxin. *Infect Immun*, **43**, 93–100.

Landesman, S. H., Rao, S. P. and Ahonkhai, V. I. (1982). Infections in children with sickle cell anemia. Special reference to pneumococcal and *Salmonella* infections. *Am J Pediatr Hematol Oncol*, **4**, 407–18.

Lazarus, G. M. and Neu, H. C. (1975). Agents responsible for infection in chronic granulomatous disease of childhood. *J Pediatr*, **86**, 415–17.

Lee, V. T. and Schneewind, O. (1999). Type III secretion machines and the pathogenesis of enteric infections caused by *Yersinia* and *Salmonella* spp. *Immunol Rev*, **168**, 241–55.

Leen, C. L., Birch, A. D., Brettle, R. P., Welsby, P. D. and Yap, P. L. (1986). Salmonellosis in patients with primary hypogammaglobulinaemia. *J Infect*, **12**, 241–5.

Lembo, A., Kalis, C., Kirschning, C. J. *et al.* (2003). Differential contribution of Toll-like receptors 4 and 2 to the cytokine response to *Salmonella enterica* serovar Typhimurium and *Staphylococcus aureus* in mice. *Infect Immun*, **71**, 6058–62.

Levine, M. M., Ferreccio, C., Black, R. E., Tacket, C. O. and Germanier, R. (1989). Progress in vaccines against typhoid fever. *Rev Infect Dis*, **11** (suppl. 3), S552–S567.

Levy, J., Espanol-Boren, T., Thomas, C. *et al.* (1997). Clinical spectrum of X-linked hyper-IgM syndrome. *J Pediatr*, **131**, 47–54.

Liang-Takasaki, C. J., Saxen, H., Makela, P. H. and Leive, L. (1983). Complement activation by polysaccharide of lipopolysaccharide: an important virulence determinant of salmonellae. *Infect Immun*, **41**, 563–9.

Lissner, C. R., Swanson, R. N. and O'Brien, A. D. (1983). Genetic control of the innate resistance of mice to *Salmonella typhimurium*: expression of the *Ity* gene in peritoneal and splenic macrophages isolated in vitro. *J Immunol*, **131**, 3006–13.

Lo, W. F., Ong, H., Metcalf, E. S. and Soloski, M. J. (1999). T-cell responses to Gram-negative intracellular bacterial pathogens: a role for $CD8^+$ T-cells in immunity to *Salmonella* infection and the involvement of MHC class Ib molecules. *J Immunol*, **162**, 5398–406.

Lo, W. F., Woods, A. S., DeCloux, A. *et al.* (2000). Molecular mimicry mediated by MHC class Ib molecules after infection with gram-negative pathogens. *Nat Med*, **6**, 215–18.

MacFarlane, A. S., Schwacha, M. G. and Eisenstein, T. K. (1999). In vivo blockage of nitric oxide with aminoguanidine inhibits immunosuppression induced by an attenuated strain of *Salmonella typhimurium*, potentiates *Salmonella* infection, and inhibits macrophage and polymorphonuclear leukocyte influx into the spleen. *Infect Immun*, **67**, 891–8.

Maskell, D. J. and Hormaeche, C. E. (1986). Genes within the major histocompatibility complex influence the response to ampicillin therapy and severity of relapse in H-2 congenic, susceptible *Ity*^{s} mice infected with virulent *Salmonella typhimurium*. *J Immunogenet*, **13**, 451–7.

Maskell, D. J., Hormaeche, C. E., Harrington, K. A., Joysey, H. S. and Liew, F. Y. (1987). The initial suppression of bacterial growth in a *Salmonella* infection is mediated by a localized rather than a systemic response. *Microb Pathog*, **2**, 295–305.

Mastroeni, P. (2002). Immunity to systemic *Salmonella* infections. *Curr Mol Med*, **2**, 393–406.

Mastroeni, P., Arena, A., Costa, G. B. *et al.* (1991). Serum TNFα in mouse typhoid and enhancement of a *Salmonella* infection by anti-TNFα antibodies. *Microb Pathog*, **11**, 33–8.

Mastroeni, P., Chabalgoity, J. A., Dunstan, S. J., Maskell, D. J. and Dougan, G. (2001). *Salmonella*: immune responses and vaccines. *Vet J*, **161**, 132–64.

Mastroeni, P., Clare, S., Khan, S. *et al.* (1999). Interleukin 18 contributes to host resistance and gamma interferon production in mice infected with virulent *Salmonella typhimurium*. *Infect Immun*, **67**, 478–83.

Mastroeni, P., Harrison, J. A., Chabalgoity, J. A. and Hormaeche, C. E. (1996). Effect of interleukin 12 neutralization on host resistance and gamma interferon production in mouse typhoid. *Infect Immun*, **64**, 189–96.

Mastroeni, P., Harrison, J. A., Robinson, J. H. *et al.* (1998). Interleukin 12 is required for control of the growth of attenuated aromatic-compound-dependent salmonellae in BALB/c mice: role of gamma interferon and macrophage activation. *Infect Immun*, **66**, 4767–76.

Mastroeni, P., Simmons, C., Fowler, R., Hormaeche, C. E. and Dougan, G. (2000a). *Igh-6*$^{-/-}$ (B-cell-deficient) mice fail to mount solid acquired resistance to oral challenge with virulent *Salmonella enterica* serovar Typhimurium and show impaired Th1 T-cell responses to *Salmonella* antigens. *Infect Immun*, **68**, 46–53.

Mastroeni, P., Skepper, J. N. and Hormaeche, C. E. (1995). Effect of anti-tumor necrosis factor alpha antibodies on histopathology of primary *Salmonella* infections [published erratum appears in *Infect Immun* (1995 Dec) **63**(12), 4966]. *Infect Immun*, **63**, 3674–82.

Mastroeni, P., Vazquez-Torres, A., Fang, F. C. *et al.* (2000b). Antimicrobial actions of the NADPH phagocyte oxidase and inducible nitric oxide synthase in experimental salmonellosis. II. Effects on microbial proliferation and host survival *in vivo*. *J Exp Med*, **192**, 237–48.

Mastroeni, P., Villarreal-Ramos, B., Demarco de Hormaeche, R. and Hormaeche, C. E. (1993a). Delayed (footpad) hypersensitivity and Arthus reactivity using protein-rich antigens and LPS in mice immunized with live attenuated *aroA Salmonella* vaccines. *Microb Pathog*, **14**, 369–79.

Mastroeni, P., Villarreal-Ramos, B. and Hormaeche, C. E. (1992). Role of T-cells, TNFα and IFNγ in recall of immunity to oral challenge with virulent salmonellae in mice vaccinated with live attenuated *aro*$^-$ *Salmonella* vaccines. *Microb Pathog*, **13**, 477–91.

(1993b). Adoptive transfer of immunity to oral challenge with virulent salmonellae in innately susceptible BALB/c mice requires both immune serum and T-cells. *Infect Immun*, **61**, 3981–4.

(1993c). Effect of late administration of anti-TNFα antibodies on a *Salmonella* infection in the mouse model. *Microb Pathog*, **14**, 473–80.

Matsui, K. and Arai, T. (1989). Protective immunity induced by porin in experimental mouse salmonellosis. *Microbiol Immunol*, **33**, 699–708.

McSorley, S. J. and Jenkins, M. K. (2000). Antibody is required for protection against virulent but not attenuated *Salmonella enterica* serovar Typhimurium. *Infect Immun*, **68**, 3344–8.

McSorley, S. J., Cookson, B. T. and Jenkins, M. K. (2000). Characterization of CD4$^+$ T-cell responses during natural infection with *Salmonella typhimurium*. *J Immunol*, **164**, 986–93.

Medzhitov, R., Preston-Hurlburt, P. and Janeway, C. A., Jr. (1997). A human homologue of the Drosophila Toll protein signals activation of adaptive immunity. *Nature*, **388**, 394–7.

Mitchell, E. K., Mastroeni, P., Kelly, A. P. and Trowsdale, J. (2004). Inhibition of cell surface MHC class II expression by *Salmonella*. *Eur J Immunol*, **34**, 2559–67.

Mittrucker, H. W., Kohler, A. and Kaufmann, S. H. (2002). Characterization of the murine T-lymphocyte response to *Salmonella enterica* serovar Typhimurium infection. *Infect Immun*, **70**, 199–203.

Mittrucker, H. W., Kohler, A., Mak, T. W. and Kaufmann, S. H. (1999). Critical role of CD28 in protective immunity against *Salmonella typhimurium*. *J Immunol*, **163**, 6769–76.

Mittrucker, H. W., Raupach, B., Kohler, A. and Kaufmann, S. H. (2000). Cutting edge: role of B lymphocytes in protective immunity against *Salmonella typhimurium* infection. *J Immunol*, **164**, 1648–52.

Mixter, P. F., Camerini, V., Stone, B. J., Miller, V. L. and Kronenberg, M. (1994). Mouse T lymphocytes that express a gamma delta T-cell antigen receptor contribute to resistance to *Salmonella* infection in vivo. *Infect Immun*, **62**, 4618–21.

Monack, D. M., Hersh, D., Ghori, N. *et al.* (2000). *Salmonella* exploits Caspase-1 to colonize Peyer's patches in a murine typhoid model. *J Exp Med*, **192**, 249–58.

Moors, M. A., Li, L. and Mizel, S. B. (2001). Activation of interleukin 1 receptor-associated kinase by gram-negative flagellin. *Infect Immun*, **69**, 4424–9.

Mouy, R., Fischer, A., Vilmer, E., Seger, R. and Griscelli, C. (1989). Incidence, severity, and prevention of infections in chronic granulomatous disease. *J Pediatr*, **114**, 555–60.

Muotiala, A. and Makela, P. H. (1990). The role of IFNγ in murine *Salmonella typhimurium* infection. *Microb Pathog*, **8**, 135–41.

(1993). Role of gamma interferon in late stages of murine salmonellosis. *Infect Immun*, **61**, 4248–53.

Murphy, J. R., Baqar, S., Munoz, C. *et al.* (1987). Characteristics of humoral and cellular immunity to *Salmonella typhi* in residents of typhoid-endemic and typhoid-free regions. *J Infect Dis*, **156**, 1005–9.

Murphy, J. R., Wasserman, S. S., Baqar, S. *et al.* (1989). Immunity to *Salmonella typhi*: considerations relevant to measurement of cellular immunity in typhoid-endemic regions. *Clin Exp Immunol*, **75**, 228–33.

Naiki, Y., Nishimura, H., Kawano, T. *et al.* (1999). Regulatory role of peritoneal NK1.1$^+$ alpha beta T-cells in IL12 production during *Salmonella* infection. *J Immunol*, **163**, 2057–63.

Nauciel, C. (1990). Role of CD4$^+$ T-cells and T-independent mechanisms in acquired resistance to *Salmonella typhimurium* infection. *J Immunol*, **145**, 1265–9.

Nauciel, C. and Espinasse-Maes, F. (1992). Role of gamma interferon and tumor necrosis factor alpha in resistance to *Salmonella typhimurium* infection. *Infect Immun*, **60**, 450–4.

Nauciel, C., Ronco, E., Guenet, J. L. and Pla, M. (1988). Role of H-2 and non-H-2 genes in control of bacterial clearance from the spleen in *Salmonella typhimurium*-infected mice. *Infect Immun*, **56**, 2407–11.

Nauciel, C., Ronco, E. and Pla, M. (1990). Influence of different regions of the H-2 complex on the rate of clearance of *Salmonella typhimurium*. *Infect Immun*, **58**, 573–4.

Netea, M. G., Kullberg, B. J., Joosten, L. A. *et al.* (2001). Lethal *Escherichia coli* and *Salmonella typhimurium* endotoxemia is mediated through different pathways. *Eur J Immunol*, **31**, 2529–38.

Nishimura, H., Hiromatsu, K., Kobayashi, N. *et al.* (1996). IL15 is a novel growth factor for murine gamma delta T-cells induced by *Salmonella* infection. *J Immunol*, **156**, 663–9.

Nishimura, H., Washizu, J., Naiki, Y. *et al.* (1999). MHC class II-dependent NK1.1^{+} gammadelta T-cells are induced in mice by *Salmonella* infection. *J Immunol*, **162**, 1573–81.

O'Brien, A. D. and Metcalf, E. S. (1982). Control of early *Salmonella typhimurium* growth in innately *Salmonella*-resistant mice does not require functional T lymphocytes. *J Immunol*, **129**, 1349–51.

O'Brien, A. D., Metcalf, E. S. and Rosenstreich, D. L. (1982). Defect in macrophage effector function confers *Salmonella typhimurium* susceptibility on C3H/HeJ mice. *Cell Immunol*, **67**, 325–33.

O'Brien, A. D., Scher, I. and Formal, S. B. (1979). Effect of silica on the innate resistance of inbred mice to *Salmonella typhimurium* infection. *Infect Immun*, **25**, 513–20.

Paul, C., Shalala, K., Warren, R. and Smith, R. (1985). Adoptive transfer of murine host protection to salmonellosis with T-cell growth factor-dependent, *Salmonella*-specific T-cell lines. *Infect Immun*, **48**, 40–3.

Paul, C. C., Norris, K., Warren, R. and Smith, R. A. (1988). Transfer of murine host protection by using interleukin 2-dependent T- lymphocyte lines. *Infect Immun*, **56**, 2189–92.

Peel, J. E., Voirol, M. J., Kolly, C., Gobet, D. and Martinod, S. (1990). Induction of circulating tumor necrosis factor cannot be demonstrated during septicemic salmonellosis in calves. *Infect Immun*, **58**, 439–42.

Picard, C., Fieschi, C., Altare, F. *et al.* (2001). Inherited Interleukin 12 deficiency: IL12B genotype and clinical phenotype of 13 patients from six kindreds. *Am J Hum Genet*, **70**, 2.

Pie, S., Matsiota-Bernard, P., Truffa-Bachi, P. and Nauciel, C. (1996). Gamma interferon and interleukin 10 gene expression in innately susceptible and resistant mice during the early phase of *Salmonella typhimurium* infection. *Infect Immun*, **64**, 849–54.

Pie, S., Truffa-Bachi, P., Pla, M. and Nauciel, C. (1997). Th1 response in *Salmonella typhimurium*-infected mice with a high or low rate of bacterial clearance. *Infect Immun*, **65**, 4509–14.

Poltorak, A., Smirnova, I., He, X. *et al.* (1998). Genetic and physical mapping of the *lps* locus: identification of the Toll-4 receptor as a candidate gene in the critical region. *Blood Cells Mol Dis*, **24**, 340–55.

Pope, M., Kotlarski, I. and Doherty, K. (1994). Induction of Lyt-2^{+} cytotoxic T lymphocytes following primary and secondary *Salmonella* infection. *Immunology*, **81**, 177–82.

Qureshi, S. T., Lariviere, L., Leveque, G. *et al.* (1999). Endotoxin-tolerant mice have mutations in Toll-like receptor 4 (Tlr4). *J Exp Med*, **189**, 615–25.

Ramarathinam, L., Niesel, D. W. and Klimpel, G. R. (1993). *Salmonella typhimurium* induces IFNγ production in murine splenocytes. Role of natural killer cells and macrophages. *J Immunol*, **150**, 3973–81.

Reisfeld, R. A., Niethammer, A. G., Luo, Y. and Xiang, R. (2004). DNA vaccines suppress tumor growth and metastases by the induction of anti-angiogenesis. *Immunol Rev*, **199**, 181–90.

Reitmeyer, J. C., Peterson, J. W. and Wilson, K. J. (1986). *Salmonella* cytotoxin: a component of the bacterial outer membrane. *Microb Pathog*, **1**, 503–10.

Rescigno, M., Urbano, M., Valzasina, B. *et al.* (2001). Dendritic cells express tight junction proteins and penetrate gut epithelial monolayers to sample bacteria. *Nat Immunol*, **2**, 361–7.

Richter-Dahlfors, A., Buchan, A. M. J. and Finlay, B. B. (1997). Murine salmonellosis studied by confocal microscopy: *Salmonella typhimurium* resides intracellularly inside macrophages and exerts a cytotoxic effect on phagocytes in vivo. *J Exp Med*, **186**, 569–80.

Robbins, J. D. and Robbins, J. B. (1984). Reexamination of the protective role of the capsular polysaccharide (Vi antigen) of *Salmonella typhi*. *J Infect Dis*, **150**, 436–49.

Robertsson, J. A., Fossum, C., Svenson, S. B. and Lindberg, A. A. (1982). *Salmonella typhimurium* infection in calves: specific immune reactivity against O-antigenic polysaccharide detectable in in vitro assays. *Infect Immun*, **37**, 728–36.

Rosenberger, C. M., Scott, M. G., Gold, M. R., Hancock, R. E. and Finlay, B. B. (2000). *Salmonella typhimurium* infection and lipopolysaccharide stimulation induce similar changes in macrophage gene expression. *J Immunol*, **164**, 5894–904.

Royle, M. C., Totemeyer, S., Alldridge, L. C., Maskell, D. J. and Bryant, C. E. (2003). Stimulation of Toll-like receptor 4 by lipopolysaccharide during cellular invasion by live *Salmonella typhimurium* is a critical but not exclusive event leading to macrophage responses. *J Immunol*, **170**, 5445–54.

Salcedo, S. P., Noursadeghi, M., Cohen, J. and Holden, D. W. (2001). Intracellular replication of *Salmonella typhimurium* strains in specific subsets of splenic macrophages in vivo. *Cell Microbiol*, **3**, 587–97.

Salerno-Goncalves, R., Pasetti, M. F. and Sztein, M. B. (2002). Characterization of $CD8^+$ effector T-cell responses in volunteers immunized with *Salmonella enterica* serovar Typhi strain Ty21a typhoid vaccine. *J Immunol*, **169**, 2196–203.

Salerno-Goncalves, R., Wyant, T. L., Pasetti, M. F. *et al.* (2003). Concomitant induction of $CD4^+$ and $CD8^+$ T-cell responses in volunteers immunized with *Salmonella enterica* serovar Typhi strain CVD 908-*htrA*. *J Immunol*, **170**, 2734–41.

Salzman, N. H., Ghosh, D., Huttner, K. M., Paterson, Y. and Bevins, C. L. (2003). Protection against enteric salmonellosis in transgenic mice expressing a human intestinal defensin. *Nature*, **422**, 522–6.

Saxen, H., Reima, I. and Makela, P. H. (1987). Alternative complement pathway activation by *Salmonella* O polysaccharide as a virulence determinant in the mouse. *Microb Pathog*, **2**, 15–28.

Schwacha, M. G. and Eisenstein, T. K. (1997). Interleukin 12 is critical for induction of nitric oxide-mediated immunosuppression following vaccination of mice with attenuated *Salmonella typhimurium*. *Infect Immun*, **65**, 4897–903.

Sebastiani, G., Leveque, G., Lariviere, L. *et al.* (2000). Cloning and characterization of the murine Toll-like receptor 5 (Tlr5) gene: sequence and mRNA expression studies in *Salmonella*-susceptible MOLF/Ei mice. *Genomics*, **64**, 230–40.

Segall, T. and Lindberg, A. A. (1993). Oral vaccination of calves with an aromatic-dependent *Salmonella dublin* (O9,12) hybrid expressing O4,12 protects against *S. dublin* (O9,12) but not against *Salmonella typhimurium* (O4,5,12). *Infect Immun*, **61**, 1222–31.

Senterfitt, V. C. and Shands, J. W., Jr. (1968). Salmonellosis in mice infected with *Mycobacterium tuberculosis* BCG. I. Role of endotoxin in infection. *J Bacteriol*, **96**, 287–92.

Sheppard, M., Webb, C., Heath, F. *et al.* (2003). Dynamics of bacterial growth and distribution within the liver during *Salmonella* infection. *Cell Microbiol*, **5**, 593–600.

Sinha, K., Mastroeni, P., Harrison, J., de Hormaeche, R. D. and Hormaeche, C. E. (1997). *Salmonella typhimurium aroA*, *htrA*, and *aroD htrA* mutants cause progressive infections in athymic (*nu/nu*) BALB/c mice. *Infect Immun*, **65**, 1566–9.

Skeen, M. J. and Ziegler, H. K. (1993). Induction of murine peritoneal gamma/delta T-cells and their role in resistance to bacterial infection. *J Exp Med*, **178**, 971–84.

Srinivasan, A., Foley, J. and McSorley, S. J. (2004). Massive number of antigen-specific CD4 T-cells during vaccination with live attenuated *Salmonella* causes interclonal competition. *J Immunol*, **172**, 6884–93.

Svenson, S. B., Nurminen, M. and Lindberg, A. A. (1979). Artificial *Salmonella* vaccines: O-antigenic oligosaccharide-protein conjugates induce protection against infection with *Salmonella typhimurium*. *Infect Immun*, **25**, 863–72.

Svensson, M., Johansson, C. and Wick, M. J. (2000). *Salmonella enterica* serovar Typhimurium-induced maturation of bone marrow-derived dendritic cells. *Infect Immun*, **68**, 6311–20.

Sztein, M. B., Tanner, M. K., Polotsky, Y., Orenstein, J. M. and Levine, M. M. (1995). Cytotoxic T lymphocytes after oral immunization with attenuated vaccine strains of *Salmonella typhi* in humans. *J Immunol*, **155**, 3987–93.

Sztein, M. B., Wasserman, S. S., Tacket, C. O. *et al.* (1994). Cytokine production patterns and lymphoproliferative responses in volunteers orally immunized with attenuated vaccine strains of *Salmonella typhi*. *J Infect Dis*, **170**, 1508–17.

Tacket, C. O., Hone, D. M., Curtiss, R., III *et al.* (1992). Comparison of the safety and immunogenicity of Δ*aroC* Δ*aroD* and Δ*cya* Δ*crp Salmonella typhi* strains in adult volunteers. *Infect Immun*, **60**, 536–41.

Takeshita, F., Leifer, C. A., Gursel, I. *et al.* (2001). Cutting edge: role of Toll-like receptor 9 in CpG DNA-induced activation of human cells. *J Immunol*, **167**, 3555–8.

Takeuchi, O., Hoshino, K., Kawai, T. *et al.* (1999). Differential roles of TLR2 and TLR4 in recognition of Gram-negative and Gram-positive bacterial cell wall components. *Immunity*, **11**, 443–51.

Tarr, P. E., Sneller, M. C., Mechanic, L. J. *et al.* (2001). Infections in patients with immunodeficiency with thymoma (Good syndrome). Report of 5 cases and review of the literature. *Medicine (Baltimore)*, **80**, 123–33.

Thatte, J., Rath, S. and Bal, V. (1993). Immunization with live versus killed *Salmonella typhimurium* leads to the generation of an IFNγ-dominant versus an IL4-dominant immune response. *Int Immunol*, **5**, 1431–6.

Totemeyer, S., Foster, N., Kaiser, P., Maskell, D. J. and Bryant, C. E. (2003). Toll-like receptor expression in C3H/HeN and C3H/HeJ mice during *Salmonella enterica* serovar Typhimurium infection. *Infect Immun*, **71**, 6653–7.

Ugrinovic, S., Menager, N., Goh, N. and Mastroeni, P. (2003). Characterization and development of T-cell immune responses in B-cell-deficient ($Igh_{-6}^{-/-}$)

mice with *Salmonella enterica* serovar Typhimurium infection. *Infect Immun*, **71**, 6808–19.

Van Amersfoort, E. S., Van Berkel, T. J. and Kuiper, J. (2003). Receptors, mediators, and mechanisms involved in bacterial sepsis and septic shock. *Clin Microbiol Rev*, **16**, 379–414.

van der Velden, A. W., Velasquez, M. and Starnbach, M. N. (2003). *Salmonella* rapidly kill dendritic cells via a Caspase-1-dependent mechanism. *J Immunol*, **171**, 6742–9.

Vazquez-Torres, A., Fantuzzi, G., Edwards, C. K., III, Dinarello, C. A. and Fang, F. C. (2001). Defective localization of the NADPH phagocyte oxidase to *Salmonella*-containing phagosomes in tumor necrosis factor p55 receptor-deficient macrophages. *Proc Natl Acad Sci USA*, **98**, 2561–5.

Vazquez-Torres, A., Jones-Carson, J., Baumler, A. J. *et al.* (1999). Extraintestinal dissemination of *Salmonella* by CD18-expressing phagocytes. *Nature*, **401**, 804–8.

Vazquez-Torres, A., Jones-Carson, J., Mastroeni, P., Ischiropoulos, H. and Fang, F. C. (2000a). Antimicrobial actions of the NADPH phagocyte oxidase and inducible nitric oxide synthase in experimental salmonellosis. I. Effects on microbial killing by activated peritoneal macrophages in vitro. *J Exp Med*, **192**, 227–36.

Vazquez-Torres, A., Xu, Y., Jones-Carson, J. *et al.* (2000b). *Salmonella* pathogenicity island 2-dependent evasion of the phagocyte NADPH oxidase. *Science*, **287**, 1655–8.

Vidal, S. M., Malo, D., Vogan, K., Skamene, E. and Gros, P. (1993). Natural resistance to infection with intracellular parasites: isolation of a candidate for Bcg. *Cell*, **73**, 469–85.

Villarreal, B., Mastroeni, P., de Hormaeche, R. D. and Hormaeche, C. E. (1992). Proliferative and T-cell specific interleukin (IL2/IL4) production responses in spleen cells from mice vaccinated with *aroA* live attenuated *Salmonella* vaccines. *Microb Pathog*, **13**, 305–15.

Villarreal-Ramos, B., Manser, J., Collins, R. A. *et al.* (1998). Immune responses in calves immunised orally or subcutaneously with a live *Salmonella typhimurium aro* vaccine. *Vaccine*, **16**, 45–54.

Vogel, S. N., Johnson, D., Perera, P. Y. *et al.* (1999). Cutting edge: functional characterization of the effect of the C3H/HeJ defect in mice that lack an *lpsn* gene: in vivo evidence for a dominant negative mutation. *J Immunol*, **162**, 5666–70.

Warren, J., Mastroeni, P., Dougan, G. *et al.* (2002). Increased susceptibility of C1q-deficient mice to *Salmonella enterica* serovar Typhimurium infection. *Infect Immun*, **70**, 551–7.

Weinberger, M. and Pizzo, P. (1992). The evaluation and management of neutropenic patients with unexplained fever. In *Infections in immunocompromised infants and children*, ed. C. C. Patrich. New York: Churchill Livingstone, pp. 338–41.

Wyant, T. L., Tanner, M. K. and Sztein, M. B. (1999). *Salmonella typhi* flagella are potent inducers of proinflammatory cytokine secretion by human monocytes. *Infect Immun*, **67**, 3619–24.

Xu, H. R., Hsu, H. S., Moncure, C. W. and King, R. A. (1993). Correlation of antibody titres induced by vaccination with protection in mouse typhoid. *Vaccine*, **11**, 725–9.

Yrlid, U., Svensson, M., Hakansson, A. *et al.* (2001). In vivo activation of dendritic cells and T-cells during *Salmonella enterica* serovar Typhimurium infection. *Infect Immun*, **69**, 5726–35.

Zasloff, M. (2002). Trypsin, for the defense. *Nat Immunol*, **3**, 508–10.

Zhang, S., Adams, L. G., Nunes, J. *et al.* (2003). Secreted effector proteins of *Salmonella enterica* serotype Typhimurium elicit host-specific chemokine profiles in animal models of typhoid fever and enterocolitis. *Infect Immun*, **71**, 4795–803.

CHAPTER 9

Interactions of *S. enterica* with phagocytic cells

Bruce D. McCollister and Andres Vazquez-Torres

9.1 INTRODUCTION

Mononuclear phagocytes associate with *S. enterica* early in the disease process before acute inflammatory abscesses are formed, as well as during later stages of the acquired immune response in which macrophages form part of well-organized granulomas (Mastroeni *et al.*, 1995; Richter-Dahlfors *et al.*, 1997). The ability to survive within macrophages is a key event in the pathogenesis of *Salmonella enterica* (Fields *et al.*, 1986). A growing body of information indicates that macrophages can serve as sites for *S. enterica* replication, even though they can be activated to exert potent anti-*S. enterica* activity. The great majority of the intimate interactions between *S. enterica* and macrophages take place inside a specialized endocytic vacuole named the phagosome. This chapter discusses the dynamic *S. enterica* phagosome as it pertains to the pathogenesis of this intracellular Gram-negative bacterium.

Immunological and genetic manipulations in animal models of infection, as well as the observation of naturally occurring genetic traits, have revealed that genetic loci encoding Nramp1, TLR4, NADPH oxidase and IFNγ play key roles in resistance to *S. enterica* infection. These host defenses are expressed directly by macrophages or, as in the case of IFNγ, up-regulate the anti-*S. enterica* activity of mononuclear phagocytes. In the following sections, we will discuss both the mechanisms by which these host defenses contribute to the anti-*S. enterica* activity of macrophages, and the virulence factors used by *S. enterica* to avoid these components of the antimicrobial arsenal of professional phagocytes.

'Salmonella' Infections: Clinical, Immunological and Molecular Aspects, ed. Pietro Mastroeni and Duncan Maskell. Published by Cambridge University Press.

9.2 INTERACTIONS OF *S. ENTERICA* WITH THE MACROPHAGE ENDOSOMAL PATHWAYS

The interaction of macrophage surface receptors with their cognate microbial ligands often results in the internalization of the microbial particle into a membrane-bound organelle called the phagosome. The young phagosome is unable to digest its content and must undergo a maturation process that is initiated through fusion and fission events with early endosomes, and continues through interactions with late endosomes and lysosomes (Vieira *et al.*, 2002). Some investigators have reported that the phagosomes containing *S. enterica*, similar to the vacuoles containing *Coxiella burnetii*, completely fuse with lysosomes (Oh *et al.*, 1996). However, more recent data suggest that *S. enterica* belongs to the category of microorganisms that are capable of blocking phagolysosomal fusion (Buchmeier and Heffron, 1991; Garvis *et al.*, 2001; Hashim *et al.*, 2000; Rathman *et al.*, 1997; Uchiya *et al.*, 1999). These studies have revealed that inhibition of phagosome-lysosome fusion is independent of the route of bacterial entry into the macrophage, requires *S. enterica* protein synthesis shortly after entry, and involves the type III secretion system encoded within the *S. enterica* pathogenicity island 2 (SPI-2) (Buchmeier and Heffron, 1991; Garvis *et al.*, 2001; Hashim *et al.*, 2000; Rathman *et al.*, 1997; Uchiya *et al.*, 1999).

Characterization of the unique niche occupied by *S. enterica* in phagocytic cells suggests that *S. enterica* selectively avoids some endosomal compartments while facilitating the interaction with others. Several groups of investigators using epithelial cells and macrophages have shown that the vacuoles containing *S. enterica* establish contact with early endosomes (Hashim *et al.*, 2000; Mukherjee *et al.*, 2000; Steele-Mortimer *et al.*, 1999). However, other investigators contend that the phagosomes-containing *S. enterica* do not have significant interactions with early endosomes (Rathman *et al.*, 1997; Uchiya *et al.*, 1999). There also seems to be some controversy as to whether phagosomes containing *S. enterica* exclude from their membranes the small GTPase rab7, which is associated with trafficking to late endosomes and lysosomes (Hashim *et al.*, 2000; Meresse *et al.*, 1999). Phagosomes containing *S. enterica* are negative for mannose 6-phosphate receptors (M6PR), and avoid interactions with NADPH oxidase- and inducible nitric oxide synthase (iNOS)-containing vesicles (Chakravortty *et al.*, 2002; Rathman *et al.*, 1997; Steele-Mortimer *et al.*, 1999; Vazquez-Torres *et al.*, 2000b). On the other hand, these phagosomes selectively acquire lysosomal-associated membrane glycoproteins from the trans Golgi network (Garcia-del Portillo and Finlay, 1995; Rathman *et al.*, 1997), suggesting that *S. enterica* has the potential to secure biomolecules in transit from the exocytic pathway. Overall, these studies illustrate the ability of *S. enterica* to block contact with selected endosomes containing antimicrobial defenses, whilst promoting interactions with other vesicles containing potential sources of nutrients.

During the maturation process, phagosomes acidify to a pH < 5 (Vieira *et al.*, 2002). The use of pharmacological inhibitors and cell biological approaches suggest that phagosomes containing *S. enterica* acquire vacuolar ATPases that acidify its lumen (Rathman *et al.*, 1996; Steele-Mortimer *et al.*, 1999). However, although phagosomes containing *S. enterica* become acidic they do so to a lesser extent than phagosomes containing inert particles (Alpuche Aranda *et al.*, 1992) and three independent lines of evidence support the idea that *S. enterica* attenuates the acidification of its phagosome. Firstly, phagosomes containing live *S. enterica* seem to recruit significantly lower amounts of vacuolar ATPases than vacuoles containing dead bacteria (Hashim *et al.*, 2000). Secondly, cathepsinD, which is normally cleaved and activated at pH 5, is maintained in its immature form within the *S. enterica* phagosome (Hashim *et al.*, 2000). Thirdly, phagosomes containing *S. enterica* acidify to a pH of 6.0 during the initial hour of infection, irrespective of whether they bear a wild-type or mutant allele of the Nramp1 divalent metal transporter (Cuellar-Mata *et al.*, 2002). Rathman *et al.* have confirmed that *S. enterica* reside in an acidic environment, although they have reported that the phagosome acidifies to a pH of 4.0–5.0 within an hour of contact (Rathman *et al.*, 1996). Therefore, the picture emerging from these studies supports the notion that the *S. enterica* phagosome acidifies, but the final pH and kinetics of acidification are still being debated.

Acidification of the phagosomal lumen increases the intracellular survival of *S. enterica* (Rathman *et al.*, 1996). The association of acidification and *S. enterica* virulence may be related to the fact that transcription and secretion of SPI-2 effectors are triggered within the acidified phagosome (Beuzon *et al.*, 1999; Cirillo *et al.*, 1998). However, acidification of the phagosome is not an absolute requirement for *S. enterica* virulence (Steele-Mortimer *et al.*, 2000). This latter observation might indicate that, in addition to acidification, low osmolarity and calcium also regulate the expression of SPI-2 virulence genes (Beuzon *et al.*, 1999; Cirillo *et al.*, 1998; Garmendia *et al.*, 2003).

Transcriptional profiling has revealed that the phagosomes containing *S. enterica* are low in phosphate and magnesium, but rich in oxygen, potassium, weak acids, gluconate and antimicrobial peptides (Eriksson *et al.*, 2003). The availability of nutrients regulates the activation of transcriptional factors that coordinate the adaptive response of *S. enterica* to the intracellular environment. For example, the low intraphagosomal magnesium concentration serves as a signal for the phosphorylation of the PhoP response regulator, which coordinates the intracellular survival of *S. enterica* and its resistance to antimicrobial peptides (Groisman and Saier, 1990).

9.3 INNATE ANTI-*S. ENTERICA* ACTIVITY OF THE NRAMP1 DIVALENT METAL TRANSPORTER

9.3.1 General

The *bcg/lsh/ity* autosomal dominant locus is on mouse chromosome 1 and regulates resistance to *S. enterica* (Plant and Glynn, 1976; Hormaeche, 1979). Positional cloning identified a candidate gene for the *bcg/lsh/ity* locus (Vidal *et al.*, 1993), which encodes a putative transmembrane glycoprotein that was initially called natural resistance associated macrophage protein-1 (*Nramp1*) and more recently renamed *Slc11a1*. A glycine to aspartate substitution at position 169 within the fourth transmembrane domain of Nramp1 results in susceptibility of mice to systemic infection with *S. enterica* (Vidal *et al.*, 1993). The definitive proof that *Nramp1* mediates resistance to intracellular pathogens such as *S. enterica* has been obtained using a murine strain in which the *Nramp1* gene has been genetically disrupted (Vidal *et al.*, 1995). The inability of mice lacking a functional *Nramp1* allele to control replication of *S. enterica* is manifested as early as 24 hours after challenge (Swanson and O'Brien, 1983), suggesting that this glycoprotein is key to the innate host response to *S. enterica*. Consistent with this view, Nramp1 is localized to late endosomes, phagosomes and lysosomes of mononuclear and polymorphonuclear phagocytes of the myeloid cell lineage (Gruenheid *et al.*, 1997). However, a recent study in humans has failed to demonstrate a genetic linkage between polymorphisms of *Nramp1* and bacteraemia in patients experiencing typhoid fever (Dunstan *et al.*, 2001). A more exhaustive investigation is needed to rule out a possible linkage between human *Nramp1* polymorphisms and the multiple clinical aspects associated with *S. enterica* pathogenesis.

Although the central role for the *Nramp1* locus in resistance to salmonellosis has been recognized for over 30 years, we are just starting to understand at the molecular level how a single amino acid substitution exerts such a profound effect on the pathophysiology of *S. enterica* infections. In the following sections we will discuss the diverse mechanisms by which Nramp1 promotes anti-*S. enterica* activity of macrophages, putting special emphasis on the effects of Nramp1 on nutrition, vesicular trafficking and overall macrophage gene expression.

9.3.2 Nutritional effects of Nramp1

Experimental data from three groups of investigators suggest that Nramp1 is a multispecific divalent metal ion symporter that deprives intracellular pathogens of divalent metals by removing Fe^{++} and Mn^{++} from the acidic luminal space of late endosomal, lysosomal and phagosomal vesicles (Atkinson and Barton, 1999; Forbes and Gros, 2003; Jabado *et al.*, 2000). Because iron and other divalent cations are cofactors of vital enzymes, it is not surprising that *S. enterica* expresses a battery of transporters that compete with the host for the trace amounts of divalent metals found in the nutritionally deprived intraphagosomal environment (Hantke, 1997; Kehres *et al.*, 2002; Tsolis *et al.*, 1996).

An alternative model has been suggested in which Nramp1 transports divalent metal ions into the phagosome (Kuhn *et al.*, 1999; Zwilling *et al.*, 1999). In support of this model is the observation that Nramp1 is a pH-dependent antiporter with the ability to flux divalent metal ions in either direction against a proton gradient (Goswami *et al.*, 2001). Within the acidic pH environment of the phagosomes, Nramp1 would transport Fe^{++} and other divalent cations into the proton rich luminal space of the *S. enterica* phagosome. The antimicrobial activity associated with Nramp1 in this model stems from the capacity of Fe^{++} to serve as a Fenton catalyst that reduces hydrogen peroxide to the potent antimicrobial hydroxyl radical.

Further experimentation is required to reconcile these apparently contradictory models defining the function that Nramp1 plays on maintaining metal homeostasis.

9.3.3 Effects of Nramp1 on vesicular trafficking

Nramp1 can promote the maturation of phagosomes containing *S. enterica* along the degradative pathway. In fact, mannose-6-phosphate receptor (M6PR) and fluid-phase markers, which are normally excluded from these phagosomes, co-localize to the vacuoles containing *S. enterica* in the presence of a functional Nramp1 (Cuellar-Mata *et al.*, 2002). These effects of Nramp1 are very likely to be associated with its function as an iron efflux pump, because treatment of susceptible *Nramp1*$^{(\mathrm{Asp169})}$ macrophages with an iron chelator increases co-localization of M6PR and fluid-phase markers to the phagosome containing *S. enterica* (Jabado *et al.*, 2003). A drop in the concentration of intraphagosomal divalent cations in turn up-regulates the expression of SPI-2 genes (Zaharik *et al.*, 2002). SPI-2 expression may allow *S. enterica* to offset the Nramp1-stimulated maturation of the phagosome.

9.3.4 Nramp1 modulates genetic expression of host cell defenses against *S. enterica*

A functional Nramp1 divalent metal ion transporter has global effects on the transcriptional profiles of phagocytic cells (Lalmanach *et al.*, 2001). Macrophages bearing a resistant *Nramp1*$^{(Gly169)}$ allele transcribe higher levels of iNOS mRNA and produce significantly more nitrite in response to IFNγ, TNFα, LPS or *S. enterica* than *Nramp1*$^{(Asp169)}$ susceptible controls (Ables *et al.*, 2001; Barrera *et al.*, 1994). The general effects of Nramp1 on gene expression seem to stem from its effects on the stabilization of mRNA transcripts (Lafuse *et al.*, 2002). Furthermore, presence of functional Nramp1 leads to increased binding of STAT1 to the promoter of interferon regulatory factor 1 that is essential for the induction of iNOS expression (Fritsche *et al.*, 2003).

9.4 OXYGEN-DEPENDENT KILLING OF *S. ENTERICA*

9.4.1 General

Macrophages make use of a wide array of oxygen-dependent mechanisms to counteract *S. enterica* growth. Macrophages consume oxygen by the activity of both NADPH and iNOS enzymatic complexes. Consequently, a battery of reactive oxygen species (ROS) and reactive nitrogen species (RNS) are generated with distinct anti-*S. enterica* activity.

9.4.2 The NADPH oxidase

The importance of the NADPH oxidase in resistance to salmonellosis is clearly demonstrated by the overt susceptibility to *S. enterica* of humans suffering from chronic granulomatous disease (CGD), and murine strains lacking a functional NADPH oxidase (Vazquez-Torres and Fang, 2001a). CGD syndrome can arise from X-linked or autosomal mutations in the membrane-bound or cytosolic subunits of the NADPH oxidase. Consequently, NADPH oxidase-deficient phagocytes fail to mount a respiratory burst and lack the high output of ROS.

NADPH oxidase catalyzes the reduction of molecular oxygen with the generation of superoxide anions. This deceptively simple reaction is catalyzed by a very complex enzymatic system comprising six subunits segregated in the cytosol and cytoplasmic membranes of resting phagocytes. Engagement of macrophage CD11b/CD18 heterodimers either by residues in the LPS O-antigen or by complement components deposited on the *S. enterica* surface activates arachidonic acid-, phospholipases A- and D-, and protein kinase C-dependent signaling cascades that lead to the phosphorylation of the p47 cytosolic subunit (Chateau and Caravano, 1997; Vazquez-Torres and Fang, 2001a). Phosphorylated p47 and the p40 and p67 cytosolic components reach the membrane and assemble with the cytochrome b_{558} heterodimer. The high degree of regulation required for the activation of the NADPH oxidase ensures that production of ROS is controlled in space and time, minimizing collateral cytotoxic effects to critical host targets.

Two models have been proposed for the activation of the NADPH oxidase in response to *S. enterica*. According to the vesicular model, assembled enzymatic complexes traffic to the *S. enterica*-containing vesicle from the cytoplasmic membrane (Vazquez-Torres *et al.*, 2000b). Mobilization of NADPH oxidase-containing vesicles to the proximity of the *S. enterica* phagosome requires a functional TNFα/TNFRp55 signaling pathway, and is antagonized by SPI-2 effector proteins (Vazquez-Torres *et al.*, 2001). In contrast, the phagosomal model proposes that membrane-bound and cytosolic subunits of the NADPH oxidase assemble on the surface of the phagosomes that contain *S. enterica*, a process that is inhibited by SPI-2 effectors (Gallois *et al.*, 2001). Irrespective of the model, the action of SPI-2 effector proteins on the NADPH oxidase constitutes a novel mechanism that significantly reduces the oxidative stress that *S. enterica* has to endure inside professional phagocytes.

9.4.3 Biological chemistry of ROS in *S. enterica* pathogenesis

The toxicity of superoxide is mainly due to the oxidation of Fe-S clusters (Imlay, 2003). By stopping NADPH oxidase enzymatic complexes from reaching the *S. enterica* phagosome, SPI-2 significantly reduces the amount of superoxide that this intracellular pathogen has to withstand within macrophages (Gallois *et al.*, 2001; Vazquez-Torres and Fang, 2001b; Vazquez-Torres *et al.*, 2001; Vazquez-Torres *et al.*, 2000b). Some superoxide derived from the activity of the NADPH oxidase must nevertheless reach the *S. enterica* phagosome, since the superoxide-consuming periplasmic Cu/Zn superoxide dismutase (SOD) encoded within the *Gifsy*-2 prophage and the *sspJ* locus play an essential role in *S. enterica* pathogenesis (De Groote *et al.*, 1997; Tsolis *et al.*, 1995; van der Straaten *et al.*, 2001).

Superoxide anions are ultimately detoxified to water and during this process several ROS are generated by successive reduction reactions. In contrast to superoxide anions, hydrogen peroxide formed by reduction of superoxide anions can readily cross cell membranes. A bimodal pattern of killing by hydrogen peroxide that has been shown to be effective against *E. coli* may also be applicable to *S. enterica* (Imlay and Linn, 1986). At the micromolar concentrations that are likely to reach the *S. enterica* phagosome, hydrogen peroxide is reduced by Fe^{++} in the Fenton reaction to give rise to hydroxyl and/or ferryl radicals. These radicals mediate killing through DNA damage. In favor of this model is the observation that, *S. enterica* strains with mutations in genes of the DNA base excision repair (BER) system are hypersusceptible to hydrogen peroxide and macrophage killing (Suvarnapunya *et al.*, 2003).

S. enterica can detoxify hydrogen peroxide via any of its catalases. However, catalases seem to be dispensable for the survival of *S. enterica* within macrophages (Buchmeier *et al.*, 1995). Small reducing molecules and DNA repair systems may protect *S. enterica* against hydrogen peroxide-associated cytotoxicity (Buchmeier *et al.*, 1995; Lundberg *et al.*, 1999).

9.4.4 Inducible nitric oxide synthase (iNOS)

Nitric oxide (NO) synthases catalyze the oxidation of the guanidino nitrogen of L-arginine for the production of NO. There are several NOS isoforms that can generate physiological amounts of this diatomic radical. However, the iNOS isoform, also known as NOS2, is the only one able to synthesize the high NO output associated with anti-*S. enterica* activity. Regulation of iNOS expression is controlled primarily at the level of transcription. The promoter of iNOS contains two NF-κB binding sites flanking several IFNγ response elements. As predicted from the promoter sequence, LPS/TLR4 and IFNγ signaling cascades act synergistically in inducing optimal iNOS expression.

NO produced from the oxidation of L-arginine is an integral part of the anti-bacterial arsenal of macrophages and RNS are critical for host resistance to *S. enterica*. *S. enterica* infection in mice induces an increase in the levels of mRNA specific for iNOS resulting in the production of the NO oxidative metabolites nitrite and nitrate and NO-Fe adducts (Alam *et al.*, 2002; Chakravortty *et al.*, 2002; Vazquez-Torres, Fang, 2001). Genetic or pharmacological inhibition of iNOS decreases resistance to systemic salmonellosis indicating a role for NO derivatives in host resistance to *S. enterica* infections (Alam *et al.*, 2002; Mastroeni *et al.*, 2000). The contribution of NO to resistance to *S. enterica* infections correlates very well with the capacity of macrophages to express iNOS, produce nitrogen oxides and exert NO-dependent antimicrobial activity (Vazquez-Torres *et al.*, 2000).

9.4.5 Biological chemistry of *NO* in the pathogenesis of *S. enterica*

NO is relatively unreactive and most of the direct effects of NO on biological systems stem from its interaction with metals such as iron and oxygen metabolites. NO for example directly reacts with the iron cofactor of Fur (ferric uptake regulator), which in *S. enterica* induces the expression of the *hmp*-encoded flavohemoprotein (Crawford and Goldberg, 1998). Zinc metalloproteins have been discovered as very exciting new targets in the anti-*S. enterica* actions of NO (Schapiro *et al.*, 2003). Mobilization of zinc from these metalloproteins most likely involves the nitrosylation of cysteine thiol groups that coordinate the zinc-finger domain. As a consequence of the mobilization of zinc from DNA-binding metalloproteins, NO arrests DNA replication, a signal that activates the SOS response. The SOS response is a genetic program for the management of damaged DNA and failed replication, and it involves nucleotide excision repair, as well as resumption of cell division and DNA replication. Interestingly, DNA replication arrest leads to indirect RecBCD exonuclease-mediated double strand chromosomal breaks. In addition, the strong oxidants N_2O_3, $NO_2^{\bullet}$; and peroxynitrite, which are formed in the reaction of NO with molecular oxygen and superoxide anion, can directly damage DNA.

Peroxynitrite is a very potent oxidant and is very efficient at killing *S. enterica* in vitro. Even though peroxynitrite can be produced by macrophages infected with *S. enterica*, this radical does not seem to exert any considerable activity against *S. enterica* in macrophages. Several reasons seem to contribute to this apparent paradox. Firstly, peroxynitrite is presumably synthesized in NADPH oxidase-containing vesicles that are excluded from the *S. enterica* phagosome by SPI-2 (Vazquez-Torres and Fang, 2001). Secondly, the strategically situated Cu/Zn SOD consumes superoxide that is required for peroxynitrite synthesis (De Groote *et al.*, 1997). Thirdly, for the most part, synthesis of the peroxynitrite precursors (superoxide and NO) is segregated in time, whereby an early NADPH oxidase-dependent oxidative burst is followed by a nitrosative phase that relies on iNOS (Vazquez-Torres *et al.*, 2000a). And fourthly, *S. enterica* expresses a peroxynitrite reductase that metabolizes peroxynitrite to nitrite (Bryk *et al.*, 2000). Therefore, the coordinated action of several *S. enterica* virulence factors, combined with the temporal segregation in the production of ROS and RNS, minimizes the oxidative stress that this intracellular pathogen has to withstand inside professional phagocytes.

9.4.6 Cell biology of iNOS in macrophages infected with *S. enterica*

Similar to the NADPH oxidase, it seems that iNOS protein is also excluded from the *S. enterica* phagosome. Two models have been put forward for the cellular distribution of iNOS in macrophages infected with *S. enterica*. The first model is based on the association of iNOS with the submembranous cortical actin cytoskeleton (Webb *et al.*, 2001). In this model, the lack of iNOS trafficking does not seem to be a consequence of a functional SPI-2 secretion system as described for the NADPH oxidase, since ingestion of inert particles does not mobilize iNOS to the nascent phagosome. An alternative model has been proposed by which trafficking of iNOS-containing vesicles is inhibited by SPI-2 (Chakravortty *et al.*, 2002). More experimentation examining diverse populations of phagocytes in different states of differentiation and activation is needed to reconcile these apparently contradictory models.

9.4.7 *S. enterica* defenses against NO stress

S. enterica uses enzymatic and nonenzymatic mechanisms to detoxify nitrosative stress encountered within macrophages. *S. enterica* expresses the *hmp*-encoded flavohaemoprotein within macrophages (Eriksson *et al.*, 2003; Stevanin *et al.*, 2002). The dioxygenase activity of Hmp protects *S. enterica* against nitrosative stress in vitro and within macrophages (Crawford and Goldberg, 1998; Stevanin *et al.*, 2002). In addition, *S. enterica* has the potential to detoxify peroxynitrite to nitrite by the enzymatic activity of peroxynitrite reductase (Bryk *et al.*, 2000). A variety of scavengers such as homocysteine, and a number of transcriptional regulators including Fur add to the coordinated efforts of *S. enterica* against nitrosative stress encountered within activated macrophages (Crawford and Goldberg, 1998; De Groote *et al.*, 1996).

9.5 ACTIVATIVATION OF MACROPHAGE ACTIVITY AGAINST *S. ENTERICA*

9.5.1 General

The response of macrophages to *S. enterica* is initiated through the direct recognition of *S. enterica* antigens by innate host cell pattern recognition receptors (PRR). Toll-like receptor 4 (TLR4) is the PRR that has been shown to contribute most dramatically to stimulate innate macrophage anti-*S. enterica* defenses (Tapping *et al.*, 2000). In addition, soluble mediators produced by myeloid and lymphoid cells can modulate the response of macrophages. Of the multiple cytokines that can activate macrophage effector functions, TNFα and IFNγ have been most clearly shown to enhance macrophage activity against *S. enterica* thus playing a pivotal role in host resistance to salmonellosis.

9.5.2 Toll like receptors (TLR) activate innate host defenses in response to *S. enterica* lipopolysaccharide

Lipopolysaccharide (LPS) is present in the outer membrane of Gram-negative bacteria and its recognition triggers changes in gene expression in macrophages (Rosenberger *et al.*, 2000). A variety of PRR are recruited to the nascent phagosome to sample its cargo, but only the subset that is engaged by cognate ligands initiates the appropriate host response (Ozinsky *et al.*, 2000). TLR4 appears to be the principal receptor for *S. enterica* LPS (Tapping *et al.*, 2000).

Hydrophobic domains of the lipid A are responsible for the majority of the biological effects associated with LPS (Schletter *et al.*, 1995). Optimal recognition of *S. enterica* lipid A acyl chains by TLR4 requires a functional MD2 protein (Muroi *et al.*, 2002). In addition to lipid A acyl chains, conserved sugars of the *S. enterica* LPS core antigen provide further structures for the activation of TLR4/MD2 signaling (Muroi and Tanamoto, 2002).

A nucleotide mutation in the *lps* locus in mice results in a proline to histidine amino acid substitution at position 712 within the cytoplasmic domain of TLR4 (Poltorak *et al.*, 1998). C3H/HeJ mice carrying the *lps* susceptibility locus are hyporesponsive to LPS-induced shock and are extremely vulnerable to a variety of Gram-negative infections, including *S. enterica* (O'Brien *et al.*, 1980; Vazquez-Torres *et al.*, 2004). These data suggest that TLR4 is important in resistance to salmonellosis. A role for this PRR in resistance to *S. enterica* is further supported by the enhanced resistance of TLR4 transgenic mice

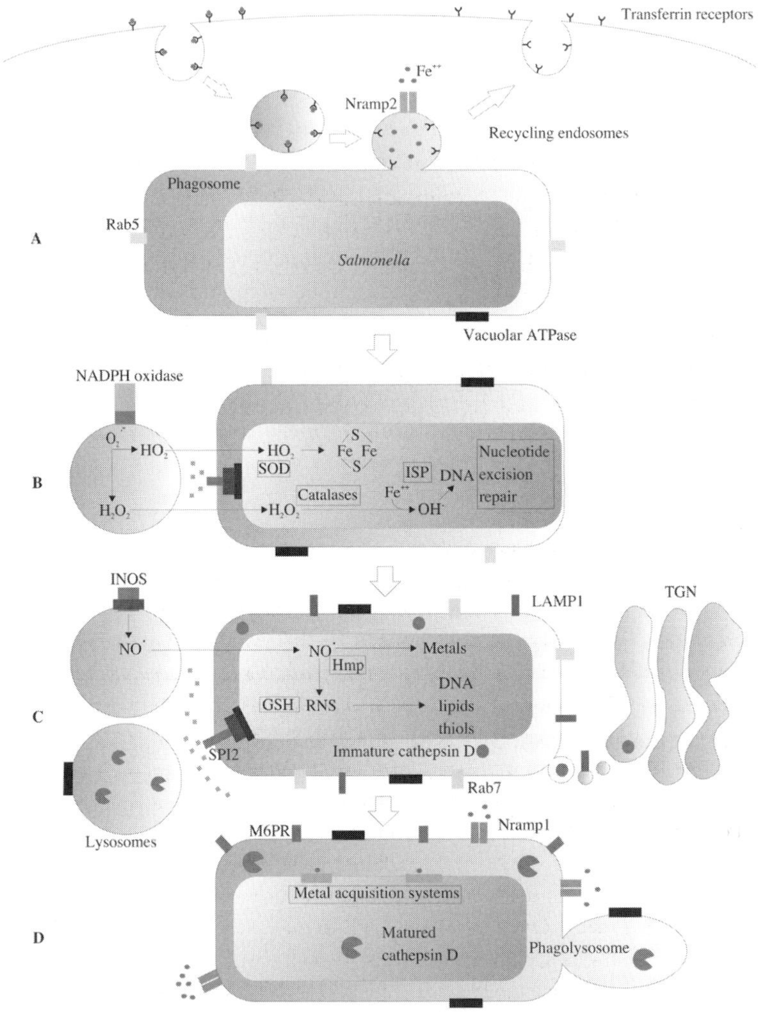

Figure 9.1. A model of the bidirectional interactions between *S. enterica* virulence factors and phagocytic host defenses. (See colour plate section.)

(A) *S. enterica* is internalized by professional phagocytes into a membrane-bound vacuole called the phagosome. The young phagosome expresses the small GTPase Rab5 that allows this vesicle to maintain a dynamic relationship with early endosomes containing transferrin (pink small spheres in figure 9.1 in color plate section). In the acidic environment of recycling endosomes, transferrin is released from its receptor and iron is pumped out to the cytosol by the Nramp2 transporter. Shortly after its formation, the *S. enterica* phagosome associates with vacuolar ATPases (black rectangles) that acidify the lumen. (B) Uptake of *S. enterica* activates the assembly of membrane-bound and cytosolic components of the NADPH oxidase in small vesicles. Effectors (green squares in to salmonellosis (Bihl *et al.*, 2003). Furthermore, some TLR4 allelic variants have been linked to increased susceptibility to *S. enterica* in chickens (Leveque *et al.*, 2003). Immunodeficient strains of mice have revealed that complexes of TLR4 with LPS-binding protein, CD14 and MD2 not only mediate the recognition of *S. enterica* LPS, but also play a critical role in resistance to this intracellular pathogen (Bernheiden *et al.*, 2001; Nagai *et al.*, 2002). The contribution of TLR4 to the host response to *S. enterica* is partially mediated by its effect on macrophages (O'Brien *et al.*, 1982). TLR4 signaling activates the translocation of NF-kB heterodimers to the nucleus (Dobrovolskaia and Vogel, 2002), and stimulation of the expression of anti-*S. enterica* genes such as iNOS and TNFα (Royle *et al.*, 2003; Vazquez-Torres *et al.*, 2004).

The role of other TLR in resistance to *S. enterica* infections is less well defined. TLR2 responds to bacterial proteins (Aliprantis *et al.*, 1999; Brightbill

color plate) of the type III secretion system encoded within the *S. enterica* pathogenicity island 2 (SPI-2) reduce the oxidative stress that *S. enterica* has to endure by blocking fusion of NADPH oxidase-containing vesicles with the phagosome. Some membrane soluble ROS such as hydroperoxyl (HO_2) and hydrogen peroxide (H_2O_2) however gain access to *S. enterica*, oxidizing iron sulfur clusters, proteins and DNA. *S. enterica* protects vital targets from the cytotoxicity associated with the NADPH oxidase by expressing an array of superoxide dismutases (SOD), catalases and scavengers such as reduced glutathione (GSH). In addition, this intracellular pathogen stores iron in an inactive form inside iron storage proteins (ISP) to prevent formation of hydroxyl radical (OH) from hydrogen peroxide, and repairs damage incurred to the DNA by nucleotide excision repair enzymes. (C) Recognition of LPS on the surface of *S. enterica* by TLR4 induces the expression of iNOS. Analogous to functional NADPH oxidase complexes, iNOS-containing vesicles are excluded from the *S. enterica* phagosome by SPI-2 effectors. NO however diffuses freely across membranes. Reactive nitrogen species (RNS) such as NO, dinitrogen trioxide, nitrogen dioxide, nitrosothiols and dinitrosyl iron complexes exert their toxicity by reacting with metal cofactors, DNA, lipids and thiol groups. *S. enterica* detoxifies NO via the enzymatic activity of the Hmp flavohaemoprotein, and uses a battery of scavengers such as GSH for the detoxification of a variety of RNS. The SPI-2 type III system prevents fusion of the *S. enterica* phagosome with lysosomes along the degradative pathway. The *S. enterica* phagosome nevertheless maintains a dynamic interaction with late endosomes and the trans Golgi network (TGN), acquiring immature cathepsin D and lysosomal-associated membrane glycoproteins (LAMP). (D) Acquisition of the Nramp1 divalent metal transporter reduces the intraphagosomal concentration of iron and manganese. *S. enterica* expresses a battery of metal acquisition systems that compete with the host for divalent metals such as iron and manganese. The Nramp1-associated diminution in intraphagosomal iron concentration hastens the maturation of the *S. enterica* phagosome, promoting its interaction with early endosomal markers, mature hydrolases and the mannose 6-phosphate receptor (M6PR).

et al., 1999; Hirschfeld *et al.*, 2000) and peptidoglycan (Flo *et al.*, 2000; Schwandner *et al.*, 1999; Takeuchi *et al.*, 2000a, b; Yoshimura *et al.*, 1999) and activation of this receptor has principally been associated with recognition of Gram-positive bacterial ligands. TLR2 is up-regulated during *Salmonella* infection in vivo (Totemeyer *et al.*, 2003) and is involved in triggering cytokine production from cultured macrophages and in mice exposed to *Salmonella*. In fact, TNFα and IL6 responses are reduced in mice lacking both TLR2 and TLR4 as compared to mice lacking only TLR4 (Lembo *et al.*, 2003).

TLR9 mediates the host inflammatory response to bacterial DNA (Bauer *et al.*, 2001; Hemmi *et al.*, 2000; Takeshita *et al.*, 2001). There is no information available yet to suggest that animals or patients with a functionally defective TLR9 have altered susceptibility to *Salmonella* infection. TLR5 is involved in the host response to *Salmonella* infections by responding to bacterial flagellin protein (Hayashi *et al.*, 2001; Moors *et al.*, 2001; Sebastiani *et al.*, 2000). MOLF/Ei mice have defective TLR5 function that might be at the basis of their increased susceptibility to *Salmonella* infections (Sebastiani *et al.*, 2000). However, a polymorphism in the TLR5 gene that introduces a premature stop codon (TLR5(392stop)) is not associated with susceptibility to typhoid fever in humans (Dunstan *et al.*, 2005).

9.5.3 Activation of macrophage anti-*S. enterica* activity by TNFα

Experimental evidence strongly implicates TNFα in resistance to *S. enterica* (Mastroeni *et al.*, 1991; Mastroeni *et al.*, 1995; Nakano *et al.*, 1990; Tite *et al.*, 1991; Vazquez-Torres *et al.*, 2001). TNFα affects multiple arms of the innate and the acquired immune response to *S. enterica* (Everest *et al.*, 1998; Nauciel and Espinasse-Maes, 1992; Tite *et al.*, 1991). Initially, TNFα was shown to be a critical factor for the formation of macrophage-rich granulomas in salmonellosis (Mastroeni *et al.*, 1995). Subsequent studies have shown that TNFα also enhances the anti-*S. enterica* activity of macrophages (Ables *et al.*, 2001; Vazquez-Torres *et al.*, 2001). TNFα/TNFp55 receptor signaling is essential for the effective localization of NADPH oxidase-harboring vesicles to the vicinity of the *S. enterica* phagosome (Vazquez-Torres *et al.*, 2001). In addition, TNFα signaling through its p55 surface receptor plays a critical role in the induction of high NO synthesis (Ables *et al.*, 2001; Everest *et al.*, 1998).

9.5.4 Activation of macrophage anti-*S. enterica* activity by IFNγ

The notion that IFNγ plays a critical role in host resistance to *S. enterica* is conclusively supported by the fact that mutations in the IFNγ signaling cascade predispose humans and experimental animals to salmonellosis (de Jong *et al.*, 1998; Hess *et al.*, 1996). Furthermore, the administration of IFNγ reduces the severity of *S. enterica* infections and partially corrects the recurrent infections in CGD patients (Ezekowitz *et al.*, 1988; Muotiala and Makela, 1990). During the course of *S. enterica* infection, IFNγ is initially produced by NK cells, and subsequently synthesized by antigen-specific αβ T-helper 1 lymphocytes (Boehm *et al.*, 1997; Mittrucker and Kaufmann, 2000). IFNγ participates in an array of innate and acquired host defense mechanisms, which include the direct stimulation of macrophage anti-*S. enterica* activity (Boehm *et al.*, 1997; Kagaya *et al.*, 1989; Muotiala and Makela, 1993). In fact, IFNγ impairs replication of *S. enterica* in macrophages by inducing NADPH oxidase-dependent bacterial filamentation and iNOS-mediated antimicrobial activity (Rosenberger and Finlay, 2002; Vazquez-Torres *et al.*, 2000a).

9.6 CONCLUSIONS

By exploiting the intracellular niche of host cells, *S. enterica*, in part, bypasses competition for nutrients and space with other bacteria, and avoids exposure to antibiotics produced by the microflora that colonize the gastrointestinal mucosa. But living inside the intracellular environment of professional phagocytes does not come free of cost (Figure 9.1). Professional phagocytes recognize the presence of *S. enterica* and activate an array of oxygen-dependent and -independent antimicrobial defenses that range from the utilization of cytotoxic ROS and RNS, and lysosomal hydrolases, to the expression of transport systems that limit the availability of essential nutrients. *S. enterica* in turn expresses a number of virulence factors that detoxify host cytotoxic molecules or repair the damage inflicted by these defenses. In addition, *S. enterica* avoids the challenges encountered inside eukaryotic cells by diverting the trafficking of its phagosome away from the degradative endocytic pathway. A better understanding of the relation between this intracellular pathogen and its host phagocytic cells will thus shed important information on *S. enterica* pathogenesis and will provide novel strategies to harness the cellular responses that protect against this common infection.

9.7 ACKNOWLEDGEMENTS

We would like to thank Jessica Jones-Carson and Olivia Steele-Mortimer for helpful discussions on the manuscript. The work in Dr. Vazquez-Torres laboratory is supported by AI054959 from the National Institutes of Health and by the Schweppe Foundation.

REFERENCES

Ables, G. P., Takamatsu, D., Noma, H. *et al.* (2001). The roles of *Nramp1* and TNFα genes in nitric oxide production and their effect on the growth of *Salmonella typhimurium* in macrophages from *Nramp1* congenic and tumor necrosis factor-alpha$^{-/-}$ mice. *J Interferon Cytokine Res*, **21**, 53–62.

Alam, M. S., Akaike, T., Okamoto, S. *et al.* (2002). Role of nitric oxide in host defense in murine salmonellosis as a function of its antibacterial and anti-apoptotic activities. *Infect Immun*, **70**, 3130–42.

Aliprantis, A. O., Yang, R. B., Mark, M. R. *et al.* (1999). Cell activation and apoptosis by bacterial lipoproteins through Toll-like receptor-2. *Science*, **285**, 736–9.

Alpuche Aranda, C. M., Swanson, J. A., Loomis, W. P. and Miller, S. I. (1992). *Salmonella typhimurium* activates virulence gene transcription within acidified macrophage phagosomes. *Proc Natl Acad Sci USA*, **89**, 10079–83.

Atkinson, P. G. and Barton, C. H. (1999). High level expression of Nramp1^{G169} in RAW264.7 cell transfectants: analysis of intracellular iron transport. *Immunology*, **96**, 656–62.

Barrera, L. F., Kramnik, I., Skamene, E. and Radzioch, D. (1994). Nitrite production by macrophages derived from BCG-resistant and -susceptible congenic mouse strains in response to IFNγ and infection with BCG. *Immunology*, **82**, 457–64.

Bauer, S., Kirschning, C. J., Hacker, H. *et al.* (2001). Human TLR9 confers responsiveness to bacterial DNA via species-specific CpG motif recognition. *Proc Natl Acad Sci USA*, **98**, 9237–42.

Bernheiden, M., Heinrich, J. M., Minigo, G. *et al.* (2001). LBP, CD14, TLR4 and the murine innate immune response to a peritoneal *Salmonella* infection. *J Endotoxin Res*, **7**, 447–50.

Beuzon, C. R., Banks, G., Deiwick, J., Hensel, M. and Holden, D. W. (1999). pH-dependent secretion of SseB, a product of the SPI-2 type III secretion system of *Salmonella typhimurium*. *Mol Microbiol*, **33**, 806–16.

Bihl, F., Salez, L., Beaubier, M. *et al.* (2003). Overexpression of Toll-like receptor 4 amplifies the host response to lipopolysaccharide and provides a survival advantage in transgenic mice. *J Immunol*, **170**, 6141–50.

Boehm, U., Klamp, T., Groot, M. and Howard, J. C. (1997). Cellular responses to interferon-gamma. *Annu Rev Immunol*, **15**, 749–95.

Brightbill, H. D., Libraty, D. H., Krutzik, S. R. *et al.* (1999). Host defense mechanisms triggered by microbial lipoproteins through Toll-like receptors. *Science*, **285**, 732–6.

Bryk, R., Griffin, P. and Nathan, C. (2000). Peroxynitrite reductase activity of bacterial peroxiredoxins. *Nature*, **407**, 211–15.

Buchmeier, N. A. and Heffron, F. (1991). Inhibition of macrophage phagosome–lysosome fusion by *Salmonella typhimurium*. *Infect Immun*, **59**, 2232–8.

Buchmeier, N. A., Libby, S. J., Xu, Y. *et al.* (1995). DNA repair is more important than catalase for *Salmonella* virulence in mice. *J Clin Invest*, **95**, 1047–53.

Chakravortty, D., Hansen-Wester, I. and Hensel, M. (2002). *Salmonella* pathogenicity island 2 mediates protection of intracellular *Salmonella* from reactive nitrogen intermediates. *J Exp Med*, **195**, 1155–66.

Chateau, M. T. and Caravano, R. (1997). The oxidative burst triggered by *Salmonella typhimurium* in differentiated U937 cells requires complement and a complete bacterial lipopolysaccharide. *FEMS Immunol Med Microbiol*, **17**, 57–66.

Cirillo, D. M., Valdivia, R. H., Monack, D. M. and Falkow, S. (1998). Macrophage-dependent induction of the *Salmonella* pathogenicity island 2 type III secretion system and its role in intracellular survival. *Mol Microbiol*, **30**, 175–88.

Crawford, M. J. and Goldberg, D. E. (1998). Regulation of the *Salmonella typhimurium* flavohemoglobin gene. A new pathway for bacterial gene expression in response to nitric oxide. *J Biol Chem*, **273**, 34028–32.

Cuellar-Mata, P., Jabado, N., Liu, J. *et al.* (2002). Nramp1 modifies the fusion of *Salmonella typhimurium*-containing vacuoles with cellular endomembranes in macrophages. *J Biol Chem*, **277**, 2258–65.

De Groote, M. A., Ochsner, U. A., Shiloh, M. U. *et al.* (1997). Periplasmic superoxide dismutase protects *Salmonella* from products of phagocyte NADPH-oxidase and nitric oxide synthase. *Proc Natl Acad Sci USA*, **94**, 13997–4001.

De Groote, M. A., Testerman, T., Xu, Y., Stauffer, G. and Fang, F. C. (1996). Homocysteine antagonism of nitric oxide-related cytostasis in *Salmonella typhimurium*. *Science*, **272**, 414–17.

de Jong, R., Altare, F., Haagen *et al.* (1998). Severe mycobacterial and *Salmonella* infections in interleukin-12 receptor-deficient patients. *Science*, **280**, 1435–8.

Dobrovolskaia, M. A. and Vogel, S. N. (2002). Toll receptors, CD14, and macrophage activation and deactivation by LPS. *Microbes Infect*, **4**, 903–14.

Dunstan, S. J., Ho, V. A., Duc, C. M. *et al.* (2001). Typhoid fever and genetic polymorphisms at the natural resistance-associated macrophage protein 1. *J Infect Dis*, **183**, 1156–60.

Dunstan, S. J., Hawn, T. R., Hue, N. T. *et al.* (2005). Host susceptibility an clinical outcomes in Toll-like receptor 5-deficient patients with typhoid fever in Vietnam. *J Infect Dis*, **191**, 1068–71.

Eriksson, S., Lucchini, S., Thompson, A., Rhen, M. and Hinton, J. C. (2003). Unravelling the biology of macrophage infection by gene expression profiling of intracellular *Salmonella enterica*. *Mol Microbiol*, **47**, 103–18.

Everest, P., Roberts, M. and Dougan, G. (1998). Susceptibility to *Salmonella typhimurium* infection and effectiveness of vaccination in mice deficient in the tumor necrosis factor alpha p55 receptor. *Infect Immun*, **66**, 3355–64.

Ezekowitz, R. A., Dinauer, M. C., Jaffe, H. S., Orkin, S. H. and Newburger, P. E. (1988). Partial correction of the phagocyte defect in patients with X-linked chronic granulomatous disease by subcutaneous interferon gamma. *N Engl J Med*, **319**, 146–51.

Fields, P. I., Swanson, R. V., Haidaris, C. G. and Heffron, F. (1986). Mutants of *Salmonella typhimurium* that cannot survive within the macrophage are avirulent. *Proc Natl Acad Sci USA*, **83**, 5189–93.

Flo, T. H., Halaas, O., Lien, E. *et al.* (2000). Human Toll-like receptor 2 mediates monocyte activation by *Listeria monocytogenes*, but not by group B streptococci or lipopolysaccharide. *J Immunol*, **164**, 2064–9.

Forbes, J. R. and Gros, P. (2003). Iron, manganese, and cobalt transport by Nramp1 (Slc11a1) and Nramp2 (Slc11a2) expressed at the plasma membrane. *Blood*, **102**, 1884–92.

Fritsche, G., Dlaska, M., Barton, H. *et al.* (2003). Nramp1 functionality increases inducible nitric oxide synthase transcription via stimulation of IFNγ regulatory factor 1 expression. *J Immunol*, **171**, 1994–8.

Gallois, A., Klein, J. R., Allen, L. A., Jones, B. D. and Nauseef, W. M. (2001). *Salmonella* pathogenicity island 2-encoded type III secretion system mediates exclusion of NADPH oxidase assembly from the phagosomal membrane. *J Immunol*, **166**, 5741–8.

Garcia-del Portillo, F. and Finlay, B. B. (1995). Targeting of *Salmonella typhimurium* to vesicles containing lysosomal membrane glycoproteins bypasses compartments with mannose 6-phosphate receptors. *J Cell Biol*, **129**, 81–97.

Garmendia, J., Beuzon, C. R., Ruiz-Albert, J. and Holden, D. W. (2003). The roles of SsrA-SsrB and OmpR-EnvZ in the regulation of genes encoding the *Salmonella typhimurium* SPI-2 type III secretion system. *Microbiology*, **149**, 2385–96.

Garvis, S. G., Beuzon, C. R. and Holden, D. W. (2001). A role for the PhoP/Q regulon in inhibition of fusion between lysosomes and *Salmonella*-containing vacuoles in macrophages. *Cell Microbiol*, **3**, 731–44.

Goswami, T., Bhattacharjee, A., Babal, P. *et al.* (2001). Natural-resistance-associated macrophage protein 1 is an H^+/bivalent cation antiporter. *Biochem J*, **354**, 511–19.

Groisman, E. A. and Saier, M. H., Jr. (1990). *Salmonella* virulence: new clues to intramacrophage survival. *Trends Biochem Sci*, **15**, 30–3.

Gruenheid, S., Pinner, E., Desjardins, M. and Gros, P. (1997). Natural resistance to infection with intracellular pathogens: the Nramp1 protein is recruited to the membrane of the phagosome. *J Exp Med*, **185**, 717–30.

Hantke, K. (1997). Ferrous iron uptake by a magnesium transport system is toxic for *Escherichia coli* and *Salmonella typhimurium*. *J Bacteriol*, **179**, 6201–4.

Hashim, S., Mukherjee, K., Raje, M., Basu, S. K. and Mukhopadhyay, A. (2000). Live *Salmonella* modulate expression of Rab proteins to persist in a specialized compartment and escape transport to lysosomes. *J Biol Chem*, **275**, 16281–8.

Hayashi, F., Smith, K. D., Ozinsky, A. *et al.* (2001). The innate immune response to bacterial flagellin is mediated by Toll-like receptor 5. *Nature*, **410**, 1099–103.

Hemmi, H., Takeuchi, O., Kawai, T. *et al.* (2000). A Toll-like receptor recognizes bacterial DNA. *Nature*, **408**, 740–5.

Hess, J., Ladel, C., Miko, D. and Kaufmann, S. H. (1996). *Salmonella typhimurium aroA*$^-$ infection in gene-targeted immunodeficient mice: major role of CD4$^+$ TCR-alpha beta cells and IFNγ in bacterial clearance independent of intracellular location. *J Immunol*, **156**, 3321–6.

Hirschfeld, M., Ma, Y., Weis, J. H., Vogel, S. N. and Weis, J. J. (2000). Cutting edge: repurification of lipopolysaccharide eliminates signaling through both human and murine Toll-like receptor 2. *J Immunol*, **165**, 618–22.

Hormaeche, C. E. (1979). Genetics of natural resistance to *Salmonella* in mice *Immunology*, **37**, 319–27.

Imlay, J. A. (2003). Pathways of oxidative damage. *Annu Rev Microbiol*, **57**, 395–418.

Imlay, J. A. and Linn, S. (1986). Bimodal pattern of killing of DNA-repair-defective or anoxically grown *Escherichia coli* by hydrogen peroxide. *J Bacteriol*, **166**, 519–27.

Jabado, N., Cuellar-Mata, P., Grinstein, S. and Gros, P. (2003). Iron chelators modulate the fusogenic properties of *Salmonella*-containing phagosomes. *Proc Natl Acad Sci USA*, **100**, 6127–32.

Jabado, N., Jankowski, A., Dougaparsad, S. *et al.* (2000). Natural resistance to intracellular infections: natural resistance-associated macrophage protein 1 (Nramp1) functions as a pH-dependent manganese transporter at the phagosomal membrane. *J Exp Med*, **192**, 1237–48.

Kagaya, K., Watanabe, K. and Fukazawa, Y. (1989). Capacity of recombinant gamma interferon to activate macrophages for *Salmonella*-killing activity. *Infect Immun*, **57**, 609–15.

Kehres, D. G., Janakiraman, A., Slauch, J. M. and Maguire, M. E. (2002). SitABCD is the alkaline Mn^{2+} transporter of *Salmonella enterica* serovar Typhimurium. *J Bacteriol*, **184**, 3159–66.

Kuhn, D. E., Baker, B. D., Lafuse, W. P. and Zwilling, B. S. (1999). Differential iron transport into phagosomes isolated from the RAW264.7 macrophage cell lines transfected with $Nramp1^{Gly169}$ or $Nramp1^{Asp169}$. *J Leukoc Biol*, **66**, 113–19.

Lafuse, W. P., Alvarez, G. R. and Zwilling, B. S. (2002). Role of MAP kinase activation in *Nramp1* mRNA stability in RAW264.7 macrophages expressing $Nramp1^{Gly169}$. *Cell Immunol*, **215**, 195–206.

Lalmanach, A. C., Montagne, A., Menanteau, P. and Lantier, F. (2001). Effect of the mouse *Nramp1* genotype on the expression of IFNγ gene in early response to *Salmonella* infection. *Microbes Infect*, **3**, 639–44.

Lembo, A., Kalis, C., Kirschning, C. J. *et al.* (2003). Differential contribution of Toll-like receptors 4 and 2 to the cytokine response to *Salmonella enterica* serovar Typhimurium and *Staphylococcus aureus* in mice. *Infect Immun*, **71**, 6058–62.

Leveque, G., Forgetta, V., Morroll, S. *et al.* (2003). Allelic variation in TLR4 is linked to susceptibility to *Salmonella enterica* serovar Typhimurium infection in chickens. *Infect Immun*, **71**, 1116–24.

Lundberg, B. E., Wolf, R. E., Jr., Dinauer, M. C., Xu, Y. and Fang, F. C. (1999). Glucose 6-phosphate dehydrogenase is required for *Salmonella typhimurium* virulence and resistance to reactive oxygen and nitrogen intermediates. *Infect Immun*, **67**, 436–8.

Mastroeni, P., Arena, A., Costa, G. B. *et al.* (1991). Serum TNFα in mouse typhoid and enhancement of a *Salmonella* infection by anti-TNFα antibodies. *Microb Pathog*, **11**, 33–8.

Mastroeni, P., Skepper, J. N. and Hormaeche, C. E. (1995). Effect of anti-tumor necrosis factor alpha antibodies on histopathology of primary *Salmonella* infections. *Infect Immun*, **63**, 3674–82.

Mastroeni, P., Vazquez-Torres, A., Fang, F. C. *et al.* (2000). Antimicrobial actions of the NADPH phagocyte oxidase and inducible nitric oxide synthase in experimental salmonellosis. II. Effects on microbial proliferation and host survival *in vivo*. *J Exp Med*, **192**, 237–48.

Meresse, S., Steele-Mortimer, O., Finlay, B. B. and Gorvel, J. P. (1999). The rab7 GTPase controls the maturation of *Salmonella typhimurium*-containing vacuoles in HeLa cells. *Embo J*, **18**, 4394–403.

Mittrucker, H. W. and Kaufmann, S. H. (2000). Immune response to infection with *Salmonella typhimurium* in mice. *J Leukoc Biol*, **67**, 457–63.

Moors, M. A., Li, L. and Mizel, S. B. (2001). Activation of interleukin-1 receptor-associated kinase by gram-negative flagellin. *Infect Immun*, **69**, 4424–9.

Mukherjee, K., Siddiqi, S. A., Hashim, S. *et al.* (2000). Live *Salmonella* recruits N-ethylmaleimide-sensitive fusion protein on phagosomal membrane and promotes fusion with early endosome. *J Cell Biol*, **148**, 741–53.

Muotiala, A. and Makela, P. H. (1990). The role of IFNγ in murine *Salmonella typhimurium* infection. *Microb Pathog*, **8**, 135–41.

(1993). Role of gamma interferon in late stages of murine salmonellosis. *Infect Immun*, **61**, 4248–53.

Muroi, M. and Tanamoto, K. (2002). The polysaccharide portion plays an indispensable role in *Salmonella* lipopolysaccharide-induced activation of NF-κB through human Toll-like receptor 4. *Infect Immun*, **70**, 6043–7.

Muroi, M., Ohnishi, T. and Tanamoto, K. (2002). MD-2, a novel accessory molecule, is involved in species-specific actions of *Salmonella* lipid A. *Infect Immun*, **70**, 3546–50.

Nagai, Y., Akashi, S., Nagafuku, M. *et al.* (2002). Essential role of MD-2 in LPS responsiveness and TLR4 distribution. *Nat Immunol*, **3**, 667–72.

Nakano, Y., Onozuka, K., Terada, Y., Shinomiya, H. and Nakano, M. (1990). Protective effect of recombinant tumor necrosis factor-alpha in murine salmonellosis. *J Immunol*, **144**, 1935–41.

Nauciel, C. and Espinasse-Maes, F. (1992). Role of gamma interferon and tumor necrosis factor alpha in resistance to *Salmonella typhimurium* infection. *Infect Immun*, **60**, 450–4.

O'Brien, A. D., Metcalf, E. S. and Rosenstreich, D. L. (1982). Defect in macrophage effector function confers *Salmonella typhimurium* susceptibility on C3H/HeJ mice. *Cell Immunol*, **67**, 325–33.

O'Brien, A. D., Rosenstreich, D. L., Scher, I. *et al.* (1980). Genetic control of susceptibility to *Salmonella typhimurium* in mice: role of the LPS gene. *J Immunol*, **124**, 20–4.

Oh, Y. K., Alpuche-Aranda, C., Berthiaume, E. *et al.* (1996). Rapid and complete fusion of macrophage lysosomes with phagosomes containing *Salmonella typhimurium*. *Infect Immun*, **64**, 3877–83.

Ozinsky, A., Underhill, D. M., Fontenot, J. D. *et al.* (2000). The repertoire for pattern recognition of pathogens by the innate immune system is defined by cooperation between Toll-like receptors. *Proc Natl Acad Sci USA*, **97**, 13766–71.

Plant, J. and Glynn, A. A. (1976). Genetics of resistance to infection with *Salmonella typhimurium* in mice. *J Infect Dis*, **133**, 72–8.

Poltorak, A., He, X., Smirnova, I. *et al.* (1998). Defective LPS signaling in C3H/HeJ and C57BL/10ScCr mice: mutations in *tlr4* gene. *Science*, **282**, 2085–8.

Rathman, M., Barker, L. P. and Falkow, S. (1997). The unique trafficking pattern of *Salmonella typhimurium*-containing phagosomes in murine macrophages is independent of the mechanism of bacterial entry. *Infect Immun*, **65**, 1475–85.

Rathman, M., Sjaastad, M. D. and Falkow, S. (1996). Acidification of phagosomes containing *Salmonella typhimurium* in murine macrophages. *Infect Immun*, **64**, 2765–73.

Richter-Dahlfors, A., Buchan, A. M. J. and Finlay, B. B. (1997). Murine salmonellosis studied by confocal microscopy: *Salmonella typhimurium* resides intracellularly inside macrophages and exerts a cytotoxic effect on phagocytes *in vivo*. *J Exp Med*, **186**, 569–80.

Rosenberger, C. M. and Finlay, B. B. (2002). Macrophages inhibit *Salmonella typhimurium* replication through MEK/ERK kinase and phagocyte NADPH oxidase activities. *J Biol Chem*, **277**, 18753–62.

Rosenberger, C. M., Scott, M. G., Gold, M. R., Hancock, R. E. and Finlay, B. B. (2000). *Salmonella typhimurium* infection and lipopolysaccharide stimulation induce similar changes in macrophage gene expression. *J Immunol*, **164**, 5894–904.

Royle, M. C., Totemeyer, S., Alldridge, L. C., Maskell, D. J. and Bryant, C. E. (2003). Stimulation of Toll-like receptor 4 by lipopolysaccharide during cellular invasion by live *Salmonella typhimurium* is a critical but not exclusive event leading to macrophage responses. *J Immunol*, **170**, 5445–54.

Schapiro, J. M., Libby, S. J. and Fang, F. C. (2003). Inhibition of bacterial DNA replication by zinc mobilization during nitrosative stress. *Proc Natl Acad Sci USA*, **100**, 8496–501.

Schletter, J., Heine, H., Ulmer, A. J. and Rietschel, E. T. (1995). Molecular mechanisms of endotoxin activity. *Arch Microbiol*, **164**, 383–9.

Schwandner, R., Dziarski, R., Wesche, H., Rothe, M. and Kirschning, C. J. (1999). Peptidoglycan- and lipoteichoic acid-induced cell activation is mediated by Toll-like receptor 2. *J Biol Chem*, **274**, 17406–9.

Sebastiani, G., Leveque, G., Lariviere, L. *et al.* (2000). Cloning and characterization of the murine Toll-like receptor 5 (*tlr5*) gene: sequence and mRNA expression studies in *Salmonella*-susceptible MOLF/Ei mice. *Genomics*, **64**, 230–40.

Steele-Mortimer, O., Meresse, S., Gorvel, J. P., Toh, B. H. and Finlay, B. B. (1999). Biogenesis of *Salmonella typhimurium*-containing vacuoles in epithelial cells involves interactions with the early endocytic pathway. *Cell Microbiol*, **1**, 33–49.

Steele-Mortimer, O., St-Louis, M., Olivier, M. and Finlay, B. B. (2000). Vacuole acidification is not required for survival of *Salmonella enterica* serovar typhimurium within cultured macrophages and epithelial cells. *Infect Immun*, **68**, 5401–4.

Stevanin, T. M., Poole, R. K., Demoncheaux, E. A. and Read, R. C. (2002). Flavohemoglobin Hmp protects *Salmonella enterica* serovar Typhimurium from nitric oxide-related killing by human macrophages. *Infect Immun*, **70**, 4399–405.

Suvarnapunya, A. E., Lagasse, H. A. and Stein, M. A. (2003). The role of DNA base excision repair in the pathogenesis of *Salmonella enterica* serovar Typhimurium. *Mol Microbiol*, **48**, 549–59.

Swanson, R. N. and O'Brien, A. D. (1983). Genetic control of the innate resistance of mice to *Salmonella typhimurium: Ity* gene is expressed in vivo by 24 hours after infection. *J Immunol*, **131**, 3014–20.

Takeshita, F., Leifer, C. A., Gursel, I. *et al.* (2001). Cutting edge: role of Toll-like receptor 9 in CpG DNA-induced activation of human cells. *J Immunol*, **167**, 3555–8.

Takeuchi, O., Hoshino, K. and Akira, S. (2000). Cutting edge: TLR2-deficient and MyD88-deficient mice are highly susceptible to *Staphylococcus aureus* infection. *J Immunol*, **165**, 5392–6.

Takeuchi, O., Takeda, K., Hoshino, K. *et al.* (2000). Cellular responses to bacterial cell wall components are mediated through MyD88-dependent signaling cascades. *Int Immunol*, **12**, 113–17.

Tapping, R. I., Akashi, S., Miyake, K., Godowski, P. J. and Tobias, P. S. (2000). Toll-like receptor 4, but not Toll-like receptor 2, is a signaling receptor for *Escherichia* and *Salmonella* lipopolysaccharides. *J Immunol*, **165**, 5780–7.

Tite, J. P., Dougan, G. and Chatfield, S. N. (1991). The involvement of tumor necrosis factor in immunity to *Salmonella* infection. *J Immunol*, **147**, 3161–4.

Totemeyer, S., Foster, N., Kaiser, P., Maskell, D. J. and Bryant, C. E. (2003). Toll-like receptor expression in C3H/HeN and C3H/HeJ mice during *Salmonella enterica* serovar Typhimurium infection. *Infect Immun*, **71**, 6653–7.

Tsolis, R. M., Baumler, A. J. and Heffron, F. (1995). Role of *Salmonella typhimurium* Mn-superoxide dismutase (SodA) in protection against early killing by J774 macrophages. *Infect Immun*, **63**, 1739–44.

Tsolis, R. M., Baumler, A. J., Heffron, F. and Stojiljkovic, I. (1996). Contribution of TonB- and Feo-mediated iron uptake to growth of *Salmonella typhimurium* in the mouse. *Infect Immun*, **64**, 4549–56.

Uchiya, K., Barbieri, M. A., Funato, K. *et al.* (1999). A *Salmonella* virulence protein that inhibits cellular trafficking. *Embo J*, **18**, 3924–33.

van der Straaten, T., van Diepen, A., Kwappenberg, K. *et al.* (2001). Novel *Salmonella enterica* serovar Typhimurium protein that is indispensable for virulence and intracellular replication. *Infect Immun*, **69**, 7413–18.

Vazquez-Torres, A. and Fang, F. C. (2001a). Oxygen-dependent anti-*Salmonella* activity of macrophages. *Trends Microbiol*, **9**, 29–33.

(2001b). *Salmonella* evasion of the NADPH phagocyte oxidase. *Microbes Infect,* **3**, 1313–20.

Vazquez-Torres, A., Fantuzzi, G., Edwards, C. K. R., Dinarello, C. A. and Fang, F. C. (2001). Defective localization of the NADPH phagocyte oxidase to *Salmonella*-containing phagosomes in tumor necrosis factor p55 receptor-deficient macrophages. *Proc Natl Acad Sci USA,* **98**, 2561–5.

Vazquez-Torres, A., Jones-Carson, J., Mastroeni, P., Ischiropoulos, H. and Fang, F. C. (2000a). Antimicrobial actions of the NADPH phagocyte oxidase and inducible nitric oxide synthase in experimental salmonellosis. I. Effects on microbial killing by activated peritoneal macrophages in vitro. *J Exp Med,* **192**, 227–36.

Vazquez-Torres, A., Vallance, B. A., Bergman, M. A. *et al.* (2004). Toll-like receptor 4 dependence of innate and adaptive immunity to *Salmonella*: importance of the Kupffer cell network. *J Immunol,* **172**, 6202–8.

Vazquez-Torres, A., Xu, Y., Jones-Carson, J. *et al.* (2000b). *Salmonella* pathogenicity island 2-dependent evasion of the phagocyte NADPH oxidase. *Science,* **287**, 1655–8.

Vidal, S., Tremblay, M. L., Govoni, G. *et al.* (1995). The *Ity/Lsh/Bcg* locus: natural resistance to infection with intracellular parasites is abrogated by disruption of the *Nramp1* gene. *J Exp Med,* **182**, 655–66.

Vidal, S. M., Malo, D., Vogan, K., Skamene, E. and Gros, P. (1993). Natural resistance to infection with intracellular parasites: isolation of a candidate for *Bcg Cell,* **73**, 469–85.

Vieira, O. V., Botelho, R. J. and Grinstein, S. (2002). Phagosome maturation: aging gracefully. *Biochem J,* **366**, 689–704.

Webb, J. L., Harvey, M. W., Holden, D. W. and Evans, T. J. (2001). Macrophage nitric oxide synthase associates with cortical actin but is not recruited to phagosomes. *Infect Immun,* **69**, 6391–400.

Yoshimura, A., Lien, E., Ingalls, R. R., Tuomanen, E., Dziarski, R. and Golenbock, D. (1999). Cutting edge: recognition of Gram-positive bacterial cell wall components by the innate immune system occurs via Toll-like receptor 2. *J Immunol,* **163**, 1–5.

Zaharik, M. L., Vallance, B. A., Puente, J. L., Gros, P. and Finlay, B. B. (2002). Host–pathogen interactions: host resistance factor Nramp1 up-regulates the expression of *Salmonella* pathogenicity island-2 virulence genes. *Proc Natl Acad Sci USA,* **99**, 15705–10.

Zwilling, B. S., Kuhn, D. E., Wikoff, L., Brown, D. and Lafuse, W. (1999). Role of iron in Nramp1-mediated inhibition of mycobacterial growth. *Infect Immun,* **67**, 1386–92.

CHAPTER 10

Interactions between *Salmonella* and dendritic cells: what happens along the way?

Cecilia Johansson, Malin Sundquist and Mary Jo Wick

10.1 INTRODUCTION

Dendritic cells (DC) are efficient antigen-presenting cells and are likely to be involved in the initiation of T-cell responses to *Salmonella*. However, it is not known what type of DC initiate immune responses to *Salmonella* or where this initiation takes place. Studies on interactions between *Salmonella* and DC are emerging and are shedding light on this topic. This chapter will review how *Salmonella* interacts with DC, following the course the bacteria take after oral infection. One of the earliest sites of *Salmonella* replication is within the Peyer's patches of the gut. Thereafter, *Salmonella* can be found in the gut-draining mesenteric lymph nodes. After systemic release of bacteria or bacteria-containing cells, *Salmonella* spread to the spleen and liver and replicate further. The relevance of the interactions between *Salmonella* and DC in these organs for initiating antibacterial T-cell responses is discussed. This is preceded by a brief overview of the biology of DC.

10.2 DENDRITIC CELLS

DC originate from precursors in the bone marrow and were named because of their morphology having long, branched dendrites (Steinman and Cohn, 1973; Steinman *et al.*, 1974). DC are widely distributed in lymphoid as well as non-lymphoid tissues (Steptoe *et al.*, 2000; Vremec and Shortman, 1997; Steiniger *et al.*, 1984). DC are often defined as $CD11c^+$ (p150/95 integrin) cells, particularly in mice, and can be divided into two main subtypes originally referred to as myeloid and lymphoid DC on the basis of their

'Salmonella' Infections: Clinical, Immunological and Molecular Aspects, ed. Pietro Mastroeni and Duncan Maskell. Published by Cambridge University Press.

expression of CD11b and CD8α, respectively (Vremec and Shortman, 1997; Wu *et al.*, 1996; Inaba *et al.*, 1993). DC expressing high levels of CD11c have been further divided into CD8α^+CD4$^-$CD11b^{lo} (herein referred to as CD8α^+), CD8α^-CD4$^+$CD11b^{hi} (referred to as CD4$^+$) and CD8α^-CD4$^-$CD11b^{hi} (referred to as CD8α^-CD4$^-$) subsets (Shortman and Liu, 2002; Vremec *et al.*, 2000). A fourth CD11c^{hi} subset, CD8α^-CD4$^-$CD11b^{lo} DC, has also been identified in Peyer's patches, mesenteric lymph nodes and liver (Johansson and Wick, 2004; Iwasaki and Kelsall, 2001). These different DC subsets develop independently and all subsets can develop into mature DC (Ardavin, 2003; Iwasaki and Kelsall, 2001; Kamath *et al.*, 2000). In addition, cells expressing intermediate levels of CD11c and expressing B220 and Gr-1 but lacking CD11b, so-called plasmacytoid DC, are present in lymphoid tissues (Asselin-Paturel *et al.*, 2001; Nakano *et al.*, 2001; Björck, 2001). These cells are particularly important during viral infections.

DC have the ability to process and present antigens in association with major histocompatibility complex (MHC) class I and class II molecules. DC are the most efficient stimulators of naïve T-cells due to their ability to migrate from peripheral sites to secondary lymphoid organs and provide a high level of costimulation (Banchereau and Steinman, 1998). The functional activities of DC depend on their state of maturation. An immature DC, which has not yet encountered antigen or received signals from pro-inflammatory cytokines, is very efficient at capturing and internalising antigens. Signals that induce DC maturation include tumor necrosis factor α (TNFα), interleukin 1β (IL1-β), microbial products such as lipopolysaccharide (LPS) and unmethylated CpG DNA sequences as well as immune complexes (Regnault *et al.*, 1999; Sparwasser *et al.*, 1998; Winzler *et al.*, 1997; Sallusto *et al.*, 1995). Phenotypic and functional changes that occur during DC maturation include increased surface expression of MHC molecules, as well as adhesion and costimulatory molecules including CD80, CD86 and CD40. Production of cytokines and chemokines, chemokine receptor expression, and migration to lymphoid tissues are also modified during DC maturation.

DC have been studied extensively and different functions, such as T-cell stimulatory capacity, cytokine production and the ability to internalize antigens have been suggested for the different subsets found in various organs (Moser and Murphy, 2000). In addition, the different subsets of DC are distinctly localized in lymphoid organs (Iwasaki and Kelsall, 2001; Kirby *et al.*, 2001; Iwasaki and Kelsall, 2000; Crowley *et al.*, 1999). DC subsets also redistribute differentially within the organ in response to stimuli such as bacteria or microbial products (Kirby *et al.*, 2001; Iwasaki and Kelsall, 2000; Reis e Sousa *et al.*, 1997; De Smedt *et al.*, 1996).

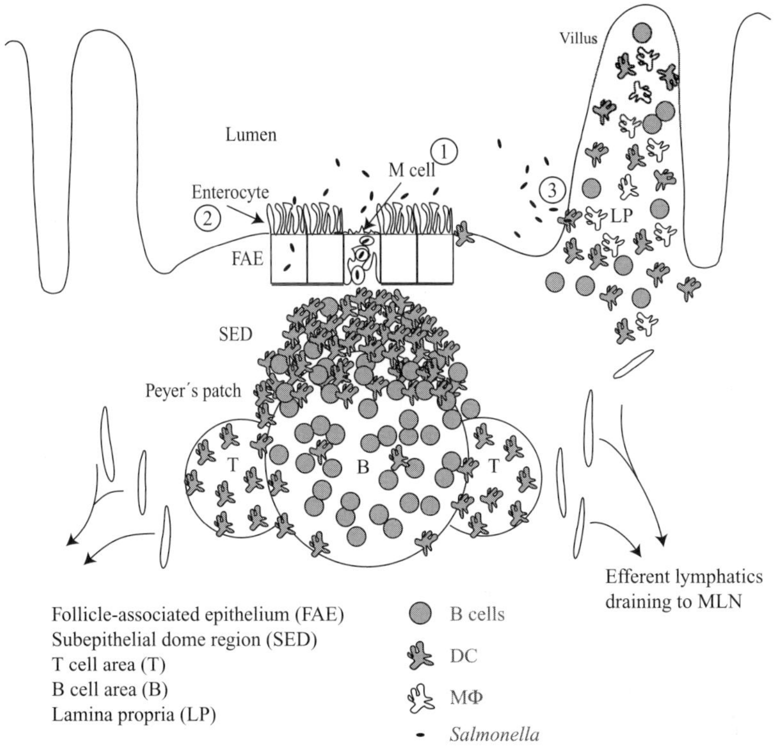

Figure 10.1. *Salmonella* enters the host by crossing the intestinal epithelium via M cells ①, enterocytes ② or via cells that penetrate the epithelium ③. Cells in the Peyer's patch or in the lamina propria can then take up bacteria. DC in the subepithelial dome (SED) consist of $CD8\alpha^{-}CD11b^{+}$ and $CD8\alpha^{-}CD11b^{-}$ subsets while $CD8\alpha^{+}CD11b^{-}$ DC are located in the T-cell areas (Iwasaki and Kelsall, 2000). Iwasaki and Kelsall (2001) also show that DC can be found in the follicle-associated epithelium (FAE).

10.3 DENDRITIC CELLS AND THE ENTRY OF *SALMONELLA* INTO THE HOST

Salmonella infections usually occur by the oral route. To establish infection in the host, *Salmonella* must penetrate the intestinal mucosa. The epithelium at the intestinal surface consists mainly of enterocytes. The small intestine contains lymphoid follicles called Peyer's patches (PP) that are overlaid by an epithelial cell layer called the follicle-associated epithelium (FAE). The FAE contains a special type of epithelial cell referred to as M cells that are specialized in transporting antigens from the lumen to cells in PP (Figure 10.1). M cells have sparse, irregular microvilli on the apical surface and a basolateral cytoplasmic invagination where cells such as lymphocytes are located (Neutra *et al.*, 1996). Some pathogens, such as *Salmonella*, *Yersinia*, *Listeria* and *Shigella*, preferentially exploit M cells as a route of infection (Jensen *et al.*, 1998; Jepson and Clark, 2001; Jones *et al.*, 1994; Vazquez-Torres and Fang, 2000) (Figure 10.1). *Salmonella* may also cross the intestinal barrier via epithelial cells of the FAE or via the villus epithelium (Figure 10.1). If the latter is used, the bacteria will encounter B cells, DC and macrophages in the lamina propria (Iwasaki and Kelsall, 2000; Kelsall and Strober, 1996; Maric *et al.*, 1996; Pavli *et al.*, 1990) (Figure 10.1). An alternative mechanism for *Salmonella* entry into the host may be via cells in PP or lamina propria that breach the epithelial barrier (Figure 10.1). For example, a ligated loop assay has shown that DC in the intestine are recruited to the epithelium in the presence of bacteria (Rescigno *et al.*, 2001). The same study showed that DC express proteins involved in the formation of tight junctions, and that dendrites from DC could penetrate the epithelial layer into the lumen without disrupting its barrier function. It has also recently been shown in vivo that DC can extend dendrites into the intestinal lumen and transport bacteria across the epithelial barrier (Niess *et al.*, 2005). Moreover, CD18-expressing phagocytes have been suggested to mediate an M cell-independent route of *Salmonella* entry (Vazquez-Torres *et al.*, 1999). Thus, DC seem capable of directly sampling bacteria in the intestinal lumen (Rescigno *et al.*, 2001; Niess *et al.*, 2005).

10.4 DENDRITIC CELL INTERACTIONS WITH *SALMONELLA* IN THE PEYER'S PATCHES

PP are important inductive sites of the gut-associated lymphoid tissue. Just a few hours after oral infection with *Salmonella*, bacteria are found in the subepithelial dome, which is located just beneath the FAE (Figure 10.1) (Nagler-Anderson, 2001; Hopkins *et al.*, 2000).

There is a dense layer of DC in the subepithelial dome, ideally localized for picking up incoming antigens (Iwasaki and Kelsall, 2000; Kelsall and Strober, 1996) (Figure 10.1). DC are also found in the T-cell areas and scattered through the B cell follicles (Iwasaki and Kelsall, 2000; Kelsall and Strober, 1996). DC in the PP are capable of processing and presenting antigens to T-cells very efficiently. For example, DC in the PP were found to be more immunostimulatory in a mixed lymphocyte reaction than splenic DC, probably due to higher expression of MHC class II molecules on PP DC compared to splenic DC (Iwasaki and Kelsall, 1999).

In PP, three different DC subsets have been identified by immunohistochemical (Iwasaki and Kelsall, 2000) and flow cytometric (Iwasaki and

Kelsall, 2001) studies. Interestingly, the DC subsets localize to different regions of PP. One population of DC, $CD11b^+CD8\alpha^-$, resides in the subepithelial dome, whereas another subset, $CD11b^-CD8\alpha^+$, is located in the T-cell areas (Iwasaki and Kelsall, 2001; Iwasaki and Kelsall, 2000). A third subset of DC that lacks expression of both CD11b and CD8α is present in the subepithelial dome, FAE and the T-cell areas (Iwasaki and Kelsall, 2001; Iwasaki and Kelsall, 2000) (Figure 10.1).

Some studies suggest that the PP DC subsets have unique abilities to induce immune responses. For instance, the Th2-inducing capacity of PP DC was found to be solely dependent on $CD11b^+$ DC (Iwasaki and Kelsall, 2001). The other subsets primed $CD4^+$ T-cells to produce IFNγ, similar to splenic DC (Iwasaki and Kelsall, 2001). Also, only $CD11b^+$ DC produce IL10 upon stimulation with soluble CD40L trimer or *Staphylococcus aureus* and IFNγ. In contrast, $CD8\alpha^+$ and $CD8\alpha^-CD11b^-$ DC produce the Th1-promoting cytokine IL12p70 in response to this bacterium (Iwasaki and Kelsall, 2001). PP $CD11b^+$ DC have also been shown to produce higher levels of IL6 than PP $CD8\alpha^+$ DC or splenic DC populations, which is important for the induction of IgA production by naive B cells (Sato *et al.*, 2003).

Taken together, these studies show that PP DC are capable of processing and presenting orally acquired antigens to T-cells. Moreover, the DC subsets seem to have unique abilities to modulate this process, for example, by influencing the Th1/Th2 differentiation of T-cells or by influencing the type of antibodies produced by B cells.

After successful penetration of the intestinal mucosa, *Salmonella* can be recovered from PP (Hopkins *et al.*, 2000). DC are abundant in the subepithelial dome of PP (Iwasaki and Kelsall, 2000; Kelsall and Strober, 1996), and *Salmonella* are found within CD11c-expressing DC in the subepithelial dome shortly after oral infection (Hopkins *et al.*, 2000). Although the subset of DC containing *Salmonella* has not been determined, it is likely that the bacteria are residing either within $CD11b^+$ DC or $CD8\alpha^-CD11b^-$ DC, both of which are located in the subepithelial dome (Iwasaki and Kelsall, 2000). Bacterial localization in $CD8\alpha^+$ DC would require the migration of these DC from the T-cell areas to the subepithelial dome in response to oral *Salmonella* infection. Alternatively, bacteria would need to enter the T-cell areas before being internalized. Although the DC subsets show different capacities to produce cytokines and induce Th2 differentiation of T-cells in vitro, it is not known how the DC subsets respond to a bacterial infection in vivo. It is obvious that *Salmonella* overrides the Th2 environment within the PP, because a strong Th1 response with IFNγ production is seen in the PP after oral *Salmonella* infection (George, 1996; Karem *et al.*, 1996; Okahashi *et al.*, 1996).

When DC encounter bacteria they typically upregulate costimulatory and MHC molecules, lose their phagocytic ability and become migratory (Sundquist *et al.*, 2003; Svensson *et al.*, 2000; Banchereau and Steinman, 1998). DC that have picked up antigen in peripheral tissues migrate via lymph or blood vessels to T-cell areas in secondary lymphoid organs, where they display processed antigens on MHC molecules. The behaviour of DC in PP may be unique, since the site where luminal antigens are acquired is not far from organized T and B cell areas. The fate of *Salmonella*-containing DC in the subepithelial dome is not known. However, $CD11b^+$ DC migrate from the subepithelial dome to the T-cell areas in response to systemic injection of a soluble extract of *Toxoplasma gondii* (Iwasaki and Kelsall, 2000). Also, orally administered microspheres are taken up by $CD11c^+$ DC in the subepithelial dome (Shreedhar *et al.*, 2003). DC containing microspheres redistribute from the subepithelial dome to the T-cell areas 24 hours after oral administration of *S. enterica* serovar Typhimurium or cholera toxin. In contrast, DC containing microspheres remain in the subepithelial dome for up to 14 days in the absence of infectious stimuli (Shreedhar *et al.*, 2003). The work of Shreedhar *et al.*, however, does not elucidate if the same microsphere-containing DC with a long half-life remain in the subepithelial dome in the absence of a microbial stimulus, or if the DC population in the subepithelial dome turns over but DC harbouring microspheres are nonetheless present. However, data available thus far suggest the possibility that DC that have engulfed *Salmonella* in the subepithelial dome migrate to the T-cell areas of PP. DC may also leave PP via lymph and migrate to other organs.

10.5 DENDRITIC CELL INTERACTIONS WITH *SALMONELLA* IN MESENTERIC LYMPH NODES

In addition to PP, mesenteric lymph nodes (MLN) are also important sites for the initiation of immune responses to gastrointestinal antigens. The structure of the MLN is similar to that of other lymph nodes, with the presence of B cell follicles and T-cell areas in the cortex. The afferent lymph empties into the subcapsular sinus and must pass through the dense lattice of lymphocytes in the cortex before the lymph is collected in the inner medulla and then exits via the efferent lymph vessel.

The same DC subsets found in PP ($CD8\alpha^+CD11b^-$, $CD8\alpha^-CD11b^+$ and $CD8\alpha^-CD11b^-$) are also found in MLN (Iwasaki and Kelsall, 2001). However, the relative abundance of the DC subsets differs in PP versus MLN (Table 10.1). Similar to their counterparts in PP, DC from MLN are able to

Table 10.1. *Relative composition of DC subsets in Peyer's patches (PP), mesenteric lymph nodes (MLN) and liver.*

Organ	$CD8\alpha^+CD11b^-$	$CD8\alpha^-CD11b^-$	$CD8\alpha^-CD11b^-$
PP	40[a]	30	30
MLN	30	40	30
Liver	10	50	40

[a] The approximate percentage of the indicated DC subset is shown.
The values are based on reference sources (Iwasaki and Kelsall, 2001; Johansson and Wick, 2004; Shortman and Liu, 2002; Vremec and Shortman, 1997).

present antigens to specific T-cells as has been shown in vitro (Johansson-Lindbom *et al.*, 2003; Wilson *et al.*, 2003; Yrlid and Wick, 2002).

Most DC that reside in MLN in the steady state are phenotypically and functionally immature, with an intermediate level of MHC class II expression and a capacity to phagocytose antigens and activate T-cells in vitro (Wilson *et al.*, 2003). However, a population of DC ($CD8\alpha^{lo}CD205^{hi}$) expressing higher levels of MHC class II molecules, but with low antigen presenting efficiency due their inability to internalise antigens, can be found in MLN (Wilson *et al.*, 2003). These cells could represent DC that have migrated to the draining lymph node after picking up antigens in the periphery. Indeed, the $CD8\alpha^{lo}CD205^{hi}$ DC can capture antigen in the skin and respiratory tract and then migrate to the subcutaneous and mediastinal lymph nodes, respectively, acquiring a mature phenotype (Henri *et al.*, 2001; Vermaelen *et al.*, 2001). Whether $CD8\alpha^{lo}CD205^{hi}$ DC have a similar function in MLN is not known.

The exact route of *Salmonella* dissemination from the intestine to MLN is not clear. Free bacteria may be present in the lymph draining the intestine. Alternatively, *Salmonella* might be carried from PP or lamina propria to MLN within DC via the lymph. In fact, DC constantly migrate from the intestine into the lymphatics leading to the MLN and the migration is accelerated by systemic injection of LPS in a TNFα-dependent manner (MacPherson *et al.*, 1995). Migration of DC from the intestine to the T-cell areas of MLN is likely to be enhanced by *Salmonella* infection.

The location where *Salmonella*-specific T-cells are primed and the identity of the responsible antigen-presenting cells are not known. However, ex vivo tests of $CD4^+$ T-cell function indicate that *Salmonella*-specific T-cells are activated both within PP and MLN (McSorley *et al.*, 2002b; VanCott *et al.*, 1998; George, 1996). For example, $CD4^+$ T-cells specific for *Salmonella* flagellin were activated to express CD69 in PP within 3 hours after oral infection, and their numbers peaked 12 hours after infection (McSorley *et al.*, 2002a). A similar activation of flagellin-specific T-cells occurred in MLN with a 3–6 hour delay in kinetics compared to PP. The activated $CD4^+$ T-cells then produced IL2 and proliferated in both PP and MLN, with the kinetics of T-cell activation within MLN lagging a few hours behind that of PP (McSorley *et al.*, 2002a). These data show that antigen-specific $CD4^+$ T-cells are activated in both PP and MLN shortly after oral *Salmonella* infection, with activation in MLN somewhat delayed relative to PP. Although the identity of the antigen-presenting cell that activates specific T-cells is not known, MLN DC are capable of phagocytosing and processing *Salmonella* for presentation on both MHC class I and MHC class II for $CD8^+$ and $CD4^+$ T-cells, respectively (Yrlid and Wick, 2002). Thus, MLN DC that contain *Salmonella* may be important in initiating the immune response to orally acquired *Salmonella*.

10.6 DENDRITIC CELL INTERACTIONS WITH *SALMONELLA* IN THE SPLEEN

The spleen is the largest lymphoid tissue of the body. It filters antigen from blood and is an important site for the induction of immune responses. It consists of distinct regions referred to as the red and white pulp (Figure 10.2). The blood drains into and is filtered by the red pulp. The white pulp consists of lymphocytes organized around arterioles in B and T-cell zones (Figure 10.2). DC in the spleen are divided into $CD8\alpha^+$, $CD4^+$ and $CD8\alpha^-CD4^-$ subsets (Shortman and Liu, 2002; Vremec *et al.*, 2000). While $CD4^+$ DC make up 50% of total splenic DC, $CD8\alpha^+$ and $CD8\alpha^-CD4^-$ each comprise 25% of DC in this organ (Shortman and Liu, 2002; Vremec *et al.*, 2000). DC are situated throughout the spleen and the $CD8\alpha^+$ subset preferentially localizes to the T-cell areas (Crowley *et al.*, 1999; Leenen *et al.*, 1998; Steinman *et al.*, 1997) (Figure 10.2). Similar to DC in PP and MLN, splenic DC are able to present antigens to specific T-cells (Yrlid and Wick, 2002). *Salmonella* probably leaves MLN in the efferent lymph, which eventually empties into the blood via the thoracic duct. When the blood is filtered, *Salmonella* or *Salmonella*-containing cells will end up in the spleen. Whether the bacteria are transported within cells or as free bacteria is not known. After oral infection, *Salmonella* reaches the spleen later than PP and MLN (McSorley *et al.*, 2002a). In addition, *Salmonella* has been shown to be internalized by splenic DC 4 hours after intraperitoneal or intravenous infection (Yrlid and Wick, 2002; Yrlid *et al.*, 2001).

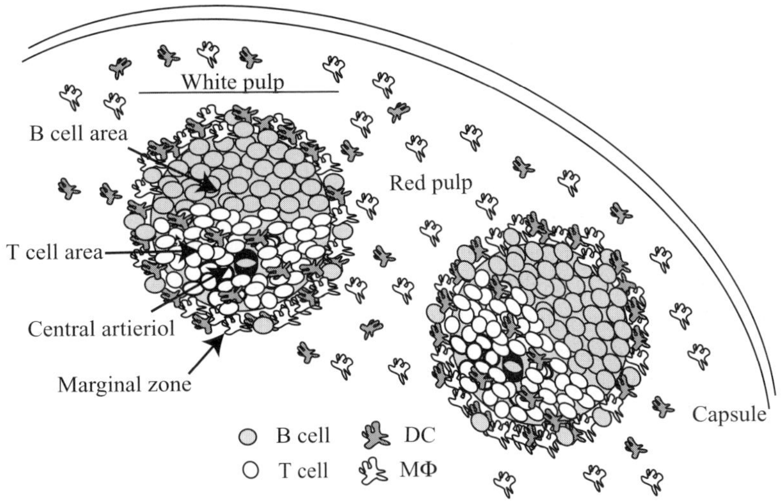

Figure 10.2. A schematic picture of a section of the spleen. The spleen consists of red and white pulp. The white pulp is composed of distinct areas enriched in T and B cells, so-called T and B cell zones, that are adjacent to each other as shown. These areas are surrounded by the marginal zone. DC are situated in the marginal zone, in the T-cell zones and in the red pulp. Macrophages (MΦ) are situated in the marginal zone and in the red pulp.

Oral *Salmonella* infection results in an increase in total DC in the spleen and redistribution within the organ (Kirby *et al.*, 2001). Moreover, these changes occur in a subset-specific fashion. For example, $CD4^+$ DC surrounding splenic B cell areas, which were present in naïve mice, were no longer detected in this location 5 days after oral *Salmonella* infection despite the presence of these cells within the follicles (Kirby *et al.*, 2001). In addition, the number of $CD8\alpha^+$ and $CD8\alpha^-CD4^-$ DC increased in the red pulp in the early stage of infection (Kirby *et al.*, 2001). Injection of other microbial stimuli, such as LPS or an extract of *Toxoplasma gondii,* also resulted in increased numbers of splenic DC (Reis e Sousa *et al.*, 1997; De Smedt *et al.*, 1996). These studies also detected the apparent redistribution of DC from the marginal zone to T-cell areas (Figure 10.2). Together these data show that subset-specific redistribution of DC occurs in response to microbial stimuli in vivo. This could potentially enhance efficient antigen presentation to specific T-cells.

The three conventional murine splenic DC subsets ($CD8\alpha^+$, $CD4^+$, $CD8\alpha^-CD4^-$) internalize *Salmonella* with approximately the same efficiency in vitro and in vivo (Yrlid and Wick, 2002). In addition, both $CD8\alpha^+$ and $CD8\alpha^-$ splenic DC (the latter including both $CD4^+$ and $CD8\alpha^-CD4^-$

subsets) either purified from naïve mice and infected with *Salmonella* in vitro, or purified from orally infected mice and co-cultured with specific T-cells *ex vivo*, can present *Salmonella* antigens to primary antigen-specific T-cells (Yrlid and Wick, 2002). Indeed, *Salmonella* infection activates splenic DC to upregulate maturation markers such as CD40 and CD86 in vivo (Yrlid *et al.*, 2001). In addition, T-cells specific for *Salmonella* have been generated in the spleen after oral infection (McSorley *et al.*, 2002a; McSorley *et al.*, 2000). Together these data suggest that splenic DC have the ability to prime naïve T-cells during *Salmonella* infection.

Cytokine production by DC and DC subsets upon exposure to *Salmonella* has been examined in vitro as well as in vivo (Wick, 2002). For example, studies of cytokine expression using RT-PCR showed that splenic DC infected with *Salmonella* have increased levels of mRNA for IL1, IL6 and IL12p40 (Marriott *et al.*, 1999). Intracellular cytokine staining and flow cytometric analyses have shown that splenic DC co-cultured with *Salmonella* produce TNFα and IL12p40 (Yrlid and Wick, 2002). While TNFα was produced by a large percentage of CD8α^- DC, relatively few CD8α^+ DC produced this cytokine following co-culture with *Salmonella*. The largest number of DC producing IL12p40, however, were amongst CD8α^+ DC. These studies also showed that bacterial contact but not internalization was required for cytokine production by DC, and a significant fraction of cytokine-producing cells were among those DC that did not contain bacteria (Yrlid and Wick, 2002). This may indicate that DC without any internalized *Salmonella* can produce cytokines and influence an immune response in vivo. The subset difference in cytokine production may have an impact on the microenvironment where the presentation of bacterial antigens to T-cells occurs. It is not known which subsets of DC present *Salmonella* antigens and stimulate T-cells in vivo.

Other studies, which quantitated cytokine-producing cells directly ex vivo using flow cytometric analysis, showed a transient increase in CD4$^+$ DC producing TNFα and a continual rise in the number of CD8α^+ DC producing this cytokine during the first 5 days after oral *Salmonella* infection (Kirby *et al.*, 2001). Despite the fact that in vitro studies have shown that splenic DC, particularly CD8α^+ DC, as well as bone marrow-derived DC can produce IL12p40 upon exposure to *Salmonella*, no significant increase in the number of IL12p40$^+$ splenic DC was detected during the early stages of an oral *Salmonella* infection (Kirby *et al.*, 2001). This could be due to the existence of different parameters that regulate the interactions between DC and *Salmonella* in vitro and in vivo. For example, a large fraction of DC (10%) will engulf *Salmonella* under in vitro culture conditions (Yrlid and Wick, 2002; Svensson and Wick, 1999) whilst relatively few splenic DC

(<1%) contain bacteria after oral infection (Johansson and Wick, unpublished observation). Alternatively, signals required to induce IL12 production by DC early after oral infection, such as IFNγ or CD40 interaction with CD40L on activated T-cells, may be lacking (Cella *et al.*, 1996; Kelsall *et al.*, 1996; Koch *et al.*, 1996; Maldonado-Lopez *et al.*, 2001; Schulz *et al.*, 2000). This could be particularly apparent when quantitating IL12^{+} DC early during primary infection (Kirby *et al.*, 2001). Thus, DC are potential sources of cytokines important in resistance against *Salmonella* infection, and individual DC subsets might have a differential capacity to produce TNFα and IL12p40 in response to *Salmonella*. However, the capacity of the different DC subsets to skew T-cell responses during *Salmonella* infection requires further investigation.

10.7 DENDRITIC CELL INTERACTIONS WITH *SALMONELLA* IN THE LIVER

The liver consists mainly of hepatocytes. However, nonparenchymal cells such as T-cells, a minor fraction of B cells, NK cells and NKT-cells are also present in this organ (Wick *et al.*, 2002). The liver also contains cells with antigen presenting ability including sinusoidal endothelial cells, Kupffer cells (resident macrophages), and DC (Wick *et al.*, 2002).

The liver has the largest number of DC compared to other parenchymal organs such as the kidney, heart or pancreas (Steptoe *et al.*, 2000). However, the relatively large size of the liver results in the lowest density of DC among the parenchymal organs (Steptoe *et al.*, 2000). Within the liver, DC are located mainly in the portal areas (O'Connell *et al.*, 2000; Shurin *et al.*, 1997; Woo *et al.*, 1994; Steiniger *et al.*, 1984) (Figure 10.3) and have been proposed to enter the liver via blood-lymph translocation in the sinusoids, perhaps via binding to Kupffer cells (Kudo *et al.*, 1997). Phenotypic analysis reveals that liver DC consist mostly of CD8α^{-} DC, particularly CD8α^{-}CD4^{-} DC, with few CD8α^{-}CD4^{+} and CD8α^{+} DC (Johansson and Wick, 2004; Lian *et al.*, 2003; O'Connell *et al.*, 2000) (Table 10.1). Similar to PP and MLN, the liver also contains CD8α^{-}CD11b^{-} DC (Johansson and Wick, 2004). Freshly isolated DC from the livers of mice, as well as DC cultured from progenitors in the liver, have an immature phenotype and express very low levels of CD40, CD80 and CD86 (Johansson and Wick, 2004; Lian *et al.*, 2003; Morelli *et al.*, 2000; O'Connell *et al.*, 2000). However, studies in rats have shown that the DC migrating from the liver via the lymphatics are either already mature immunostimulatory cells or they rapidly mature upon leaving the liver (Sato *et al.*, 1998; Matsuno *et al.*, 1996).

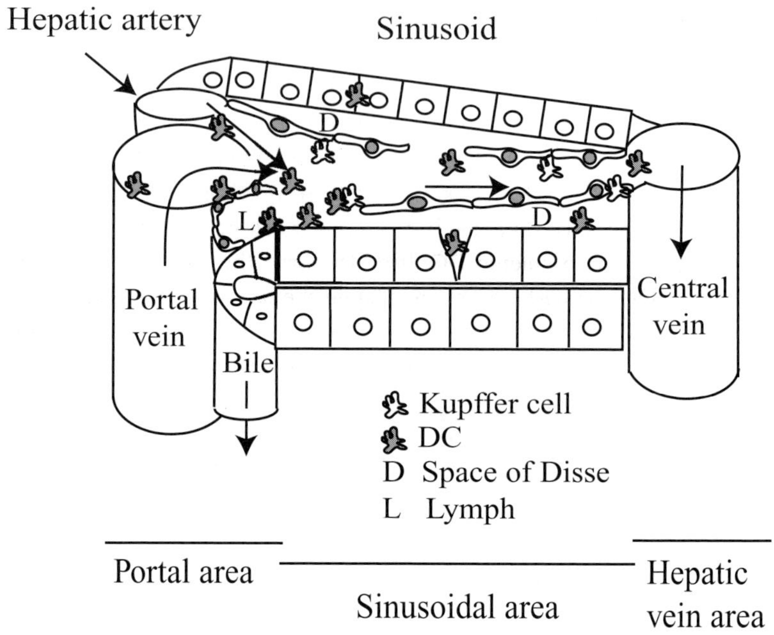

Figure 10.3. A sinusoid in the liver. Venous blood from the gastrointestinal tract and pancreas enter the liver via the portal vein. Oxygen-rich blood from the circulation enters the liver through the hepatic artery. Thus, blood and its contents, including cells and bacteria, pass through the liver and enter hepatic sinusoids before leaving via the central vein or through efferent lymphatics.

Large volumes of blood pass through the liver, both from the intestinal tract via the portal vein and from the circulation via the hepatic artery. Thus, the liver is a target organ of orally acquired and blood-borne antigens including bacteria such as *Salmonella*. Orally acquired *Salmonella* could reach the liver after exiting the gut via the lymph, which reaches the thoracic duct and then joins the blood circulation. It is also possible that *Salmonella* leave the intestine in the blood and thus access the liver (Vazquez-Torres and Fang, 2000; Vazquez-Torres *et al.*, 1999). Indeed, *Salmonella* has been found in $CD18^+$ cells in the liver (Sheppard *et al.*, 2003; Richter-Dahlfors *et al.*, 1997) and in $CD11c^+$ cells purified from the liver after intravenous infection (Johansson and Wick, 2004).

Little is know about the role of liver DC during bacterial infection, but *Salmonella* can be recovered from the liver of orally infected mice as early as 2–3 days post infection. In addition, *Salmonella* was found in $CD11c^+$ cells, both $CD8\alpha^-$ and $CD8\alpha^+$, in the liver 4 hours after intravenous infection (Johansson and Wick, 2004). DC purified from the liver have been shown to process and present *Salmonella*-derived peptides on both MHC class I and MHC class II molecules after a short co-incubation with *S. enterica* serovar Typhimurium (Johansson and Wick, 2004). In addition, purified liver DC with a $CD11c^{+}CD8\alpha^{+}$ or $CD11c^{+}CD8\alpha^{-}$ phenotype have been shown to produce IL12p40 and TNFα after a 2 hour co-incubation with *Salmonella* (Johansson and Wick, 2004). However, only about 3% of $IL12p40^{+}$ and 10% of TNFα positive cells contained bacteria in both of these DC subsets.

Thus, similar to splenic DC, liver DC have the capacity to produce cytokines after *Salmonella* exposure despite the lack of intracellular bacteria, and have the potential to interact with bacteria-specific T-cells during the immune responses to *Salmonella* (Johansson and Wick, 2004; Yrlid and Wick, 2002). However, despite the fact that DC containing *Salmonella* and T-cells specific for *Salmonella* are both found in the liver, little is known about whether these two cell populations interact within the liver or liver-draining lymph nodes. (Johansson and Wick, 2004; Kursar *et al.*, 2002; Pope *et al.*, 2001).

10.8 CONCLUSIONS

DC are a cell type important for the induction of immune responses against antigens including bacteria (Jung *et al.*, 2002; Banchereau and Steinman, 1998). These cells are located in most tissues of the body including lymphoid and non-lymphoid organs. They are therefore able to encounter pathogenic bacteria, such as *Salmonella*, at peripheral infection sites. DC location in lymphoid organs including PP, MLN and the spleen as well as the liver and liver-draining lymph nodes, enhances their ability to participate in initiating and maintaining a protective immune response against pathogens. The ability of DC to phagocytose *Salmonella* and process and present bacterial antigens to T-cells is central to the induction of an immune response. In addition, cytokine and chemokine production by DC after interaction with *Salmonella* could influence the nature of the ensuing T-cell response to one dominated by IFNγ-producing Th1 T-cells, thus allowing bacterial clearance.

Many aspects of the interaction between *Salmonella* and DC remain to be elucidated. For example, the role of DC in the dissemination of *Salmonella* from the intestine to deeper organs needs to be clarified, although progress is being made (Niess *et al.*, 2005). Moreover, the role of different DC subsets in inducing or maintaining the T-cell response is not yet known. However, what is known about the biology and function of DC in general, and what is known thus far about their role in various infection models, indicates that DC are key players in the host-pathogen battle that occurs during *Salmonella* infection.

10.9 ACKNOWLEDGEMENTS

Work in Professor Wick's laboratory is supported by the Swedish Research Council, the Swedish Cancer foundation, the Swedish Foundation for Strategic Research and Sahlgrenska Academy at Göteborg University.

REFERENCES

Ardavin, C. (2003). Origin, precursors and differentiation of mouse dendritic cells. *Nat Rev Immunol*, **3**, 582–90.

Asselin-Paturel, C., Boonstra, A., Dalod, M. *et al.* (2001). Mouse type I IFNproducing cells are immature APCs with plasmacytoid morphology. *Nat Immunol*, **2**, 1144–50.

Banchereau, J. and Steinman, R. M. (1998). Dendritic cells and the control of immunity. *Nature*, **392**, 245–52.

Björck, P. (2001). Isolation and characterization of plasmacytoid dendritic cells from Flt3 ligand and granulocyte–macrophage colony-stimulating factor-treated mice. *Blood*, **98**, 3520–26.

Cella, M., Scheidegger, D., Palmer-Lehmann, K. *et al.* (1996). Ligation of CD40 on dendritic cells triggers production of high levels of interleukin-12 and enhances T-cell stimulatory capacity: T-T help via APC activation. *J Exp Med*, **184**, 747–52.

Crowley, M. T., Reilly, C. R. and Lo, D. (1999). Influence of lymphocytes on the presence and organization of dendritic cell subsets in the spleen. *J Immunol*, **163**, 4894–900.

De Smedt, T., Pajak, B., Muraille, E. *et al.* (1996). Regulation of dendritic cell numbers and maturation by lipopolysaccharide *in vivo*. *J Exp Med*, **184**, 1413–24.

George, A. (1996). Generation of gamma interferon responses in murine Peyer's patches following oral immunization. *Infect Immun*, **64**, 4606–11.

Henri, S., Vremec, D., Kamath, A. *et al.* (2001). The dendritic cell populations of mouse lymph nodes. *J Immunol*, **167**, 741–8.

Hopkins, S. A., Niedergang, F., Corthesy-Theulaz, I. E. and Kraehenbuhl, J. P. (2000). A recombinant *Salmonella typhimurium* vaccine strain is taken up and survives within murine Peyer's patch dendritic cells. *Cell Microbiol*, **2**, 59–68.

Inaba, K., Inaba, M., Deguchi, M. *et al.* (1993). Granulocytes, macrophages, and dendritic cells arise from a common major histocompatibility complex class II-negative progenitor in mouse bone marrow. *Proc Natl Acad Sci USA*, **90**, 3038–42.

Iwasaki, A. and Kelsall, B. L. (1999). Freshly isolated Peyer's patch, but not spleen, dendritic cells produce interleukin 10 and induce the differentiation of T-helper type 2 cells. *J Exp Med*, **190**, 229–39.

(2000). Localization of distinct Peyer's patch dendritic cell subsets and their recruitment by chemokines macrophage inflammatory protein (MIP)-3α, MIP-3β, and secondary lymphoid organ chemokine. *J Exp Med*, **191**, 1381–94.

(2001). Unique functions of $CD11b^+$, $CD8\alpha^+$, and double-negative Peyer's patch dendritic cells. *J Immunol*, **166**, 4884–90.

Jensen, V. B., Harty, J. T. and Jones, B. D. (1998). Interactions of the invasive pathogens *Salmonella typhimurium*, *Listeria monocytogenes*, and *Shigella flexneri* with M cells and murine Peyer's patches. *Infect Immun*, **66**, 3758–66.

Jepson, M. A. and Clark, M. A. (2001). The role of M cells in *Salmonella* infection. *Microbes Infect*, **3**, 1183–90.

Johansson, C. and Wick, M. (2004). Liver dendritic cells present bacterial antigens and produce cytokines upon *Salmonella* encounter. *J Immunol*, **172**, 2496–503.

Johansson-Lindbom, B., Svensson, M., Wurbel, M. A. *et al.* (2003). Selective generation of gut tropic T-cells in gut-associated lymphoid tissue (GALT): requirement for GALT dendritic cells and adjuvant. *J Exp Med*, **198**, 963–9.

Jones, B. D., Ghori, N. and Falkow, S. (1994). *Salmonella typhimurium* initiates murine infection by penetrating and destroying the specialized epithelial M cells of the Peyer's patches. *J Exp Med*, **180**, 15–23.

Jung, S., Unutmaz, D., Wong, P. *et al.* (2002). In vivo depletion of $CD11c^+$ dendritic cells abrogates priming of $CD8^+$ T-cells by exogenous cell-associated antigens. *Immunity*, **17**, 211–20.

Kamath, A. T., Pooley, J., O'Keeffe, M. A. *et al.* (2000). The development, maturation, and turnover rate of mouse spleen dendritic cell populations. *J Immunol*, **165**, 6762–70.

Karem, K. L., Kanangat, S. and Rouse, B. T. (1996). Cytokine expression in the gut associated lymphoid tissue after oral administration of attenuated *Salmonella* vaccine strains. *Vaccine*, **14**, 1495–502.

Kelsall, B. L. and Strober, W. (1996). Distinct populations of dendritic cells are present in the subepithelial dome and T-cell regions of the murine Peyer's patch. *J Exp Med*, **183**, 237–47.

Kirby, A. C., Yrlid, U., Svensson, M. and Wick, M. J. (2001). Differential involvement of dendritic cell subsets during acute *Salmonella* infection. *J Immunol*, **166**, 6802–11.

Koch, F., Stanzl, U., Jennewein, P. *et al.* (1996). High level IL12 production by murine dendritic cells: upregulation via MHC class II and CD40 molecules and downregulation by IL4 and IL10. *J Exp Med*, **184**, 741–6.

Kudo, S., Matsuno, K., Ezaki, T. and Ogawa, M. (1997). A novel migration pathway for rat dendritic cells from the blood: hepatic sinusoids-lymph translocation. *J Exp Med*, **185**, 777–84.

Kursar, M., Bonhagen, K., Kohler, A. *et al.* (2002). Organ-specific $CD4^+$ T-cell response during *Listeria monocytogenes* infection. *J Immunol*, **168**, 6382–7.

Leenen, P. J., Radosevic, K., Voerman, J. S. *et al.* (1998). Heterogeneity of mouse spleen dendritic cells: *in vivo* phagocytic activity, expression of macrophage markers, and subpopulation turnover. *J Immunol*, **160**, 2166–73.

Lian, Z. X., Okada, T., He, X. S. *et al.* (2003). Heterogeneity of dendritic cells in the mouse liver: identification and characterization of four distinct populations. *J Immunol*, **170**, 2323–30.

MacPherson, G. G., Jenkins, C. D., Stein, M. J. and Edwards, C. (1995). Endotoxin-mediated dendritic cell release from the intestine. Characterization of released dendritic cells and TNFα dependence. *J Immunol*, **154**, 1317–22.

Maldonado-Lopez, R., Maliszewski, C., Urbain, J. and Moser, M. (2001). Cytokines regulate the capacity of $CD8\alpha^+$ and $CD8\alpha^-$ dendritic cells to prime Th1/Th2 cells *in vivo*. *J Immunol*, **167**, 4345–50.

Maric, I., Holt, P. G., Perdue, M. H. and Bienenstock, J. (1996). Class II MHC antigen (Ia)-bearing dendritic cells in the epithelium of the rat intestine. *J Immunol*, **156**, 1408–14.

Marriott, I., Hammond, T. G., Thomas, E. K. and Bost, K. L. (1999). *Salmonella* efficiently enter and survive within cultured $CD11c^+$ dendritic cells initiating cytokine expression. *Eur J Immunol*, **29**, 1107–15.

Matsuno, K., Ezaki, T., Kudo, S. and Uehara, Y. (1996). A life stage of particle-laden rat dendritic cells *in vivo*: their terminal division, active phagocytosis, and translocation from the liver to the draining lymph. *J Exp Med*, **183**, 1865–78.

McSorley, S. J., Cookson, B. T. and Jenkins, M. K. (2000). Characterization of $CD4^+$ T-cell responses during natural infection with *Salmonella typhimurium*. *J Immunol*, **164**, 986–93.

McSorley, S. J., Asch, S., Costalonga, M., Reinhardt, R. L. and Jenkins, M. K. (2002a). Tracking *Salmonella*-specific CD4 T-cells *in vivo* reveals a local mucosal response to a disseminated infection. *Immunity*, **16**, 365–77.

McSorley, S. J., Ehst, B. D., Yu, Y. and Gewirtz, A. T. (2002b). Bacterial flagellin is an effective adjuvant for $CD4^+$ T-cells *in vivo*. *J Immunol*, **169**, 3914–19.

Morelli, A. E., O'Connell, P. J., Khanna, A. *et al.* (2000). Preferential induction of Th1 responses by functionally mature hepatic ($CD8\alpha^-$ and $CD8\alpha^+$) dendritic cells: association with conversion from liver transplant tolerance to acute rejection. *Transplantation,* **69**, 2647–57.

Moser, M. and Murphy, K. M. (2000). Dendritic cell regulation of TH1–TH2 development. *Nat Immunol,* **1**, 199–205.

Nagler-Anderson, C. (2001). Man the barrier! Strategic defences in the intestinal mucosa. *Nat Rev Immunol,* **1**, 59–67.

Nakano, H., Yanagita, M. and Gunn, M. D. (2001). $CD11c^+B220^+Gr\text{-}1^+$ cells in mouse lymph nodes and spleen display characteristics of plasmacytoid dendritic cells. *J Exp Med,* **194**, 1171–8.

Neutra, M. R., Frey, A. and Kraehenbuhl, J. P. (1996). Epithelial M cells: gateways for mucosal infection and immunization. *Cell,* **86**, 345–8.

Niess, J. H., Brand, S., Gu, X. *et al.* (2005). CX_3CR1-mediated dendritic cell access to the intestinal lumen and bacterial clearance. *Science,* **307**, 254–8.

O'Connell, P. J., Morelli, A. E., Logar, A. J. and Thomson, A. W. (2000). Phenotypic and functional characterization of mouse hepatic $CD8\alpha^+$ lymphoid-related dendritic cells. *J Immunol,* **165**, 795–803.

Okahashi, N., Yamamoto, M., Vancott, J. L. *et al.* (1996). Oral immunization of interleukin-4 (IL4) knockout mice with a recombinant *Salmonella* strain or cholera toxin reveals that $CD4^+$ Th2 cells producing IL6 and IL10 are associated with mucosal immunoglobulin A responses. *Infect Immun,* **64**, 1516–25.

Pavli, P., Woodhams, C. E., Doe, W. F. and Hume, D. A. (1990). Isolation and characterization of antigen-presenting dendritic cells from the mouse intestinal lamina propria. *Immunology,* **70**, 40–7.

Pope, C., Kim, S. K., Marzo, A. *et al.* (2001). Organ-specific regulation of the CD8 T-cell response to *Listeria monocytogenes* infection. *J Immunol,* **166**, 3402–9.

Regnault, A., Lankar, D., Lacabanne, V. *et al.* (1999). $Fc\gamma$ receptor-mediated induction of dendritic cell maturation and major histocompatibility complex class I-restricted antigen presentation after immune complex internalization. *J Exp Med,* **189**, 371–80.

Reis e Sousa, C. R., Hieny, S., Scharton-Kersten, T. *et al.* (1997). *In vivo* microbial stimulation induces rapid CD40 ligand-independent production of interleukin 12 by dendritic cells and their redistribution to T-cell areas. *J Exp Med,* **186**, 1819–29.

Rescigno, M., Urbano, M., Valzasina, B. *et al.* (2001). Dendritic cells express tight junction proteins and penetrate gut epithelial monolayers to sample bacteria. *Nat Immunol,* **2**, 361–7.

Richter-Dahlfors, A., Buchan, A. M. J. and Finlay, B. B. (1997). Murine salmonellosis studied by confocal microscopy: *Salmonella typhimurium* resides intracellularly inside macrophages and exerts a cytotoxic effect on phagocytes *in vivo*. *J Exp Med*, **186**, 569–80.

Sallusto, F., Cella, M., Danieli, C. and Lanzavecchia, A. (1995). Dendritic cells use macropinocytosis and the mannose receptor to concentrate macromolecules in the major histocompatibility complex class II compartment: downregulation by cytokines and bacterial products. *J Exp Med*, **182**, 389–400.

Sato, A., Hashiguchi, M., Toda, E. *et al.* (2003). $CD11b^{+}$ Peyer's patch dendritic cells secrete IL6 and induce IgA secretion from naive B cells. *J Immunol*, **171**, 3684–90.

Sato, T., Yamamoto, H., Sasaki, C. and Wake, K. (1998). Maturation of rat dendritic cells during intrahepatic translocation evaluated using monoclonal antibodies and electron microscopy. *Cell Tissue Res*, **294**, 503–14.

Schulz, O., Edwards, D. A., Schito, M. *et al.* (2000). CD40 triggering of heterodimeric IL12 p70 production by dendritic cells in vivo requires a microbial priming signal. *Immunity*, **13**, 453–62.

Sheppard, M., Webb, C., Heath, F. *et al.* (2003). Dynamics of bacterial growth and distribution within the liver during *Salmonella* infection. *Cell Microbiol*, **5**, 593–600.

Shortman, K. and Liu, Y. J. (2002). Mouse and human dendritic cell subtypes. *Nature Rev Immunol*, **2**, 151–61.

Shreedhar, V. K., Kelsall, B. L. and Neutra, M. R. (2003). Cholera toxin induces migration of dendritic cells from the subepithelial dome region to T- and B-cell areas of Peyer's patches. *Infect Immun*, **71**, 504–9.

Shurin, M. R., Pandharipande, P. P., Zorina, T. D. *et al.* (1997). FLT3 ligand induces the generation of functionally active dendritic cells in mice. *Cell Immunol*, **179**, 174–84.

Sparwasser, T., Koch, E. S., Vabulas, R. M. *et al.* (1998). Bacterial DNA and immunostimulatory CpG oligonucleotides trigger maturation and activation of murine dendritic cells. *Eur J Immunol*, **28**, 2045–54.

Steiniger, B., Klempnauer, J. and Wonigeit, K. (1984). Phenotype and histological distribution of interstitial dendritic cells in the rat pancreas, liver, heart, and kidney. *Transplantation*, **38**, 169–74.

Steinman, R. M. and Cohn, Z. A. (1973). Identification of a novel cell type in peripheral lymphoid organs of mice. I. Morphology, quantitation, tissue distribution. *J Exp Med*, **137**, 1142–62.

Steinman, R. M., Lustig, D. S. and Cohn, Z. A. (1974). Identification of a novel cell type in peripheral lymphoid organs of mice. 3. Functional properties *in vivo*. *J Exp Med*, **139**, 1431–45.

Steinman, R. M., Pack, M. and Inaba, K. (1997). Dendritic cells in the T-cell areas of lymphoid organs. *Immunol Rev*, **156**, 25–37.

Steptoe, R. J., Patel, R. K., Subbotin, V. M. and Thomson, A. W. (2000). Comparative analysis of dendritic cell density and total number in commonly transplanted organs: morphometric estimation in normal mice. *Transpl Immunol*, **8**, 49–56.

Sundquist, M., Johansson, C. and Wick, M. J. (2003). Dendritic cells as inducers of antimicrobial immunity *in vivo*. *Apmis*, **111**, 715–24.

Svensson, M. and Wick, M. J. (1999). Classical MHC class I peptide presentation of a bacterial fusion protein by bone marrow-derived dendritic cells. *Eur J Immunol*, **29**, 180–8.

Svensson, M., Johansson, C. and Wick, M. J. (2000). *Salmonella enterica* serovar Typhimurium-induced maturation of bone marrow-derived dendritic cells. *Infect Immun*, **68**, 6311–20.

VanCott, J. L., Chatfield, S. N., Roberts, M. *et al.* (1998). Regulation of host immune responses by modification of *Salmonella* virulence genes. *Nat Med*, **4**, 1247–52.

Vazquez-Torres, A. and Fang, F. C. (2000). Cellular routes of invasion by enteropathogens. *Curr Opin Microbiol*, **3**, 54–9.

Vazquez-Torres, A., Jones-Carson, J., Baumler, A. J. *et al.* (1999). Extraintestinal dissemination of *Salmonella* by CD18-expressing phagocytes. *Nature*, **401**, 804–8.

Vermaelen, K. Y., Carro-Muino, I., Lambrecht, B. N. and Pauwels, R. A. (2001). Specific migratory dendritic cells rapidly transport antigen from the airways to the thoracic lymph nodes. *J Exp Med*, **193**, 51–60.

Vremec, D. and Shortman, K. (1997). Dendritic cell subtypes in mouse lymphoid organs: cross-correlation of surface markers, changes with incubation, and differences among thymus, spleen, and lymph nodes. *J Immunol*, **159**, 565–73.

Vremec, D., Pooley, J., Hochrein, H., Wu, L. and Shortman, K. (2000). CD4 and CD8 expression by dendritic cell subtypes in mouse thymus and spleen. *J Immunol* **164**, 2978–86.

Wick, M. J. (2002). The role of dendritic cells during *Salmonella* infection. *Curr Opin Immunol*, **14**, 437–43.

Wick, M. J., Leithäuser, F. and Reimann, J. (2002). The hepatic immune system. *Crit Rev Immunol*, **22**, 47–103.

Wilson, N. S., El-Sukkari, D., Belz, G. T. *et al.* (2003). Most lymphoid organ dendritic cell types are phenotypically and functionally immature. *Blood*, **102**, 2187–94.

Winzler, C., Rovere, P., Rescigno, M. *et al.* (1997) Maturation stages of mouse dendritic cells in growth factor-dependent long-term cultures. *J Exp Med*, **185**, 317–28.

Woo, J., Lu, L., Rao, A. S., Li, Y. *et al.* (1994). Isolation, phenotype, and allostimulatory activity of mouse liver dendritic cells. *Transplantation*, **58**, 484–91.
Wu, L., Li, C. L. and Shortman, K. (1996). Thymic dendritic cell precursors: relationship to the T lymphocyte lineage and phenotype of the dendritic cell progeny. *J Exp Med*, **184**, 903–11.
Yrlid, U. and Wick, M. J. (2002). Antigen presentation capacity and cytokine production by murine splenic dendritic cell subsets upon *Salmonella* encounter. *J Immunol*, **169**, 108–16.
Yrlid, U., Svensson, M., Hakansson, A. *et al.* (2001). *In vivo* activation of dendritic cells and T-cells during *Salmonella enterica* serovar Typhimurium infection. *Infect Immun*, **69**, 5726–35.

CHAPTER 11

Immunity to *Salmonella* in domestic (food animal) species

Paul Wigley, Paul Barrow and Bernardo Villarreal-Ramos

11.1 INTRODUCTION

Salmonellosis in domestic animal species is important in terms of animal welfare and productivity. Infection may lead to decreased yields of milk, eggs or meat, and in certain cases loss of livestock. Salmonellosis in domestic species is also important for public health as the major reservoir and source of food-borne human infections.

A number of *Salmonella enterica* serovars can induce a systemic typhoid-like disease in healthy adults of a restricted range of host animal species. Other serovars colonize the intestine of the host and in some cases may induce severe enteritis. The severity of the disease will be dependent on the virulence and dose of the challenge and immune status of the host. Thus, some *S. enterica* strains that would normally induce enteritis in adult hosts are able to induce systemic disease in immuno-compromised hosts. Immunity to *S. enterica* is dependent on the nature of the disease that different serovars induce in different hosts. Thus, mucosal immunity is more likely to be important in protecting against serovars that induce enteritis, whereas systemic immunity would be more important in protecting against serovars that induce systemic disease.

Our understanding of the interaction of the host's immune system with different *S. enterica* serovars is still rudimentary. Effective control of salmonellosis affecting domestic host species requires a greater understanding of immunological mechanisms during such infections. This will provide the basis from which rational control measures, such as more effective vaccines,

'Salmonella' Infections: Clinical, Immunological and Molecular Aspects, ed. Pietro Mastroeni and Duncan Maskell. Published by Cambridge University Press.

vaccination strategies, diagnostic tools or other non-immunological tools may be developed. The current understanding of immunity to *S. enterica* in domestic species is briefly outlined in this chapter.

11.2 INNATE IMMUNITY

11.2.1 Chickens

11.2.1.1 Early interactions in the gastrointestinal tract

The innate immune system plays a crucial role in protecting chickens from disease caused by *S. enterica* infection, as well as in initiating the adaptive immune response.

A number of cell types, including enterocytes, macrophages and heterophils are likely to recognize *S. enterica* present in the gut or in the early stages of the invasion process, through surface receptors, including Toll-like receptors (TLR). As with mammals, chickens appear to have a range of different TLRs, including TLR4 and TLR2 (Boyd *et al.*, 2001; Fukui *et al.*, 2001; Leveque *et al.*, 2003).

11.2.1.2 Heterophils: avian polymorphonuclear phagocytes

S. enterica infection results in the induction of pro-inflammatory cytokines and chemokines including interleukin-6 (IL6) and IL8, (Kaiser *et al.*, 2000; Withanage, 2004). The production of these mediators leads to an influx of heterophils (polymorphonuclear cells) and macrophages into the intestine. In contrast, in vitro or in vivo infection with *S. enterica* serovar Gallinarum or Pullorum causes little activation, in some cases even down-regulation, of pro-inflammatory cytokines and chemokines (Kaiser *et al.*, 2000), and coincides with little heterophil infiltration into the gut (Henderson *et al.*, 1999). These differences suggest that the development of systemic disease by typhoid serotypes may involve inhibition of a heterophil influx, thus allowing invasion by large numbers of *S. enterica*. In fact, heterophils are crucial in the control of disease following *S. enterica* serovar Enteritidis infection of chickens (Kogut *et al.*, 1994). Depletion of heterophils with 5-fluorouracil, prior to oral infection with *S. enterica* serovar Enteritidis, leads to the development of severe systemic salmonellosis with high bacterial numbers in the spleen and liver closely resembling fowl typhoid. The susceptibility of very young chicks to *S. enterica* serovar Enteritidis or serovar Typhimurium infection, may be a consequence of relatively poor heterophil function in the first few days after hatching (Wells *et al.*, 1998).

Heterophils are very efficient killers of phagocytosed bacteria, killing in excess of 90% of *S. enterica* within an hour in vitro (Wells *et al.*, 1998). Heterophils differ significantly from mammalian neutrophils in some areas of their biology. They have no myeloperoxidase activity and the antimicrobial activity of heterophils appears to be less reliant on oxidative mechanisms and more reliant on antimicrobial peptides such as defensins (Maxwell and Robertson, 1998). Heterophils express cytokines including IL18 and the chemokine IL8 following *S. enterica* infection (Kogut *et al.*, 2003). Such responses may activate inflammatory responses and may play a critical signaling role in the development of adaptive Th1 immunity and clearance of infection (Kogut *et al.*, 2003a, b). It also appears that activated T-cells may prime heterophils for phagocytosis and killing of *S. enterica*, (Kogut *et al.*, 2003).

11.2.1.3 Role of macrophages in innate immunity to *S. enterica*

The biology of chicken and mammalian macrophages is similar, both cell types being activated by IFNγ to express MHC class II molecules, and being capable of antimicrobial activity (Qureshi, 2003). Chicken macrophages produce reactive nitrogen species (RNS) and reactive oxygen species (ROS) following infection with *S. enterica* (Wigley *et al.*, 2002). *S. enterica* requires the *Salmonella* pathogenicity island 2 (SPI-2) type III secretion system (TTSS) for survival in chicken macrophages (Jones *et al.*, 2001; Wigley *et al.*, 2002). Chicken macrophage biology during infection with *S. enterica* serovars Typhimurium and Enteritidis has not been investigated to any great extent, other than reports of changes in their numbers in the gastrointestinal or reproductive tracts following infection (Desmidt *et al.*, 1997; Van Immerseel *et al.*, 2002a; Withanage *et al.*, 1998). Interactions with, and survival within, macrophages are key stages of *S. enterica* serovar Gallinarum infections in chickens (Barrow *et al.*, 1994; Jones *et al.*, 2001). Mutation in the SPI-2 TTSS leads to attenuation in *S. enterica* serovar Gallinarum (Jones *et al.*, 2001) suggesting the importance of survival within macrophages for the virulence of this serovar. *S. enterica* serovar Pullorum is capable of long term persistence in splenic macrophages (Wigley *et al.*, 2001) and this is also dependent on the SPI-2 TTSS (Wigley *et al.*, 2002). However, although host-specific *Salmonella* serovars display a requirement for survival in macrophages in vivo, they do not appear to have any increased ability to survive within these cells in vitro (Chadfield *et al.*, 2003).

Resistance to salmonellosis in mice is linked to the presence of a functional *Nramp1* gene expressed in macrophages (Eckmann *et al.*, 1996; Gautier *et al.*, 1998; Plant and Glynn, 1974; Vidal *et al.*, 1993). In chickens, *Nramp1* plays a significant, albeit minor, role in resistance to systemic salmonellosis (Hu *et al.*, 1997). Conversely, the autosomal dominant *SAL1* gene (Mariani *et al.*, 2001) is a major determinant of susceptibility to systemic disease, and determines increased levels of oxidative burst in macrophages of chicken lines resistant to *S. enterica* infection (wigley *et al.*, 2002). Other autosomal dominant genes contribute to resistance to intestinal infection and are not related to MHC or *Nramp1* (Barrow *et al.*, 2004).

11.2.2 Cattle

11.2.2.1 Early interactions in the gastrointestinal tract

Inoculation of ligated bovine ileal loops with *S. enterica* induces the migration of peripheral blood granulocytes to the site of infection within 15 minutes of infection. The level of granulocyte infiltration is related to the ability of *S. enterica* strains to invade the intestinal epithelium, which in turn is related to the level of damage induced by the infection (Paulin *et al.*, 2002; Watson *et al.*, 1995). Following inoculation of ileal loops, bacteria can translocate from the gut lumen to mesenteric lymph nodes and, in some host and pathogen combinations, to the systemic periphery (Paulin *et al.*, 2002). Most of the bacteria detected in afferent lymph are extracellular (Paulin *et al.*, unpublished).

11.2.2.2 Interaction between *S. enterica* and macrophages or dendritic cells (DC)

S. enterica can be found intracellularly in MHC class II$^+$ antigen presenting cells both in the afferent lymph and intestinal mucosa (Paulin *et al.*, unpublished data). The early interaction between *S. enterica* and macrophages or DC is likely to affect the growth rate of the bacteria in the tissues, and to influence the profile of antigen specific acquired immunity in the later stages of the infection. The responses of bovine macrophages and DC to *S. enterica* are influenced by multiple parameters including the number and viability of the bacteria to which they are exposed (Norimatsu *et al.*, 2004; Norimatsu *et al.*, 2003). Exposure to live *S. enterica* serovar Typhimurium increases expression of MHC class I, MHC class II, CD40, CD80 and CD86 molecules on the cell surface of DC. In contrast, besides a marginal upregulation of CD40, macrophages do not exhibit detectable changes in their expression of cell surface molecules. Both macrophages and DC exposed to *S. enterica* serovar Typhimurium upregulate mRNA transcription of

TNFα, IL1β, IL6 and iNOS. Upregulation of mRNA transcripts for GM-CSF and IL12p40 occurs only in DC whilst only macrophages up-regulate IL10 (Norimatsu *et al.*, 2003). In general, live *S. enterica* vaccines induce better protection than killed ones. Although this may be due to a number of reasons, it is possible that different responses of DC following exposure to live and dead bacteria will determine the profile and the protective ability of the antigen specific response to *S. enterica*. Interestingly, live *S. enterica* induce greater upregulation of cell surface CD40 and CD86 on DC than killed bacteria (Norimatsu *et al.*, 2004). Similarly, upregulation of transcription of IL6, IL12 and GM-CSF is greater in DC exposed to live bacteria than in DC exposed to killed bacteria.

Macrophages from cattle breeds identified as resistant to challenge with *Brucella abortus* were tested for their ability to control the growth of bacteria in vitro. Results from these experiments indicated that while macrophages from resistant cattle were able to control the growth of *S. enterica* serovar Dublin, this was not the case when *S. enterica* serovar Typhimurium was used (Qureshi *et al.*, 1996). Thus, it would seem that in cattle there are differences in the genetic resistance to different *S. enterica* serovars.

11.2.3 Pigs

Little information is available on the mechanisms of host resistance that operate in the early stages of *S. enterica* infections of pigs.

IL8-induced infiltration of neutrophils can be seen into the lamina propria and villi of the intestine after oral administration of a *S. enterica* serovar Infantis mutant. Neutrophil numbers and activity are linked to increased resistance to systemic salmonellosis in the progeny of selectively bred boars (van Diemen *et al.*, 2002).

S. enterica infection can suppress some antimicrobial functions of porcine neutrophils. For example, neutrophils obtained from pigs infected with *S. enterica* showed reduced ability to uptake *S. enterica* micro-organisms (Stabel *et al.*, 2002).

11.2.4 Sheep

Little is known of the early response to *S. enterica* infection in sheep. Infection with *S. enterica* serovar Abortusovis in lambs is characterized by neutrophil infiltration that precedes the development of germinal centres (Fontaine *et al.*, 1994).

11.3 ADAPTIVE IMMUNITY

11.3.1 Chickens

11.3.1.1 B-cells and humoral immunity

The nature and kinetics of the humoral immune response to *S. enterica* infection of the chicken have been investigated extensively (Zhang-Barber *et al.*, 1999). Overall, infection with or vaccination against *S. enterica* serovar Typhimurium or Enteritidis leads to increased levels of specific IgY and IgA which may also be used for serological monitoring of flocks (Barrow *et al.*, 1992; Brito *et al.*, 1993; Hassan and Curtiss, 1990; Hassan and Curtiss, 1994). Secreted antibodies against *S. enterica* have also been found in the reproductive tract following experimental infection of that site (Withanage *et al.*, 1999). Contrasting results have been reported on the role of B-cells in the clearance of *S. enterica* from the gut. In fact, ablation of B-cells by chemical bursectomy using cyclophosphamide indicates that B-cells play a minimal role in the clearance of *S. enterica* from the intestine (Corrier *et al.*, 1991). However, bursectomy experiments using hormones and cyclophosphamide have indicated a requirement for B-cells in bacterial clearance from the gut (Desmidt *et al.*, 1998). As a consequence of this contradictory data, we do not yet have a clear understanding of the role of B-cells and antibody in the clearance of primary *Salmonella* infection in the chicken. However, the relative efficacy of killed-*Salmonella* vaccines in the chicken, discussed later in this chapter, indicates that antibody plays some role in protective immunity.

11.3.1.2 T-cells and cell mediated immunity

Few studies on the role of T-lymphocytes in immunity against *S. enterica* have been carried out in chickens.

Changes in the numbers of lymphocytes in the tissues can be observed in chickens infected with *S. enterica*. Infections with virulent and attenuated strains of *S. enterica* serovar Enteritidis in young chicks leads to an influx of T-cells into the ileum and the ceca around 20–24 hours post infection, with a later influx of B-cells (Van Immerseel *et al.*, 2002a; Van Immerseel *et al.*, 2002b). Reproductive tract infection with *S. enterica* serovar Enteritidis leads to a surge in the numbers of T-cells in the ovaries and oviduct reaching a peak at 10 days post infection, whilst a peak in B-cells occurs at 14 days post infection. Increased percentages of $CD4^+$ and $TCR1^+$ ($TCR\gamma\delta$) $CD8^+$ T-cells were observed in peripheral blood during infection of one-day-old chicks with *S. enterica* serovar Typhimurium.

T-cell responses to *S. enterica* antigens can be detected in infected animals. For example, there are reports of increased delayed type hypersensitivity in response to *Salmonella* antigens (Hassan and Curtiss, 1994; Lee *et al.*, 1983) and changes in the distribution of T- and B-cell subsets during infection (Berndt and Methner, 2001). T-cell proliferation can be detected in cells isolated from the spleen following infection with *S. enterica* serovar Typhimurium (Beal *et al.*, 2004a), or following vaccination with killed or live vaccines (Babu *et al.*, 2003). The highest levels of proliferation are found around three weeks post infection, a time that correlates with clearance of the transient systemic infection found with some invasive *S. enterica* serovar Typhimurium strains (Beal *et al.*, 2004a, b).

11.3.2 Cattle

11.3.2.1 B-cells and humoral immunity

Cattle exposed to *S. enterica* produce antibody responses against the bacterium. The age of the calves, the type of infection (*e.g.* systemic or enteric) and the route of infection all influence production of antibodies against *S. enterica* antigens. Calves of 1–2 weeks of age mount weaker responses than older calves (Da Roden *et al.*, 1992). Calves of different age groups mount responses to whole bacterial antigen preparations more readily than to LPS. The lack of response to LPS is more noticeable in young calves and the response to LPS increases with the age of the animals. In addition, repeated inoculation of killed antigen preparations trigger stronger antibody responses to whole-cell antigen preparations than inoculation with live attenuated bacteria. It has also been shown that inoculation of cattle with live *S. typhimurium aroA, aroD* vaccine strains via different routes influenced the type of antibodies generated and the antigens recognized by the antibody response (Villarreal-Ramos *et al.*, 1998). Using Western blot, it was found that different animals mounted different responses (animal to animal variation) and that oral and subcutaneous inoculation of live bacteria induced both systemic and mucosal antibody responses.

The role of antibodies in protection against salmonellosis in cattle has not been clearly defined. Jones and colleagues (Jones *et al.*, 1991) showed that calves fed with the colostrum of dams vaccinated with killed salmonellae were protected against virulent challenge with *S. enterica* serovar Typhimurium. However, the level of protection did not correlate with the level of antibodies in serum against O or H antigens. No studies were undertaken to determine the presence of mucosal antibodies against *S. enterica* in the intestinal tract. Experiments have also been carried out to determine the role of immune responses to LPS in protection against challenge (Segall and Lindberg, 1993). Calves vaccinated with live attenuated *S. enterica* serovar Dublin expressing serovar Typhimurium LPS developed antibodies against both serovar Dublin and Typhimurium LPS. However, upon challenge, animals were protected against serovar Dublin but not against serovar Typhimurium. From these experiments, it can be concluded that the presence of antibodies to *S. enterica* LPS in sera of cattle cannot be predictive of protection against challenge. Therefore, the presence of serum or milk antibodies to *S. enterica* LPS, at least in bovine salmonellosis, will have to be confined, at best, to being a diagnostic tool (Galland *et al.*, 2000). The experiments of Segall and Lindberg (1993) indicated that immune responses to a serovar specific non-LPS component are needed for the expression of protective immunity to *S. enterica* in cattle.

11.3.2.2 T-cells and cell mediated immunity

Inoculation of cattle with live attenuated *S. enterica* serovar Dublin or serovar Typhimurium induces the development of antigen-specific cell mediated immune responses that can be detected in peripheral blood (Lindberg and Andersson, 1983; Villarreal-Ramos *et al.*, 1998). These responses are assumed to be mainly from $CD4^+$ T-cells since they are measured in vitro after exposure of immune cells to antigens added exogenously. The role of $CD8^+$ T-cells in immunity to *Salmonella* in domestic species is largely unknown. It has been shown in mice that protective immune responses elicited by vaccination are dependent on the presence of $CD4^+$ and $CD8^+$ T-cells and that depletion of any of these cell subsets before challenge results in increased susceptibility to challenge with virulent *S. enterica* (Mastroeni *et al.*, 1992). To determine whether a similar situation occurs in cattle, animals were vaccinated with an *aroA aroD* mutant of *S. enterica* serovar Typhimurium and inoculated with $CD8^+$-depleting or control antibodies prior to challenge with virulent *S. enterica* serovar Typhimurium. Compared to unimmunized controls, vaccinated cattle were protected against challenge and no differences in response to the virulent challenge were detected between cattle that had been depleted of $CD8^+$ T-cells and non-depleted vaccinated controls (Villarreal-Ramos, unpublished data).

11.3.3 Pigs

Studies in pigs infected with *S. enterica* have shown antibody responses against the bacterium, and serology is increasingly being used as a means of monitoring infection (Christensen *et al.*, 1999; Nielsen *et al.*, 1995).

There is also evidence for the activation of T-cell responses in infected pigs. The production of IgG and IgM antibodies specific for LPS and outer membrane proteins is usually associated with lymphocyte proliferation (Gray *et al.*, 1996; Srinand *et al.*, 1995). Furthermore, protection of orally vaccinated gnotobiotic pigs against challenge with virulent *S. enterica* serovar Typhimurium (Dlabac *et al.*, 1997) coincides with expression of TGFβ by enterocytes (Trebichavsky *et al.*, 2003) and infiltration of TCRα/β^{+} T-cells in the tissues. Early stimulation of IL18 and IFNγ, usually involved in the initiation of T-cell immunity, can be seen in the gut of pigs infected with *S. enterica* (Splichal *et al.*, 2002).

11.3.4 Sheep

11.3.4.1 B-cells and antibody responses

Sheep exposed to *S. enterica* produce antibody specific for the bacterium. For example, inoculation of sheep with an *aroA* mutant of *S. enterica* serovar Typhimurium induces antigen specific IgM and IgG in serum accompanied by strong B-cell proliferative responses (Brennan *et al.*, 1995; Mukkur *et al.*, 1995).

Virulent strains of *S. enterica* serovar Abortusovis, as well as the vaccine Rv6 strain, elicit antibody responses in sheep, with the virulent strain inducing a greater total amount of IgM and IgG1 than the vaccine strain. On the other hand, the Rv6 vaccine induced a greater amount of IgG2 than the virulent strain (Gohin *et al.*, 1997). It remains to be established whether humoral immunity is protective in sheep salmonellosis.

11.3.4.2 T-cells and cell mediated immunity

Inoculation of sheep with the *aroA* mutant strain *S. enterica* serovar Typhimurium S25/1 induces weak antigen specific T-cell proliferative responses. Conversely, B-cells from these immunized animals proliferate strongly and produce IgM in response to *S. enterica* (Brennan *et al.*, 1995). Cytokine mRNA specific for IL2, IL10, IL1β and IL8 is upregulated in lymph nodes of sheep inoculated with *S. enterica* serovar Abortusovis, together with reduction in CD4^{+} T-cells and an increase in B-cells in the prescapular lymph node (Doucet, Bernard, 1997). This seems to indicate that the sheep humoral immune system responds efficiently to *S. enterica* infection, whereas cell mediated immunity to the bacterium is rather inefficient in these animals.

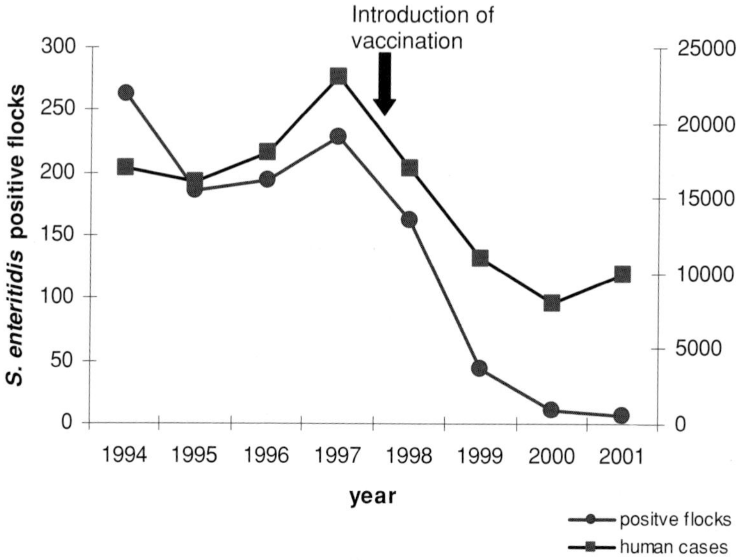

Figure 11.1. Vaccination of British flocks with Salenvac® led to a decrease in the incidence of *S. enterica* serovar Enteritidis; this in turn was accompanied by a decrease in the number of food poisoning cases due to *S. enterica* serovar Enteritidis.

11.4 VACCINES AGAINST *S. ENTERICA* INFECTIONS

11.4.1 Vaccines for chickens

Vaccination of laying hens against *S. enterica* serovar Enteritidis has been increasingly employed in the control of transmission of infection to eggs. In the UK, introduction of the killed vaccine, Salenvac, produced by Intervet, in the late 1990s led to a reduction in the incidence of serovar Enteritidis in flocks. It is believed that the decreased levels of *S. enterica* in chickens has led to a reduction in the number of food poisoning cases due to serovar Enteritidis (Figure 11.1) (Anonymous, 2000). The Salenvac vaccine is an iron-restricted bacterin that has been shown experimentally to partially protect chickens against intravenous challenge with *S. enterica* serovar Enteritidis. The vaccine also reduced subsequent egg batch contamination from 25% in unvaccinated birds to 13% in the vaccinated group (Woodward *et al.*, 2002). A bivalent killed vaccine against *S. enterica* serovars Enteritidis and Typhimurium has now been introduced for layers. Following oral infection, this vaccine protected against intestinal colonization, though significant numbers of vaccinated birds still shed *Salmonella* (Clifton-Hadley *et al.*, 2002). Live vaccines, generated through chemical mutagenesis or in vitro subculturing, have now been licensed for use in parts of Europe. One of these vaccines, the TAD E vaccine (Lohmann Animal Health), a mutant of *S. enterica* serovar Enteritidis, is believed to protect *via* induction of specific immunity and competitive exclusion of pathogenic strains. Another vaccine generated in a similar way is the TAD T vaccine, which is a *S. typhimurium* mutant, which is currently used in chickens. Recently, the *S. enterica* serovar Gallinarum 9R vaccine (Smith, 1956), produced by Intervet and commercially known as Nobilis SG9R, has been licensed for immunization against serovar Enteritidis (Barrow, 1991). A number of field trials have shown that the serovar Gallinarum 9R vaccine confers cross-protecion against serovar Enteritidis and that it is safe in terms of animal and human health (Feberwee *et al.*, 2001). All the live vaccines currently licensed in Europe are non-defined mutants or killed vaccines which have largely been developed in an empirical manner. However, *S. enterica* strains with defined mutations can also elicit protective immunity. *S. enterica* strains with attenuating mutations in genes involved in the synthesis of aromatic amino acids (e.g. *aroA*) protect chickens against infection with virulent *S. enterica* (Barrow, 1991; Cooper *et al.*, 1994a, b; Zhang-Barber *et al.*, 1999). Mutants with deletions in *cya* and *cyd* genes, incuding Megan Vacc1 (a serovar Typhimurium *cya cyd* mutant) and Megan Egg (a serovar Enteritidis *cya cyd* mutant) have also been developed and licensed for use in chickens in the USA (Curtiss and Hassan, 1996; Hassan and Curtiss, 1990; Hassan and Curtiss, 1994). The *cya cyd* mutants elicit greater T-cell responses to *S. enterica* antigens compared to killed vaccines (Babu *et al.*, 2003). *S. enterica* mutants in the gene encoding DNA adenine methylase (*dam*) are attenuated for virulence in mice (Heithoff *et al.*, 2001), and confer some, albeit limited, protection (Dueger *et al.*, 2001). A limited competitive exclusion effect has also been observed with *dam* mutants of *S. enterica* in young chicks (Dueger *et al.*, 2003).

11.4.2 Vaccines for cattle

Live attenuated vaccines against salmonellosis have been tested in cattle. Some of the *S. enterica* strains that are protective in murine models of infection surprisingly failed to confer protection in cattle. For example, mutants of *S. enterica* serovar Typhimurium auxotrophic for aromatic derivatives and undefined mutants of *S. enterica* serovar Dublin showed variable levels of protection in cattle, and there are no simple explanations for the differences in efficacy between the vaccine strains (Smith *et al.*, 1984; Smith, 1965). The *aroA* SL3261 strain of serovar Typhimurium, widely used and protective in murine studies, fails to confer protection in cattle. Equally puzzling, is the fact that *aro* mutants derived from one strain, but using two different methodologies for their generation, differ in their ability to confer protection to cattle. The *aro* mutant *S. enterica* serovar Typhimurium SL3261 was derived by transposon mutagenesis from the virulent strain SL1344, which in turn was derived from the virulent strain 4/74 (Hoiseth and Stocker, 1981; Smith *et al.*, 1984). Jones and colleagues used precise in vitro mutagenesis techniques to generate an *aroA, aroD* double mutant from the strain 4/74. Seven-day old calves vaccinated orally or parenterally with the *aroA, aroD* derivative were protected against virulent challenge (Jones *et al.*, 1991). The reasons for the differences between the protective ability of strain SL3261 and the *aroA, aroD* derivatives must be subtle and remain to be explained. Interestingly, both vaccine strains are capable of conferring protection in mice (Hoiseth and Stocker, 1981).

S. enterica strains that are safe in mice can retain virulence in cattle. For example, *purE* and *htrA* derivatives of *S. typhimurium* are attenuated in mice but virulent for cattle when delivered orally (Villarreal-Ramos *et al.*, 2000). Calves inoculated orally with *purE* or *htrA* derivatives succumb with acute salmonellosis. This precludes the use of these vaccines as potential measures to prevent salmonellosis in cattle, and strongly highlights the need for more research and vaccine testing in domestic species.

More recently, the *S. enterica* serovar Typhimurium *dam* mutant has been shown to confer protection against virulent challenge in cattle (Dueger *et al.*, 2003).

11.4.3 Vaccines for pigs

Live vaccines against salmonellosis are currently licenced for use in pigs. One of these vaccines is the undefined strain SC-54 of *S. enterica* serovar Choleraesuis that was generated through passage in pig neutrophils (Kramer *et al.*, 1992). The SC-54 *S. enterica* serovar Choleraesuis vaccine strain is highly attenuated for pigs and generates protection as detected by reduced clinical signs or decreased numbers of the wild type challenge strain in vaccinated animals (Kennedy *et al.*, 1999; Roof and Doitchinoff, 1995). A series of defined *S. enterica* serovar Choleraesuis strains which contain deletions in *cya* and *crp* genes and lack the virulence plasmid, generated a high degree of protection (Kennedy *et al.*, 1999). Other live vaccines are being evaluated, including undefined mutants of *S. enterica* serovar Typhimurium generated by chemical mutagenesis (Springer *et al.*, 2001). The *cya crp* double mutants elicited strong serological responses but apparently no cell mediated immunity. An undefined mutant of *S. enterica* serovar Typhimurium, generated by chemical mutagenesis (Springer *et al.*, 2001), conferred a degree of protection against colonization when delivered orally and parenterally. A *S. enterica* serovar Typhimurium *aroA* mutant conferred protection against enteritis induced by challenge with virulent bacteria following parenteral vaccination (Lumsden and Wilkie, 1992).

11.4.4 Vaccines for sheep

Undefined vaccines of *S. enterica* serovar Abortusovis (Pardon *et al.*, 1990) and *aroA* derivatives of *S. enterica* serovar Typhimurium (Mukkur and Walker, 1992) have been tested with success in sheep providing good protection for up to 6 months after oral or intramuscular inoculation.

11.5 LIVE *SALMONELLA* VACCINES AS VECTORS FOR THE DELIVERY OF HETEROLOGOUS ANTIGENS IN DOMESTIC SPECIES

Live attenuated salmonellae have been used successfully to elicit immune responses against heterologous antigens in murine models and in some domestic species. Although live attenuated *S. enterica* have not been as successful in delivering recombinant antigens in all host species, live attenuated *aro* mutants are capable of eliciting immune responses to recombinant antigens in cattle. An *aroA* mutant of *S. enterica* serovar Dublin expressing the p67 sporozoite antigen of *Theileria parva* was used to immunize cattle against East Coast Fever (Gentschev *et al.*, 1998). Cattle immunized intramuscularly three times with the *S. enterica* vaccine expressing p67 showed high levels of antibody production against the p67 antigen and two out of three calves were protected against challenge with *T. parva*. On the other hand, cattle immunized orally with the same construct showed low or negligible production of antibody against the guest antigen and only one out of three animals showed a degree of protection against challenge with *T. parva*. Immune responses against the fragment C of tetanus toxin (TetC) were generated in calves vaccinated orally or subcutaneously with a recombinant *S. enterica* serovar Typhimurium *aroA aroD* mutant carrying a plasmid encoding TetC (Villarreal-Ramos *et al.*, 1998; Villarreal-Ramos *et al.*, 2000). Oral immunization resulted in the generation of mucosal but not serum antibodies against TetC; subcutaneous inoculation resulted in the induction of both serum and mucosal antibodies.

S. enterica serovar Gallinarum 9R vaccine has been used as a vector to express and deliver TetC in Rhode Island Red chickens (Wigley and Barrow, unpublished observations); the vector efficiently induced circulating

IgG against TetC. Recently, *S. enterica* serovar Gallinarum and serovar Typhimurium have been used to express and deliver *Eimeria* antigens (Pogonka *et al.*, 2003; Vermeulen, 1998). *S. enterica* serovar Typhimurium constructs expressing the SO7 and TA4 *E. tenella* antigens efficiently induced antibodies against the guest antigens (Pogonka *et al.*, 2003). Although immune responses have been elicited, no attempts were reported to test the efficacy of protection induced by these constructs.

Therefore, live attenuated *S. enterica* recombinant strains show promise as multivalent vaccines for the delivery of antigens to the immune system of domestic animals.

11.6 PROTECTION INDUCED BY LIVE *S. ENTERICA* VACCINES BY NON-IMMUNE AND NON-SPECIFIC IMMUNE MECHANISMS

11.6.1 Protection by non-immune mechanisms

Besides their ability to confer protection against challenge through the generation of antigen specific immune responses, live vaccines are able to confer protection against colonization and enteritis through other mechanisms. These mechanisms include competitive exclusion, possibly based on competition for nutrients, space or production of inhibiting factors. Other mechanisms are the development of a colonization-inhibition effect as exemplified by the oral administration of live vaccines to young poultry, which results in extensive colonization (within 24 hours) that inhibits subsequent colonization by homologous challenge (Barrow *et al.*, 1987; Berchieri, Barrow, 1990). This inhibition is genus-specific and thought to be different to that obtained by use of intestinal flora preparations. The genus-specific inhibition is a microbiological effect that is thought to be similar to the down-regulation of bacterial growth observed in stationary-phase broth cultures under anaerobic or micro-aerophilic conditions (Barrow *et al.*, 1996; Berchieri and Barrow, 1991; Zhang-Barber *et al.*, 1997). The genus-specific inhibition is not the result of immune stimulation or the result of bacteriophage or bacteriocin activity (Barrow *et al.*, 1987) and shows a degree of serovar specificity (Martin *et al.*, 2002).

11.6.2 Protection by non-specific immune mechanisms

Administration of live vaccines, especially at high doses, is likely to result in large numbers of attenuated bacteria in the intestine. This would lead to the stimulation of non-specific immune mechanisms such as neutrophil/heterophil infiltration resulting in a degree of resistance to invasion by virulent micro-organisms. In the case of chickens this is a relatively short period of resistance against invasive infection (Van Immerseel *et al.*, 2002b), and in pigs the activation of non-specific immune mechanisms induces a profound resistance to enteritis caused by reinfection with virulent bacteria (Foster *et al.*, 2003).

11.7 CONCLUSIONS

Murine infection models have been useful in elucidating the fundamental mechanisms of protection against salmonellosis. However, the vast spectrum of diseases induced by different *S. enterica* serotypes in domestic animal species requires specific research approaches in individual target species. It is likely that the nature of protective immunity would be different in different types of *S. enterica* infections (e.g. systemic *vs* enteric infections). The range of infections and immune responses can be determined by the host specificity of the infecting bacterial strains as well as by intrinsic features of the immune system of different animals.

The understanding of the mechanisms of immunity to *S. enterica* in targeted species has important implications for the development of vaccines that are safe for the animal and do not pose a threat to humans if they enter the food chain.

Vaccination elicits an antibody response that may interfere with the use of serology for immunological detection of infectious diseases. The improvement of our understanding of immunity to *S. enterica* in domestic animals can facilitate the definition of correlates of protection and can improve our ability to distinguish between vaccinated and infected individuals.

REFERENCES

Anonymous (2000). *Salmonella in Livestock 1999*. London: Veterinary Laboratory Agency / Ministry of Agriculture Fisheries and Food.

Babu, U., Scott, M., Myers, M. J. *et al.* (2003). Effects of live attenuated and killed *Salmonella* vaccine on T-lymphocyte mediated immunity in laying hens. *Vet Immunol Immunopathol*, **91**, 39–44.

Barrow, P. (1991). Serological analysis for antibodies to *S. enteritidis*. *Vet Rec*, **128**, 43–4.

Barrow, P. A., Berchieri, A., Jr. and al-Haddad, O. (1992). Serological response of chickens to infection with *Salmonella gallinarum-S. pullorum* detected by enzyme-linked immunosorbent assay. *Avian Dis*, **36**, 227–36.

Barrow, P. A., Bunstead, N., Marston, K., Lovell, M. A. and Wigley, P. (2004). Faecal shedding and intestinal colonization of *Salmonella enterica* in inbred chickens: the effect of host-genetic background. *Epidemiol Infect*, **132**, 117–26.

Barrow, P. A., Huggins, M. B. and Lovell, M. A. (1994). Host specificity of *Salmonella* infection in chickens and mice is expressed *in vivo* primarily at the level of the reticuloendothelial system. *Infect Immun*, **62**, 4602–10.

Barrow, P. A., Lovell, M. A. and Barber, L. Z. (1996). Growth suppression in early stationary phase nutrient broth cultures of *Salmonella typhimurium* and *Escherichia coli* is genus specific and not regulated by sigma S. *J Bacteriol*, **178**, 3072–6.

Barrow, P. A., Tucker, J. F. and Simpson, J. M. (1987). Inhibition of colonization of the chicken alimentary tract with *Salmonella typhimurium* Gram-negative facultatively anaerobic bacteria. *Epidemiol Infect*, **98**, 311–22.

Beal, R. K., Powers, C., Wigley, P., Barrow, P. A. and Smith, A. L. (2004a). Temporal dynamics of the cellular, humoral and cytokine responses in chickens during primary and secondary infection with *Salmonella enterica* serovar Typhimurium. *Avian Pathol*, **33**, 25–33.

Beal, R. K., Wigley, P., Powers, C. *et al.* (2004b). Age at primary infection with *Salmonella enterica* serovar Typhimurium in the chicken influences persistence of infection and subsequent immunity to re-challenge. *Vet Immunol Immunopathol*, **100**, 151–64.

Berchieri, A., Jr. and Barrow, P. A. (1990). Further studies on the inhibition of colonization of the chicken alimentary tract with *Salmonella typhimurium* by pre-colonization with an avirulent mutant. *Epidemiol Infect*, **104**, 427–41.

Berndt, A. and Methner, U. (2001). Gamma/delta T-cell response of chickens after oral administration of attenuated and non-attenuated *Salmonella typhimurium* strains. *Vet Immunol Immunopathol*, **78**, 143–61.

Boyd, Y., Goodchild, M., Morroll, S. and Bumstead, N. (2001). Mapping of the chicken and mouse genes for Toll-like receptor 2 (TLR2) to an evolutionarily conserved chromosomal segment. *Immunogenetics*, **52**, 294–8.

Brennan, F. R., Oliver, J. J. and Baird, G. D. (1995). *In vitro* studies with lymphocytes from sheep orally inoculated with an aromatic-dependent mutant of *Salmonella typhimurium*. *Res Vet Sci*, **58**, 152–7.

Brito, J. R., Hinton, M., Stokes, C. R. and Pearson, G. R. (1993). The humoral and cell mediated immune response of young chicks to *Salmonella typhimurium* and *S. Kedougou*. *Br Vet J*, **149**, 225–34.

Chadfield, M. S., Brown, D. J., Aabo, S., Christensen, J. P. and Olsen, J. E. (2003). Comparison of intestinal invasion and macrophage response of *Salmonella* Gallinarum and other host-adapted *Salmonella enterica* serovars in the avian host. *Vet Microbiol*, **92**, 49–64.

Christensen, J., Baggesen, D. L., Soerensen, V. and Svensmark, B. (1999). *Salmonella* level of Danish swine herds based on serological examination of meat-juice samples and *Salmonella* occurrence measured by bacteriological follow-up. *Prev Vet Med*, **40**, 277–92.

Clifton-Hadley, F. A., Breslin, M., Venables, L. M. *et al.* (2002). A laboratory study of an inactivated bivalent iron restricted *Salmonella enterica* serovars Enteritidis and Typhimurium dual vaccine against Typhimurium challenge in chickens. *Vet Microbiol*, **89**, 167–79.

Cooper, G. L., Venables, L. M., Woodward, M. J. and Hormaeche, C. E. (1994a). Invasiveness and persistence of *Salmonella enteritidis, Salmonella typhimurium*, and a genetically defined *S. enteritidis aroA* strain in young chickens. *Infect Immun*, **62**, 4739–46.

(1994b). Vaccination of chickens with strain CVL30, a genetically defined *Salmonella enteritidis aroA* live oral vaccine candidate. *Infect Immun*, **62**, 4747–54.

Corrier, D. E., Elissalde, M. H., Ziprin, R. L. and DeLoach, J. R. (1991). Effect of immunosuppression with cyclophosphamide, cyclosporin, or dexamethasone on *Salmonella* colonization of broiler chicks. *Avian Dis*, **35**, 40–5.

Curtiss, R., III and Hassan, J. O. (1996). Nonrecombinant and recombinant avirulent *Salmonella* vaccines for poultry. *Vet Immunol Immunopathol*, **54**, 365–72.

Da Roden, L., Smith, B. P., Spier, S. J. and Dilling, G. W. (1992). Effect of calf age and *Salmonella* bacterin type on ability to produce immunoglobulins directed against *Salmonella* whole cells or lipopolysaccharide. *Am J Vet Res*, **53**, 1895–9.

Desmidt, M., Ducatelle, R. and Haesebrouck, F. (1997). Pathogenesis of *Salmonella enteritidis* phage type four after experimental infection of young chickens. *Vet Microbiol*, **56**, 99–109.

Desmidt, M., Ducatelle, R., Mast, J. *et al.* (1998). Role of the humoral immune system in *Salmonella enteritidis* phage type four infection in chickens. *Vet Immunol Immunopathol*, **63**, 355–67.

Dlabac, V., Trebichavsky, I., Rehakova, Z. *et al.* (1997). Pathogenicity and protective effect of rough mutants of *Salmonella* species in germ-free piglets. *Infect Immun*, **65**, 5238–43.

Doucet, F. and Bernard, S. (1997). *In vitro* cellular responses from sheep draining lymph node cells after subcutaneous inoculation with *Salmonella abortusovis*. *Vet Res*, **28**, 165–78.

Dueger, E. L., House, J. K., Heithoff, D. M. and Mahan, M. J. (2001). *Salmonella* DNA adenine methylase mutants elicit protective immune responses to homologous and heterologous serovars in chickens. *Infect Immun*, **69**, 7950–4.

(2003). *Salmonella* DNA adenine methylase mutants prevent colonization of newly hatched chickens by homologous and heterologous serovars. *Int J Food Microbiol*, **80**, 153–9.

Eckmann, L., Fierer, J. and Kagnoff, M. F. (1996). Genetically resistant (*Ity*ʳ) and susceptible (*Ity*ˢ) congenic mouse strains show similar cytokine responses following infection with *Salmonella dublin*. *J Immunol*, **156**, 2894–900.

Feberwee, A., Hartman, E. G., de Wit, J. J. and de Vries, T. S. (2001). The spread of *Salmonella gallinarum* 9R vaccine strain under field conditions. *Avian Dis*, **45**, 1024–9.

Fontaine, J. J., Pepin, M., Pardon, P., Marly, J. and Parodi, A. L. (1994). Comparative histopathology of draining lymph node after infection with virulent or attenuated strains of *Salmonella abortusovis* in lambs. *Vet Microbiol*, **39**, 61–9.

Foster, N., Lovell, M. A., Marston, K. L. *et al.* (2003). Rapid protection of gnotobiotic pigs against experimental salmonellosis following induction of polymorphonuclear leukocytes by avirulent *Salmonella enterica*. *Infect Immun*, **71**, 2182–91.

Fukui, A., Inoue, N., Matsumoto, M. *et al.* (2001). Molecular cloning and functional characterization of chicken Toll-like receptors. A single chicken Toll covers multiple molecular patterns. *J Biol Chem*, **276**, 47143–9.

Galland, J. C., House, J. K., Hyatt, D. R. *et al.* (2000). Prevalence of *Salmonella* in beef feeder steers as determined by bacterial culture and ELISA serology. *Vet Microbiol*, **76**, 143–51.

Gautier, A. V., Lantier, I. and Lantier, F. (1998). Mouse susceptibility to infection by the *Salmonella abortusovis* vaccine strain Rv6 is controlled by the *Ity/Nramp* 1 gene and influences the antibody but not the complement responses. *Microb Pathog*, **24**, 47–55.

Gentschev, I., Glaser, I., Goebel, W. *et al.* (1998). Delivery of the p67 sporozoite antigen of *Theileria parva* by using recombinant *Salmonella dublin*: secretion of the product enhances specific antibody responses in cattle. *Infect Immun*, **66**, 2060–144.

Gohin, I., Olivier, M., Lantier, I., Pepin, M. and Lantier, F. (1997). Analysis of the immune response in sheep efferent lymph during *Salmonella abortusovis* infection. *Vet Immunol Immunopathol*, **60**, 111–30.

Gray, J. T., Stabel, T. J. and Fedorka-Cray, P. J. (1996). Effect of dose on the immune response and persistence of *Salmonella choleraesuis* infection in swine. *Am J Vet Res*, **57**, 313–19.

Hassan, J. O. and Curtiss, R., III (1990). Control of colonization by virulent *Salmonella typhimurium* by oral immunization of chickens with avirulent Δ*cya* Δ*crp S. typhimurium*. *Res Microbiol*, **141**, 839–50.

(1994). Development and evaluation of an experimental vaccination program using a live avirulent *Salmonella typhimurium* strain to protect immunized chickens against challenge with homologous and heterologous *Salmonella* serotypes. *Infect Immun*, **62**, 5519–27.

Heithoff, D. M., Enioutina, E. Y., Daynes, R. A., Sinsheimer, R. L., Low, D. A. and Mahan, M. J. (2001). *Salmonella* DNA adenine methylase mutants confer cross-protective immunity. *Infect Immun*, **69**, 6725–30.

Henderson, S. C., Bounous, D. I. and Lee, M. D. (1999). Early events in the pathogenesis of avian salmonellosis. *Infect Immun*, **67**, 3580–6.

Hoiseth, S. K. and Stocker, B. A. (1981). Aromatic-dependent *Salmonella typhimurium* are non-virulent and effective as live vaccines. *Nature*, **291**, 238–9.

Hu, J., Bumstead, N., Barrow, P. *et al.* (1997). Resistance to salmonellosis in the chicken is linked to NRAMP1 and TNC. *Genome Res*, **7**, 693–704.

Jones, M. A., Wigley, P., Page, K. L., Hulme, S. D. and Barrow, P. A. (2001). *Salmonella enterica* serovar Gallinarum requires the *Salmonella* pathogenicity island 2 type III secretion system but not the *Salmonella* pathogenicity island 1 type III secretion system for virulence in chickens. *Infect Immun*, **69**, 5471–6.

Jones, P. W., Dougan, G., Hayward, C. *et al.* (1991). Oral vaccination of calves against experimental salmonellosis using a double *aro* mutant of *Salmonella typhimurium*. *Vaccine*, **9**, 29–34.

Kaiser, P., Rothwell, L., Galyov, E. E. *et al.* (2000). Differential cytokine expression in avian cells in response to invasion by *Salmonella typhimurium*, *Salmonella enteritidis* and *Salmonella gallinarum*. *Microbiology*, **146** (Pt 12), 3217–26.

Kennedy, M. J., Yancey, R. J., Jr., Sanchez, M. S. *et al.* (1999). Attenuation and immunogenicity of Δ*cya* Δ*crp* derivatives of *Salmonella choleraesuis* in pigs. *Infect Immun*, **67**, 4628–36.

Kogut, M. H., Rothwell, L. and Kaiser, P. (2003a). Differential regulation of cytokine gene expression by avian heterophils during receptor-mediated phagocytosis of opsonized and nonopsonized *Salmonella enteritidis*. *J Interferon Cytokine Res*, **23**, 319–27.

Kogut, M. H., Rothwell, L. and Kaiser, P. (2003b). Priming by recombinant chicken interleukin-2 induces selective expression of IL8 and IL18 mRNA in chicken heterophils during receptor-mediated phagocytosis of opsonized and nonopsonized *Salmonella enterica* serovar enteritidis. *Mol Immunol*, **40**, 603–10.

Kogut, M. H., Tellez, G. I., McGruder, E. D. *et al.* (1994). Heterophils are decisive components in the early responses of chickens to *Salmonella enteritidis* infections. *Microb Pathog*, **16**, 141–51.

Kramer, T. T., Roof, M. B. and Matheson, R. R. (1992). Safety and efficacy of an attenuated strain of *Salmonella choleraesuis* for vaccination of swine. *Am J Vet Res*, **53**, 444–8.

Lee, G. M., Jackson, G. D. and Cooper, G. N. (1983). Infection and immune responses in chickens exposed to *Salmonella typhimurium*. *Avian Dis*, **27**, 577–83.

Leveque, G., Forgetta, V., Morroll, S. *et al.* (2003). Allelic variation in TLR4 is linked to susceptibility to *Salmonella enterica* serovar Typhimurium infection in chickens. *Infect Immun*, **71**, 1116–24.

Lindberg, A. A. and Andersson, J. A. (1983). *Salmonella typhimurium* infection in calves: cell-mediated and humoral immune reactions before and after challenge with live virulent bacteria in calves given live or inactivated vaccines. *Infection and Immunity*, **41**, 751–7.

Lumsden, J. S. and Wilkie, B. N. (1992). Immune response of pigs to parenteral vaccination with an aromatic-dependent mutant of *Salmonella typhimurium*. *Can J Vet Res*, **56**, 296–302.

Mariani, P., Barrow, P. A., Cheng, H. H. *et al.* (2001). Localization to chicken chromosome 5 of a novel locus determining salmonellosis resistance. *Immunogenetics*, **53**, 786–91.

Martin, G., Methner, U., Rychlik, I. and Barrow, P. A. (2002). [Specificity of inhibition between *Salmonella* strains.] *Dtsch Tierarztl Wochenschr*, **109**, 154–7.

Mastroeni, P., Villarreal-Ramos, B. and Hormaeche, C. E. (1992). Role of T-cells, TNFα and IFNγ in recall of immunity to oral challenge with virulent salmonellae in mice vaccinated with live attenuated *aro Salmonella* vaccines. *Microbial Pathogenesis*, **13**, 477–91.

Maxwell, M. H. and Robertson, G. (1998). The avian heterophil leucocyte: a review. *World's Poultry Science Journal*, **54**, 155–78.

Mukkur, T. K. and Walker, K. H. (1992). Development and duration of protection against salmonellosis in mice and sheep immunised with live aromatic-dependent *Salmonella typhimurium*. *Res Vet Sci*, **52**, 147–53.

Mukkur, T. K., Walker, K. H., Baker, P. and Jones, D. (1995). Systemic and mucosal intestinal antibody response of sheep immunized with aromatic-dependent live or killed *Salmonella typhimurium*. *Comp Immunol Microbiol Infect Dis*, **18**, 27–39.

Nielsen, B., Baggesen, D., Bager, F., Haugegaard, J. and Lind, P. (1995). The serological response to *Salmonella* serovars Typhimurium and Infantis in experimentally infected pigs. The time course followed with an indirect anti-LPS ELISA and bacteriological examinations. *Vet Microbiol*, **47**, 205–18.

Norimatsu, M., Chance, V., Dougan, G., Howard, C. J. and Villarreal-Ramos, B. (2004). Live *Salmonella enterica* serovar Typhimurium (*S. Typhimurium*) elicit dendritic cell responses that differ from those induced by killed *S. typhimurium*. *Vet Immunol Immunopathol*, **98**, 193–201.

Norimatsu, M., Harris, J., Chance, V. *et al.* (2003). Differential response of bovine monocyte-derived macrophages and dendritic cells to infection with *Salmonella typhimurium* in a low-dose model in vitro. *Immunology*, **108**, 55–61.

Pardon, P., Marly, J., Lantier, F. and Sanchis, R. (1990). Vaccinal properties of *Salmonella abortusovis* mutants for streptomycin: screening with an ovine model. *Ann Rech Vet*, **21**, 57–67.

Paulin, S. M., Watson, P. R., Benmore, A. R. *et al.* (2002). Analysis of *Salmonella enterica* serotype-host specificity in calves: avirulence of *S. enterica* serotype Gallinarum correlates with bacterial dissemination from mesenteric lymph nodes and persistence in vivo. *Infect Immun*, **70**, 6788–97.

Plant, J. and Glynn, A. A. (1974). Natural resistance to *Salmonella* infection, delayed hypersensitivity and *Ir* genes in different strains of mice. *Nature*, **248**, 345–7.

Pogonka, T., Klotz, C., Kovacs, F. and Lucius, R. (2003). A single dose of recombinant *Salmonella typhimurium* induces specific humoral immune responses against heterologous *Eimeria tenella* antigens in chicken. *Int J Parasitol*, **33**, 81–8.

Qureshi, M. A. (2003). Avian macrophage and immune response: an overview. *Poult Sci*, **82**, 691–8.

Roof, M. B. and Doitchinoff, D. D. (1995). Safety, efficacy, and duration of immunity induced in swine by use of an avirulent live *Salmonella choleraesuis*-containing vaccine. *Am J Vet Res*, **56**, 39–44.

Segall, T. and Lindberg, A. A. (1993). Oral vaccination of calves with an aromatic-dependent *Salmonella dublin* (O9,12) hybrid expressing O4,12 protects against *S. dublin* (O9,12) but not against *Salmonella typhimurium* (O4, 5,12). *Infect Immun*, **61**, 1222–31.

Smith, B. P., Reina-Guerra, M., Stocker, B. A., Hoiseth, S. K. and Johnson, E. (1984). Aromatic-dependent *Salmonella dublin* as a parenteral modified live vaccine for calves. *Am J Vet Res*, **45**, 2231–5.

Smith, H. W. (1956). The use of live vaccines in experimental *Salmonella gallinarum* infection in chickens with observations on their interference effect. *J Hyg (Lond)*, **54**, 419–32.

(1965). The immunization of mice, calves and pigs against *Salmonella dublin* and *Salmonella choleraesuis* infections. *J Hyg (Lond)*, **63**, 117–35.

Splichal, I., Trebichavsky, I., Muneta, Y. and Mori, Y. (2002). Early cytokine response of gnotobiotic piglets to *Salmonella enterica* serotype Typhimurium. *Vet Res*, **33**, 291–7.

Springer, S., Lindner, T., Steinbach, G. and Selbitz, H. J. (2001). Investigation of the efficacy of a genetically-stabile live *Salmonella typhimurium* vaccine for use in swine. *Berl Munch Tierarztl Wochenschr*, **114**, 342–5.

Srinand, S., Robinson, R. A., Collins, J. E. and Nagaraja, K. V. (1995). Serologic studies of experimentally induced *Salmonella choleraesuis* var kunzendorf infection in pigs. *Am J Vet Res*, **56**, 1163–8.

Stabel, T. J., Fedorka-Cray, P. J. and Gray, J. T. (2002). Neutrophil phagocytosis following inoculation of *Salmonella choleraesuis* into swine. *Vet Res Commun*, **26**, 103–9.

Trebichavsky, I., Splichal, I., Splichalova, A., Muneta, Y. and Mori, Y. (2003). Systemic and local cytokine response of young piglets to oral infection with *Salmonella enterica* serotype Typhimurium. *Folia Microbiol (Praha)*, **48**, 403–7.

van Diemen, P. M., Kreukniet, M. B., Galina, L., Bumstead, N. and Wallis, T. S. (2002). Characterisation of a resource population of pigs screened for resistance to salmonellosis. *Vet Immunol Immunopathol*, **88**, 183–96.

Van Immerseel, F., De Buck, J., De Smet, I., Mast, J., Haesebrouck, F. and Ducatelle, R. (2002a). Dynamics of immune cell infiltration in the caecal lamina propria of chickens after neonatal infection with a *Salmonella enteritidis* strain. *Dev Comp Immunol*, **26**, 355–64.

(2002b). The effect of vaccination with a *Salmonella enteritidis aroA* mutant on early cellular responses in caecal lamina propria of newly-hatched chickens. *Vaccine*, **20**, 3034–41.

Vermeulen, A. N. (1998). Progress in recombinant vaccine development against coccidiosis. A review and prospects into the next millennium. *Int J Parasitol*, **28**, 1121–30.

Vidal, S. M., Malo, D., Vogan, K., Skamene, E. and Gros, P. (1993). Natural resistance to infection with intracellular parasites: isolation of a candidate for Bcg. *Cell*, **73**, 469–85.

Villarreal-Ramos, B., Manser, J., Collins, R. A. *et al.* (1998). Immune responses in calves immunised orally or subcutaneously with a live *Salmonella typhimurium* aro vaccine. *Vaccine*, **16**, 45–54.

(2000). Susceptibility of calves to challenge with *Salmonella typhimurium* 4/74 and derivatives harbouring mutations in *htrA* or *purE*. *Microbiology*, **146**, 2775–83.

Watson, P. R., Paulin, S. M., Bland, A. P., Jones, P. W. and Wallis, T. S. (1995). Characterization of intestinal invasion by *Salmonella typhimurium* and

Salmonella dublin and effect of a mutation in the *invH* gene. *Infect Immun*, **63**, 2743–54.

Wells, L. L., Lowry, V. K., DeLoach, J. R. and Kogut, M. H. (1998). Age-dependent phagocytosis and bactericidal activities of the chicken heterophil. *Dev Comp Immunol*, **22**, 103–9.

Wigley, P., Berchieri, A., Jr., Page, K. L., Smith, A. L. and Barrow, P. A. (2001). *Salmonella enterica* serovar Pullorum persists in splenic macrophages and in the reproductive tract during persistent, disease-free carriage in chickens. *Infect Immun*, **69**, 7873–9.

Wigley, P., Hulme, S. D., Bumstead, N. and Barrow, P. A. (2002a). *In vivo* and *in vitro* studies of genetic resistance to systemic salmonellosis in the chicken encoded by the SAL1 locus. *Microbes Infect*, **4**, 1111–20.

Wigley, P., Jones, M. A. and Barrow, P. A. (2002b). *Salmonella enterica* serovar Pullorum requires the *Salmonella* pathogenicity island 2 type III secretion system for virulence and carriage in the chicken. *Avian Pathol*, **31**, 501–6.

Withanage, G. S., Sasai, K., Fukata, T. *et al.* (1998). T-lymphocytes, B-lymphocytes, and macrophages in the ovaries and oviducts of laying hens experimentally infected with *Salmonella enteritidis*. *Vet Immunol Immunopathol*, **66**, 173–84.

Withanage, G. S., Sasai, K., Fukata, T., Miyamoto, T. and Baba, E. (1999). Secretion of *Salmonella*-specific antibodies in the oviducts of hens experimentally infected with *Salmonella enteritidis*. *Vet Immunol Immunopathol*, **67**, 185–93.

Withanage, G. S. K., Kaiser, P., Wigley, P., Powers, C., Mastroeni, P., Brooks, H., Barrow, P., Smith, A., Maskell, D., and McConnell, I. (2004). Rapid expression of chemokines and pro-inflammatory cytokines in newly hatched chickens infected with *Salmonella enterica* serovar Typhimurium. *Infect Immun*, **72**, 2152–9.

Woodward, M. J., Gettinby, G., Breslin, M. F., Corkish, J. D. and Houghton, S. (2002). The efficacy of Salenvac, a *Salmonella enterica* subsp. Enterica serotype Enteritidis iron-restricted bacterin vaccine, in laying chickens. *Avian Pathol*, **31**, 383–92.

Zhang-Barber, L., Turner, A. K. and Barrow, P. A. (1999). Vaccination for control of *Salmonella* in poultry. *Vaccine*, **17**, 2538–45.

Zhang-Barber, L., Turner, A. K., Martin, G. *et al.* (1997). Influence of genes encoding proton-translocating enzymes on suppression of *Salmonella typhimurium* growth and colonization. *J Bacteriol*, **179**, 7186–90.

CHAPTER 12

Newer vaccines against typhoid fever and gastrointestinal salmonelloses

Richard A. Strugnell and Odilia L. C. Wijburg

12.1 INTRODUCTION

Typhoid and paratyphoid fever result from infection with *Salmonella enterica* serovars Typhi and Paratyphi respectively. Humans are the only reservoir of these infections that are spread by the fecal-oral route. The control and near elimination of this disease in Western countries has been achieved largely because of improved sanitation, surveillance, contact tracing and successful therapy. In locations where this infrastructure does not exist, vaccines can be used as one measure to control the incidence of typhoid fever (Tarr *et al.*, 1999) and possibly even to contribute to the eventual eradication of the disease.

Non-typhoidal *S. enterica* infections are a major public health problem world-wide and reduction of the incidence of these diseases presents quite different challenges to reducing the incidence of typhoid fever. In fact, these diseases have several animal reservoirs and, in humans, a large number of different *S. enterica* serovars cause gastroenteritis probably making vaccines very difficult to realize and/or use commercially.

This chapter will outline the current development of vaccines that address disease caused by *S. enterica*, with an emphasis on typhoid fever.

12.2 TYPHOID VACCINES

12.2.1 A short history of typhoid vaccines

Typhoid fever vaccine development began shortly after the so-called "Golden Age" of microbiology, at the end of the 19th Century. The initial

'Salmonella' Infections: Clinical, Immunological and Molecular Aspects, ed. Pietro Mastroeni and Duncan Maskell. Published by Cambridge University Press.

discovery of the typhoid bacillus was made by Eberth in 1880 and it was first isolated from stools by Pfeiffer in 1896 (Warren and Hornick, 1979). Fecal-oral spread was demonstrated amongst soldiers in camps through observations by the Typhoid Commission of 1898, headed by Walter Reed, and is well described by a key member of the Commission, Victor Vaughan (Vaughan, 1926). The discoveries relating to human faeces and flies and the spread of typhoid led to the first organised implementation of typhoid vaccination that coincided with a major reduction in incidence of disease. However, it is difficult to ascertain whether this significant reduction was directly attributable to the vaccine, or to improved sanitation and use of fly nets.

Vaughan's memoirs make for interesting reading:

> In 1898 the death rate in the (US) army from typhoid fever per hundred thousand was eight hundred and seventy-nine; in 1899 it was one hundred and seven and continued with some fluctuations to fall, reaching nineteen in 1907, the lowest point before the introduction of typhoid vaccination. With compulsory vaccination, there were no cases in 1913; three per hundred thousand in 1914; none in 1915; three in each of 1916 and 1917. In France in 1918 and 1919, it was five and three-tenths and seven and one-tenth respectively. In 1898, a civilian enlisting in the army was quite sure to have typhoid fever; one in every five did. In 1917, the civilian who wished to escape typhoid fever could find no safer place than the army.
>
> In 1910, twenty per cent of our troops were vaccinated against typhoid fever. In 1911, thirty per cent; in 1912, this procedure was made compulsory. This method of increasing resistance to this infection was first practised by Sir Almroth Wright of the Army Medical School at Netley, but England ignored the discovery of its wise son and went through the Boer War without this protection. The losses from this disease in South Africa were even greater than ours in the Spanish-American War. Ours was the first army in the world in which vaccination against typhoid fever was made compulsory and therefore universal.

The early typhoid vaccines were inactivated pure cultures of the pathogen (bacterins) injected at high doses. Whether typhoid vaccination was first used in humans by the British pathologist Almoth Wright or by Robert Koch's disciple, Richard Pfeiffer, has been the subject of some recent debate (Groschel and Hornick, 1981). Regardless of their origins, these heat-, phenol- or otherwise chemically inactivated vaccines were used on a largely *ad hoc* basis for the next 50 years.

Major conflicts such as World War I often provided occasions for the introduction of mass vaccination programs against typhoid fever, but such programs lacked scientific rigour in testing whether the vaccines were effective. Heat/phenol and acetone-inactivated *S. enterica* serovar Typhi vaccines were first subjected to formal efficacy analysis in large field studies conducted in the 1960s by the WHO and others (Warren and Hornick, 1979). The largest of these studies on 160,000 polish children, who were divided in groups receiving either the vaccine or the placebo, revealed an efficacy level of 87% for the whole cell acetone-killed *S. enterica* serovar Typhi vaccine (Anonymous, 1966). Similar but smaller studies in other geographic areas revealed efficacies for whole cell acetone-killed vaccines of between 50 and 94% (Warren and Hornick, 1979). Whole cell killed typhoid vaccines are still in use today (Panchanathan *et al.*, 2001). However, the side-effects associated with this type of vaccine, most notably fever and injection site-reactions, probably relating to their endotoxin content and the relatively low efficacy in some field studies, have prompted efforts over the past 40 years to develop new typhoid vaccines. These new vaccines are designed to elicit long-lasting, protective immune responses, with minimal side-effects.

12.2.2 Performance criteria for developing an optimal typhoid vaccine

From observations of human disease and animal models and based on experience to date with existing approved vaccines, an optimal human typhoid vaccine should:

- be efficacious in large controlled field trials
- contain the Vi surface polysaccharide antigen
- induce long lasting antibody responses against Vi
- induce $CD4^+$ and $CD8^+$ T-cell responses
- be safe and attenuated in immunocompetent and immunocompromised individuals
- be administered orally (probably as a live, attenuated vaccine)
- be formulated such that the vaccine is not cold chain-dependent
- have low level, if any, reactogenicity
- not induce bacteremia

12.2.3 How do the approved typhoid vaccines and those typhoid vaccines under development satisfy these criteria?

The approved typhoid vaccines, and those vaccines under development are examined against these criteria in Table 12.1.

Table 12.1. *Criteria against which typhoid vaccines are examined*

S. typhi Vaccine Options [mutations]	Efficacy in human trials[1]	Clinical Studies (completed)[2]	Contains Vi[3]	Induce Vi antibodies[4]	Induce $CD4^+$/ $CD8^+$ T-cells[5]	Attenuated in HIV^+[6]	Oral delivery[7]	Cold chain dependent[8]	Reactogenic and/ or Bacteremic[9]
Acetone-killed bacteria	50–90%, m73%	Approved	+/−	+/−	NT	+	−	No	R+, B−
Ty21a [cryptic]	c. 50%, m51%	Approved	−	−	NT	probably	+	Yes	R−, B−
Vi	60–90%, m55%	Approved	+++	+	−	+	−	No	R−, B−
Conjugated Vi	90%	Field Tests	+++	+	−	+	−	No	R−, B−
CVD908 [*aroC, aroD*]	immunogenic	Phase II	+ (low)	−	+	NT	+	ND	R−, B+
CVD908HtrA [*aroC, aroD, htrA*]	immunogenic	Phase II	+ (low)	−	+	NT	+	ND	R−, B−
VCD908 HtrA Vi^c [*aroC, aorD, htrA*]	Pre-clinical	Pre-Clinical	++	NT	NT	NT	+	ND	NT
Ty800 [*phoP/phoQ*]	immunogenic	Phase I	NR	NT	NT	NT	+	ND	R−, B−/+
ZH9 [*aroC, ssaV*]	immunogenic	Phase I	NR	NT	NT	NT	+	No	R−, B−
WT05 [*aroC, ssaV*] (*S.* Typhimurium)	immunogenic	Phase I	NR	NT	NT	NT	+	ND	R−, B−
χ4073 [*cya, crp, cdt*]	immunogenic	Phase II	NR	NT	NT	NT	+	ND	R−/+, B?

m = meta analysis, 3 years after immunisation, NT – not tested, NR – not reported, ND – not yet done

1. Efficacy in published controlled trails, 3 of these vaccines have been subject to meta-analysis using the Cochrane database
2. Stage of clinical assessment in humans
3. Level of Vi present in the vaccine
4. Antibody produced by vaccine recipients against Vi
5. Induces measured *Salmonella*-specific T-cell responses in humans
6. Is the vaccine safe in HIV^+ individuals?
7. Delivered orally
8. Cold chain dependent for activity; a number of vaccines have not yet been formulated for use in the field
9. Is the vaccine measurably reactogenic (R) and does the vaccine cause a bacteremia (B)?

12.2.3.1 Whole cell acetone-killed vaccines

These vaccines are known to be efficacious but reactogenic, and are still produced by some government-controlled, national vaccine manufacturers. They are the best trialled of the typhoid vaccines (Warren and Hornick, 1979), yet have been superseded in many settings by the much less reactogenic oral Ty21a vaccine, or the injected Vi vaccine. Acetone-killed vaccines usually contain only low levels of Vi.

12.2.3.2 *S. enterica* serovar Typhi Ty21a strain as a live attenuated vaccine

This oral vaccine requires multiple doses and a cold chain. This approved vaccine is severely attenuated, lacks significant levels of Vi expression (Levine *et al.*, 1987; Tacket *et al.*, 1991) and induces protection at a level frequently around 50%. The bacterium appears to have an epimerase (*galE*) mutation but there is no evidence that this is responsible for attenuation, since defined GalE mutants of *S. enterica* serovar Typhi were not uniformly attenuated in human studies (Hone *et al.*, 1988). Ty21a is sold as Vivotif by Bernabiotech.

12.2.3.3 Vi and Vi-conjugate vaccines

The purified Vi surface antigen of *S. enterica* serovar Typhi has proved to be an effective vaccine. Vi was discovered by Felix and Pitt in 1934. The use of Vi as a vaccine appears to have been first explored by Landy in the early 1950s. Initial problems such as relatively short (<3 years) duration of protection, and only moderate seroconversion rates appear to have been overcome by conjugating the Vi vaccine to protein carriers (Klugman *et al.*, 1996). The original Vi vaccines induced significant protection in the absence of a protein carrier, suggesting that the polysaccharide alone was capable of inducing B-cell proliferation, maturation and antibody secretion. The addition of non-cognate T-helper cell determinants, provided by conjugation of Vi to carrier proteins, has improved the immunogenicity of the Vi vaccine and the longevity of the antibody response (Mai *et al.*, 2003). The Vi conjugate vaccine is currently in field trials. Vi vaccine is sold as Typherix by GSK and as Typhim Vi by Aventis Pasteur.

12.2.3.4 *S. enterica* serovar Typhi CVD908 and derivatives

Hoiseth and Stocker, in the early 1980s, first showed that a mutant of *S. enterica* serovar Typhimurium that was unable to biosynthetize aromatic amino acids was attenuated (Hoiseth and Stocker, 1981). Based on these findings, live attenuated *S. enterica* serovar Typhi vaccines have been developed at the Centre for Vaccine Development in Baltimore. These vaccines carry mutations in genes (*aroC, aroD*) that encode enzymes involved in the biosynthesis of aromatic aminoacids, para-amino-benzoic acid (PABA) and dihydrobenzoate. The presence of two mutations minimizes the risk of reversion to virulence (Chatfield *et al.*, 1992a; Dougan *et al.*, 1988; Hone *et al.*, 1991). *S. enterica* serotype Typhi CVD908, was immunogenic in clinical trials but the level of antibody generated to Vi was low. The vaccine also gave rise to a low grade bacteremia (Tacket *et al.*, 1997) that could be abolished by addition of a mutation in the *htrA* gene (Chatfield *et al.*, 1992b; Johnson *et al.*, 1991; Tacket *et al.*, 1997). More recently, the level of Vi expression in CVD908-*htrA* has been improved by the introduction of transgenic Vi expression from a strong constitutive promoter (Wang *et al.*, 2000), yielding CVD908-*htrA* Vi^c.

12.2.3.5 S. enterica serovar Typhi Ty800

S. enterica serovar Typhi Ty800 is a live attenuated vaccine with mutations in the *phoP* and *phoQ* genes (Hohmann *et al.*, 1996). The PhoP/PhoQ system is a global two-component regulator of gene expression and is essential to the intracellular survival of *Salmonella*. *PhoP/phoQ* mutants of *S. enterica* serovar Typhimurium are attenuated (Miller *et al.*, 1989) and immunogenic (Miller *et al.*, 1990) in animal models. Attenuation is presumably mediated through an inability by the bacterium to adapt to the intracellular environment (Alpuche Aranda *et al.*, 1992). The vaccine is being developed by Avant Immunotherapeutics.

12.2.3.6 *S. enterica* serovar Typhi ZH9 and WT05

These new typhoid vaccines have been through limited dose-ranging testing in humans (Hindle *et al.*, 2002). The vaccines carry a mutation in aromatic biosynthesis (i.e. *aroC*) together with a mutation in SPI2, in a gene known as *ssaV* (Hensel *et al.*, 1995; Shea *et al.*, 1996). ZH9 is a derivative of *S. enterica* serovar Typhi Ty2 and WT05 is mutant of *S. enterica* serovar Typhimurium TML. The vaccines have been formulated to be cold-chain independent, for reconstitution from a freeze-dried product. Both vaccines appeared to be safe and immunogenic in humans. The typhoid vaccine has been through a multicentre Phase I/II clinical trial in the US (http://www.microscience.com/portfolio.asp).

12.2.3.7 *S. enterica* serovar Typhi χ4073

This vaccine was developed following the observation that *S. enterica* serovar Typhimurium carrying mutations in the *cya* gene (involved in the biosynthesis of cAMP) and *crp* gene (encoding cAmp receptor protein) were attenuated in mice and other species (Curtiss and Kelly, 1987). A further mutation in *cdt*, a gene involved in the colonisation of deep tissues was introduced in these vaccines to reduce their reactogenicity for humans (Nardelli-Haefliger *et al.*, 1996; Tacket *et al.*, 1992; Tacket *et al.*, 1997). The clinical development of χ4073 has been halted due to the fact that other vaccines (e.g. Ty800) are more immunogenic and less reactogenic (R. Curtiss, personal communication).

12.3 VACCINES FOR USE AGAINST NON-TYPHOIDAL SALMONELLOSES IN HUMANS

The most likely path to vaccine-mediated prevention of non-typhoidal salmonelloses in humans will be through prevention of transmission of zoonotic disease from the relevant veterinary species. It is likely that vaccines will be fundamental to the control then potential eradication of *S. enterica* serovars such as Enteriditis and Typhimurium from poultry (see below).

The safety and immunogenicity of the WT05 live attenuated *S. enterica* serovar Typhimurium vaccine has now been formally tested in humans (Table 12.1) (Hindle *et al.*, 2002). The WT05 vaccine was well tolerated but appeared to induce variable immune responses and required a high dose (Hindle *et al.*, 2002).

The considerable variety in *S. enterica* serovars responsible for zoonotic disease in humans would argue against the development of any single vaccine that is serovar-specific in its protection. Conceptually, it may be possible to develop broadly cross-protective *Salmonella* vaccines that are dependent on T-cell responses for efficacy. Such a vaccine would be predicated on the belief that T-cells are the mediators of immunity in gastroenteritis. Proteins containing dominant T-cell antigens (e.g. flagella) might be delivered in adjuvants that are known to stimulate Th1-type responses, such as Montanide or ISCOM adjuvants. An alternative approach would be to present individual T-cell epitopes in an appropriate adjuvant. A synthetic string of T-cell determinants, known as a polytope, has been used to generate MHC restricted cellular immune responses, mediated through $CD8^+$ T-cells (Thomson *et al.*, 1996) and/or $CD4^+$ T-cells (Thomson *et al.*, 1998). The effects of HLA polymorphism are countered by providing a series of epitopes, each of which binds to a different HLA allele (Ramsay *et al.*, 1999). Before such a vaccine was developed it would be necessary to show that the $CD8^+$ T-cells, that have been detected in a number of human vaccination trials, are important in resistance to re-infection by non-typhoidal *S. enterica* serovars and/or that a CD4$^+$ T-cell-directed polytope would be efficiently processed and presented in vivo and strongly recognised by naïve and effector T-cells.

The large *S. enterica* genome would seem to provide ample opportunity to define a series of epitopes that are conserved across serovars and which are recognised by the different MHC Class I and Class II alleles. The challenge would then be to present such epitopes, possibly as a polytope, in an immunogenic form. The demand for such a vaccine would be limited unless the experience in Malawi, *i.e.* that non-typhoidal salmonelloses are highly virulent in people with low CD4$^+$ T-cell counts (Gordon *et al.*, 2002; Gordon *et al.*, 2001), extends through the developing world HIV$^+$ community. Unfortunately, T-cell-mediated vaccine protection elicited by 'minimised' T-cell vaccines would be problematic in this patient cohort since effector activity would be dependent on CD4$^+$ T-cells, the same cells depleted during HIV infection. The capacity of polytope-based vaccines to afford protection in this setting will be established in near-term HIV vaccine programs. As discussed, the most effective means of controlling non-typhoidal salmonelloses using vaccines will probably be through the eradication of the bacteria from the food chain.

12.4 VACCINES FOR USE IN VETERINARY SPECIES

S. enterica can infect poultry, cattle, pigs and sheep. In these animal species, *Salmonella* infections cause severe disease leading to commercial losses and/or cause contamination of the food chain. The ubiquitous interaction between *Salmonella* and domestic animals, especially poultry, and the subsequent impact on food-borne disease has led to considerable commercial interest in the development of vaccines for use in animals. These bacterin and live vaccines include Megan Vac, Nobilis Salenvac, vacT, Zoosaloral, and X3985 and are most typically used to prevent horizontal spread of infection in intensive poultry rearing, and the vertical transmission of the pathogen from layers through eggs.

Megan Health developed a vaccine, Megan Vac 1, for use in chickens that is sprayable or deliverable in water. Laying hens vaccinated with a higher dose of Megan Vac 1 were protected from internal organ infection and had significantly reduced contamination of the gastrointestinal tract, caeca and eggs after challenge with wild-type *S. enterica* serovar Enteritidis (http://www.usaha.org/reports/reports01/r01sal.html). Megan Vac 1 was co-developed with Lohmann Animal Health who have produced live attenuated vaccines for both *S. enterica* serovar Enteriditis and *S. enterica* serovar Typhimurium.

Salmonella vaccines can be used at various points in poultry production and bacterin type vaccines are known to elicit cross protection in chickens. Moreover, vaccination of hens can confer protection against the growth of *S. enterica* serovar Enteriditis in eggs. Live attenuated mutants of *S. enterica* serovar Typhimiurm confer protection to chickens (Hassan and Curtiss, 1994). The advantage of live vaccines is their relatively simple administration either through the use of sprayers, where the droplet size has been optimised to 100um, or through drinking water (Barbezange *et al.*, 2000). The size optimisation at 100um reduces deep inhalation and deposits the vaccine on the surface of the birds.

Nobilis Salenvac (Intervet), an inactivated, alum adjuvanted whole cell vaccine for *S. enterica* serovar Enteritidis, reduces both horizontal and vertical transmission of the bacterium. The vaccine is grown in iron-restricted conditions during manufacture which are held to provide for a more 'in vivo' like presentation of antigens in the killed antigen preparation (Woodward *et al.*, 2002). Salenvac is usually given to broiler chickens by intramuscular injection. The same manufacturers have extended the immunological coverage of the Nobilus vaccine by incorporation of additional immunogens from *Salmonella enterica* vars Enteritidis PT4 and Typhimurium DT104 (vaccine known as Nobilis Salenvac T) (Feberwee *et al.*, 2000). Studies have suggested that this vaccine also affords cross-protection against other serogroup B *Salmonella* serovars, such as *S. enterica* serovars Heidelberg and Agona. Intervet have also developed a live attenuated vaccine for preventing vaccine (Nobilis SG 9R) which contains a rough LPS *Salmonella* (Feberwee *et al.*, 2001).

Live *Salmonella* vaccines used in poultry have been compared for safety and the types of mutations typically engineered into commercial vaccines usually lead to transient carriage and only short term persistence (Barbezange *et al.*, 2000). In contrast, it is unlikely that sufficiently large field trials to test the efficacy of all existing commercial vaccines will be conducted, and hence it will not be possible to determine which of the many live and bacterin vaccines on offer present the best cost-effective means of reducing *Salmonella spp.* contamination of the food chain. What has been demonstrated to-date is that vaccine approaches might be less than 100% effective at eradicating salmonellae in chickens. A 'competitive exclusion' approach, where colonisation by attenuated salmonellae or related bacteria are used to exclude colonisation by more virulent *S. enterica*, might be an effective alternative means of controlling the presence of human virulent bacteria within a poultry flock (Rabsch *et al.*, 2000).

12.5 NOVEL APPROACHES TO THE DEVELOPMENT OF *S. ENTERICA* VACCINES

New *S. enterica* vaccines might be based on purified natural or recombinant proteins or carbohydrate subunits, delivered using adjuvants, depot systems with innate immune system modulators, or vaccine vectors. The vaccines will need to be safe, effective, easy to produce, simple to deliver and cost-effective.

12.5.1 DNA vaccines

The promise of DNA vaccines (Ulmer *et al.*, 1993), first described in the early 1990s, has yet to be realised. The concept of using purified plasmid DNA as a vaccine, thereby avoiding all the issues concerned with making folded, pure recombinant or native protein subunits for use in vaccines, has considerable appeal. Unfortunately, efficacy of DNA vaccination in animals appears to decrease as the animals get larger and attempts to establish DNA vaccination as an effective alternative to multivalent subunit vaccines, in humans or domestic animals, has so far failed. The current favoured approach is to combine DNA vaccination with some form of booster vaccine comprised of either pure protein, or a protein delivered by bacterial or viral vector.

Early studies on *S. enterica* serovar Typhimurium DNA vaccines that consisted of the cloning and expression of genes encoding individual *Salmonella* protein targets (e.g. *invH*, *fliC*) were not encouraging (Allen and Dougan unpublished). Some immunogenicity was observed with an experimental OmpC-based DNA vaccine, but there have been few follow-up reports regarding the efficacy of this construct (Lopez-Macias *et al.*, 1995).

12.6 CONCLUSIONS

While *S. enterica* remains a significant pathogen of humans and animals, attempts will be made to develop new and improved vaccines. The increasing costs associated with the regulatory demands on trialling then implementing new human pharmaceuticals will need to be offset against commercial income for such vaccines, once approved. Like many major human infectious diseases, much of the mortality and morbidity from *S. enterica* disease is in those countries that can least afford new and expensive medicines. One means of confronting this issue has been to establish "private–public" partnerships, a model adopted by the Gates Foundation. In these partnerships between philanthropy and business, the engagement of larger cost-effective manufacturers combined with differential pricing allows vaccines to be made efficiently and, more importantly, affordably for 'developing' countries. The production of new veterinary vaccines will be driven by demands for increased public health, scrutiny of the food chain, and technological innovation which will allow cost-effective vaccination of intensively and extensively reared animals.

12.7 ACKNOWLEDGEMENTS

The authors are supported by The Australian Government through the Cooperative Research Centres Program and the National Health and Medical Research Council.

REFERENCES

Alpuche Aranda, C. M., Swanson, J. A., Loomis, W. P. and Miller, S. I. (1992). *Salmonella typhimurium* activates virulence gene transcription within acidified macrophage phagosomes. *Proc Natl Acad Sci USA*, **89**, 10079–83.

Anonymous (1966). Controlled field trials and laboratory studies on the effectiveness of typhoid vaccines in Poland, 1961–64. *Bull WHO*, **34**, 211–22.

Barbezange, C., Ermel, G., Ragimbeau, C., Humbert, F. and Salvat, G. (2000). Some safety aspects of *Salmonella* vaccines for poultry: *in vivo* study of the genetic stability of three *Salmonella typhimurium* live vaccines. *FEMS Microbiol Lett*, **192**, 101–6.

Chatfield, S. N., Fairweather, N., Charles, I. *et al.* (1992a). Construction of a genetically defined *Salmonella typhi* Ty2 *aroA, aroC* mutant for the engineering of a candidate oral typhoid–tetanus vaccine. *Vaccine*, **10**, 53–60.

Chatfield, S. N., Strahan, K., Pickard, D. *et al.* (1992b). Evaluation of *Salmonella typhimurium* strains harbouring defined mutations in *htrA* and *aroA* in the murine salmonellosis model. *Microb Pathog*, **12**, 145–51.

Curtiss, R. and Kelly, S. M. (1987). *Salmonella typhimurium* deletion mutants lacking adenylate cyclase and cyclic AMP receptor protein are avirulent and immunogenic. *Infect Immun*, **55**, 3035–43.

Dougan, G., Chatfield, S., Pickard, D. *et al.* (1988). Construction and characterization of vaccine strains of *Salmonella* harboring mutations in two different *aro* genes. *J Infect Dis*, **158**, 1329–35.

Feberwee, A., de Vries, T. S., Elbers, A. R. and de Jong, W. A. (2000). Results of a *Salmonella enteritidis* vaccination field trial in broiler-breeder flocks in The Netherlands. *Avian Dis*, **44**, 249–55.

Feberwee, A., Hartman, E. G., de Wit, J. J. and de Vries, T. S. (2001). The spread of *Salmonella gallinarum* 9R vaccine strain under field conditions. *Avian Dis*, **45**, 1024–9.

Gordon, M. A., Banda, H. T., Gondwe, M. *et al.* (2002). Non-typhoidal *Salmonella* bacteraemia among HIV-infected Malawian adults: high mortality and frequent recrudescence. *Aids*, **16**, 1633–41.

Gordon, M. A., Walsh, A. L., Chaponda, M. *et al.* (2001). Bacteraemia and mortality among adult medical admissions in Malawi – predominance of non-typhi salmonellae and *Streptococcus pneumoniae*. *J Infect*, **42**, 44–9.

Groschel, D. H. and Hornick, R. B. (1981). Who introduced typhoid vaccination: Almroth Write or Richard Pfeiffer? *Rev Infect Dis*, **3**, 1251–4.

Hassan, J. O. and Curtiss, R. (1994) Development and evaluation of an experimental vaccination program using a live avirulent *Salmonella typhimurium* strain to protect immunized chickens against challenge with homologous and heterologous *Salmonella* serotypes. *Infect Immun*, **62**, 5519–27.

Hensel, M., Shea, J. E., Gleeson, C. *et al.* (1995). Simultaneous identification of bacterial virulence genes by negative selection. *Science*, **269**, 400–3.

Hindle, Z., Chatfield, S. N., Phillimore, J. *et al.* (2002). Characterization of *Salmonella enterica* derivatives harboring defined *aroC* and *Salmonella* pathogenicity island 2 type III secretion system (*ssaV*) mutations by immunization of healthy volunteers. *Infect Immun*, **70**, 3457–67.

Hohmann, E. L., Oletta, C. A., Killeen, K. P. and Miller, S. I. (1996). *phoP/phoQ*-deleted *Salmonella typhi* (Ty800) is a safe and immunogenic single-dose typhoid fever vaccine in volunteers. *J Infect Dis*, **173**, 1408–14.

Hoiseth, S. K. and Stocker, B. A. (1981). Aromatic-dependent *Salmonella typhimurium* are non-virulent and effective as live vaccines. *Nature*, **291**, 238–9.

Hone, D. M., Attridge, S. R., Forrest, B. *et al.* (1988). A *galE via* (Vi antigen-negative) mutant of *Salmonella typhi* Ty2 retains virulence in humans. *Infect Immun*, **56**, 1326–33.

Hone, D. M., Harris, A. M., Chatfield, S., Dougan, G. and Levine, M. M. (1991). Construction of genetically defined double *aro* mutants of *Salmonella typhi*. *Vaccine*, **9**, 810–16.

Johnson, K., Charles, I., Dougan, G. *et al.* (1991). The role of a stress-response protein in *Salmonella typhimurium* virulence. *Mol Microbiol*, **5**, 401–7.

Klugman, K. P., Koornhof, H. J., Robbins, J. B. and Le Cam, N. N. (1996). Immunogenicity, efficacy and serological correlate of protection of *Salmonella typhi* Vi capsular polysaccharide vaccine three years after immunization. *Vaccine*, **14**, 435–8.

Levine, M. M., Ferreccio, C., Black, R. E. and Germanier, R. (1987). Large-scale field trial of Ty21a live oral typhoid vaccine in enteric-coated capsule formulation. *Lancet*, **1**, 1049–52.

Lopez-Macias, C., Lopez-Hernandez, M. A., Gonzalez, C. R., Isibasi, A. and Ortiz-Navarrete, V. (1995). Induction of antibodies against *Salmonella typhi* OmpC porin by naked DNA immunization. *Ann N Y Acad Sci*, **772**, 285–8.

Mai, N. L., Phan, V. B., Vo, A. H. *et al.* (2003). Persistent efficacy of Vi conjugate vaccine against typhoid fever in young children. *N Engl J Med*, **349**, 1390–1.

Miller, S. I., Kukral, A. M. and Mekalanos, J. J. (1989). A two-component regulatory system (*phoP phoQ*) controls *Salmonella typhimurium* virulence. *Proc Natl Acad Sci USA*, **86**, 5054–8.

Miller, S. I., Mekalanos, J. J. and Pulkkinen, W. S. (1990). *Salmonella* vaccines with mutations in the *phoP* virulence regulon. *Res Microbiol*, **141**, 817–21.

Nardelli-Haefliger, D., Kraehenbuhl, J. P., Curtiss, R. *et al.* (1996). Oral and rectal immunization of adult female volunteers with a recombinant attenuated *Salmonella typhi* vaccine strain. *Infect Immun*, **64**, 5219–24.

Panchanathan, V., Kumar, S., Yeap, W. *et al.* (2001). Comparison of safety and immunogenicity of a Vi polysaccharide typhoid vaccine with a whole-cell killed vaccine in Malaysian Air Force recruits. *Bull WHO*, **79**, 811–17.

Rabsch, W., Hargis, B. M., Tsolis, R. M. *et al.* (2000). Competitive exclusion of *Salmonella enteritidis* by *Salmonella gallinarum* in poultry. *Emerg Infect Dis*, **6**, 443–8.

Ramsay, A. J., Kent, S. J., Strugnell, R. A. *et al.* (1999). Genetic vaccination strategies for enhanced cellular, humoral and mucosal immunity. *Immunol Rev*, **171**, 27–44.

Shea, J. E., Hensel, M., Gleeson, C. and Holden, D. W. (1996). Identification of a virulence locus encoding a second type III secretion system in *Salmonella typhimurium*. *Proc Natl Acad Sci USA*, **93**, 2593–7.

Tacket, C. O., Hone, D. M., Curtiss, R., III *et al.* (1992). Comparison of the safety and immunogenicity of Δ*aroC* Δ*aroD* and Δ*cya* Δ*crp* *Salmonella typhi* strains in adult volunteers. *Infect Immun*, **60**, 536–41.

Tacket, C. O., Losonsky, G., Taylor, D. N. *et al.* (1991). Lack of immune response to the Vi component of a Vi-positive variant of the *Salmonella typhi* live oral vaccine strain Ty21a in human studies. *J Infect Dis*, **163**, 901–4.

Tacket, C. O., Sztein, M. B., Losonsky, G. A. *et al.* (1997). Safety of live oral *Salmonella typhi* vaccine strains with deletions in *htrA* and *aroC aroD* and immune response in humans. *Infect Immun*, **65**, 452–6.

Tarr, P. E., Kuppens, L., Jones, T. C. *et al.* (1999). Considerations regarding mass vaccination against typhoid fever as an adjunct to sanitation and public health measures: potential use in an epidemic in Tajikistan. *Am J Trop Med Hyg*, **61**, 163–70.

Thomson, S. A., Burrows, S. R., Misko, I. S. *et al.* (1998). Targeting a polyepitope protein incorporating multiple class II-restricted viral epitopes to the secretory/endocytic pathway facilitates immune recognition by $CD4^+$ cytotoxic T lymphocytes: a novel approach to vaccine design. *J Virol*, **72**, 2246–52.

Thomson, S. A., Elliott, S. L., Sherritt, M. A. *et al.* (1996). Recombinant polyepitope vaccines for the delivery of multiple $CD8^+$ cytotoxic T-cell epitopes. *J Immunol*, **157**, 822–6.

Ulmer, J. B., Donnelly, J. J., Parker, S. E. *et al.* (1993). Heterologous protection against influenza by injection of DNA encoding a viral protein. *Science*, **259**, 1745–9.

Vaughan, V. C. (1926). *A Doctor's Memories*. Indianapolis: The Bobbs-Merril Co.

Wang, J. Y., Noriega, F. R., Galen, J. E., Barry, E. and Levine, M. M. (2000). Constitutive expression of the Vi polysaccharide capsular antigen in attenuated *Salmonella enterica* serovar Typhi oral vaccine strain CVD 909. *Infect Immun*, **68**, 4647–52.

Warren, J. W. and Hornick, R. B. (1979). Immunization against typhoid fever. *Annu Rev Med*, **30**, 457–72.

Woodward, M. J., Gettinby, G., Breslin, M. F., Corkish, J. D. and Houghton, S. (2002). The efficacy of Salenvac, a *Salmonella enterica* subsp. Enterica serotype Enteritidis iron-restricted bacterin vaccine, in laying chickens. *Avian Pathol*, **31**, 383–92.

CHAPTER 13

S. enterica-based antigen delivery systems

José A. Chabalgoity

13.1 INTRODUCTION

Salmonella enterica has been proposed as a highly efficient vector for the delivery of heterologous molecules to the immune system of the host. For more than two decades recombinant live attenuated salmonellae expressing antigens from other pathogens have been extensively assessed as oral multivalent vaccines and tested in a great diversity of experimental models. More recently, it has been demonstrated that *S. enterica* can also be used as a vector for DNA vaccines. New emerging applications for recombinant *S. enterica* include its use in the treatment of cancer and possible applications in gene therapy.

Some distinctive features of *S. enterica* have strongly contributed to make it an attractive delivery system. Among them are features of the immunobiology of *S. enterica* infections and the genetics of the bacteria. *S. enterica* naturally enter the host by the oral route, elicit strong mucosal and systemic immune responses and are eventually cleared from the tissues leaving long lasting immunological memory. Once inside the host, *S. enterica* can be found within macrophages and dendritic cells (DC), which are professional antigen presenting cells (APC). Thus, oral administration of recombinant *S. enterica* can be an effective way of directing the expression of relevant molecules (antigens or immunomodulatory molecules) to APC.

The genetics of *S. enterica* are very similar to those of *E. coli* and the full genome sequences of several *Salmonella* species and serovars are available. Therefore, the molecular tools and techniques currently available enable the rational construction of vaccine vectors based on *S. enterica*.

'Salmonella' Infections: Clinical, Immunological and Molecular Aspects, ed. Pietro Mastroeni and Duncan Maskell. Published by Cambridge University Press.

13.2 *S. ENTERICA* EXPRESSING HETEROLOGOUS ANTIGENS AS MULTIVALENT VACCINES

Live attenuated *S. enterica* can be engineered to express antigens from other pathogens. The recombinant bacteria can be administered by systemic injection or by mucosal immunization to elicit strong local and systemic immune responses. Although the oral route has been the main mucosal route used, other routes of administration like intranasal, intra-vaginal and intra-rectal have also been attempted with some success. In many cases it has been demonstrated that recombinant strains of *S. enterica* can confer protection against salmonellosis and against the pathogen(s) from which the heterologous antigens is derived. The expression and delivery of antigens from viruses, bacteria and parasites in *S. enterica* vaccine strains has been pursued and an ever increasing number of antigens are cloned and tested in vaccination experiments with recombinant *S. enterica* strains (Kochi *et al.*, 2003; Mastroeni *et al.*, 2001). Recently the expression of mouse PrP in a *Salmonella* vaccine strain as a vaccine against prion disease has been also reported (Goñi *et al.*, 2005). The effectiveness of *Salmonella* live vector vaccines in neonatal mice has been also recently demonstrated (Capozzo *et al.*, 2004).

Recombinant salmonellae have also been tested for the development of less conventional vaccines like allergy vaccines, anticaries vaccines or infertility vaccines (Redman *et al.*, 1995; Redman *et al.*, 1996; Srinivasan *et al.*, 1995; Vrtala *et al.*, 1995). Cloning mini-libraries of a particular pathogen into *S. enterica* vaccine strains and testing protection afforded by vaccination with different clones has been used as a tool for identification of genes that code for "protection-stimulating" proteins (Brayton *et al.*, 1998). A novel concept for multivalent virus vaccines has been demonstrated recently using *S. enterica* as a vector for a plasmid containing a complete viral genome. In fact it has been demonstrated that a murine cytomegalovirus (MCMV) bacterial artificial chromosome can be stably maintained in certain strains of *S. enterica* serovar Typhimurium. Injection of the recombinant strain into mice reconstitutes a virus infection, and it was shown that virus reconstitution in vivo caused elevated titers of specific anti-MCMV antibodies, and protection against lethal virus challenge (Cicin-Sain *et al.*, 2003).

13.3 EXPRESSION SYSTEMS FOR HETEROLOGOUS ANTIGENS IN *S. ENTERICA*

13.3.1 General

One of the major targets when developing heterologous vaccines based on *S. enterica* is the sustained expression of the heterologous antigen(s) in the tissues in an immunogenic form, and at levels sufficient to prime an immune response. A number of problems related to this issue can arise in the successful construction of a multivalent vaccine. Plasmid instability, toxicity of the expressed protein for the bacterial host, poor levels of gene expression and enzymatic degradation or incorrect folding of the expressed proteins are some of the most common problems encountered.

13.3.2 Expression systems based on plasmids

The expression of heterologous antigens in *S. enterica* has been achieved in the vast majority of cases by transforming the bacterial strain with a multicopy plasmid that encodes the antigen of interest under the control of a prokaryotic promoter. Plasmids with different replicons and greatly differing copy number (from a few copies to over a hundred) have been used. In some cases, the use of high-copy plasmids can improve the immunogenicity of the vaccine (Covone *et al.*, 1998; Molina and Parker, 1990). However, lowering the copy number of the plasmid often results in increased stability of the construct and/or in reduction of the expression of a heterologous antigen to non-toxic levels (Coulson *et al.*, 1994a; Coulson *et al.*, 1994b; Jagusztyn-Krynicka *et al.*, 1993; Turner *et al.*, 1993). The metabolic burden resulting from the expression of a foreign protein may create a selective pressure for plasmid segregation. Most plasmids used have an antibiotic resistance marker for maintenance during in vitro growth, but plasmid instability may arise once the bacteria are growing inside the tissues in the absence of antibiotic pressure, thus resulting in the amplification of bacterial populations that no longer contain the plasmid. Plasmid instability during in vivo replication may result in low-level expression of the cloned antigen and poor immunogenicity. On the other hand, unregulated high-level expression of foreign antigens may be toxic for the bacteria, leading to selection of plasmid-cured strains. Several different systems aimed at improving in vivo plasmid stability, or to avoid unregulated expression of the heterologous antigen have been developed.

In vivo growth and persistence of bacterial populations that contain the plasmid in the absence of antibiotic selection can be improved by incorporating essential genes into the expression plasmid. For example, the enzyme aspartate β-semialdehyde dehydrogenase (encoded by the *asd* gene) is common to the biosynthetic pathways of several amino acids as well as diaminopimelic acid (DAP), an essential constituent of the peptidoglycan of the cell wall of Gram-negative bacteria. DAP is not present in mammals and therefore the bacteria need to synthesize this compound in order to survive in the host. A plasmid vector carrying the *asd* gene was generated and used to complement a *S. enterica* strain with an *asd* mutation, constituting a balanced lethal combination ensuring that all surviving bacterial cells would possess the recombinant plasmid expressing the *asd* gene (Galan *et al.*, 1990; Nakayama *et al.*, 1988). A variation of this system was developed to allow expression of proteins that may be toxic for the *S. enterica* carrier. In this system the foreign antigen is expressed from one plasmid under the *trc* promoter, whereas a second incompatible plasmid carries the LacI repressor. Under such circumstances, expression of the foreign antigen will be repressed when grown in vitro if antibiotic selection for both plasmids is maintained. However, growth of *S. enterica* within the host will result in loss of one of the two incompatible plasmids. The plasmid that carries the foreign antigen also encodes *asd* and therefore will be essential for the survival of the *S. enterica* strain carrying the *asd* mutation. The overall effect is that expression of the antigen is repressed in vitro (avoiding a possible deleterious effect to the carrier) and after replication in the host tissues occurs, the incompatible plasmid carrying the repressor will segregate leading to constitutive expression of the foreign antigen (Ervin *et al.*, 1993).

Another balanced-lethal system for plasmid stabilization is the one based on the *hok* and *sok* genes in which *hok* encodes a lethal pore-forming Hok protein whose expression is blocked by hybridization of an antisense *sok* mRNA to *hok* mRNA. Since *sok* mRNA is highly susceptible to degradation, its protective intracellular concentration must be maintained by constitutive transcription. Therefore, bacteria that spontaneously lose such plasmids are post-segregationally killed because existing levels of the protective *sok* mRNA rapidly drop, whereas levels of the more stable toxin-encoding *hok* mRNA quickly lead to Hok synthesis and cell death (Galen *et al.*, 1999).

The use of strong but regulated, in vivo inducible promoters has been proposed as a way to ensure that expression of recombinant proteins would be increased once the bacteria reach appropriate locations in the tissues. Several different promoters have been tested for construction of recombinant *S. enterica*, with different degrees of success. The *E. coli* nitric oxide reductase (*nirB*) promoter is activated under anaerobic conditions and has been used successfully in *S. enterica* vaccines to direct the expression of recombinant antigens derived from bacteria, viruses and parasites. A number of *nirB*-based constructs delivered by *Salmonella* were immunogenic and protective (Chabalgoity *et al.*, 1997; Chabalgoity *et al.*, 1996; Chatfield *et al.*, 1992a; Galen *et al.*, 1999; Karem *et al.*, 1995; Khan *et al.*, 1994a; Khan *et al.*, 1994b; Londono *et al.*, 1996; McSorley *et al.*, 1997). The *nirB* promoter has also been used successfully to express cytokines to be delivered by *S. enterica* vaccines

(Xu *et al.*, 1998). Another anaerobically induced promoter from the same family as *nirB* is *dmsA* which drives the expression of dimethylsulphoxide reductase. The parental *dmsA* promoter and two derivatives were used to drive the expression of fragment C of tetanus toxin in an attenuated strain of *S. enterica* serovar Typhi. Intranasal immunization of mice with one of these derivatives elicited high titres of anti-tetanus antibodies and conferred protection against challenge by tetanus toxin (Orr *et al.*, 2001).

The *pagC* promoter, which is activated upon entry of the *S. enterica* into macrophages, has also been assessed for expression of foreign antigens in *S. enterica* vaccine strains. In one study, the *pagC* promoter was compared with the *nirB* promoter for driving the expression of chimeric fimbriae containing two different epitopes from the transmissible gastroenteritis virus (TGEV). The constructs with the *pagC* promoter showed increased expression of the foreign antigen and elicited significantly higher mucosal and systemic immune responses in mice compared to similar constructs with *nirB* promoter (Chen and Schifferli, 2001).

Several other promoters that are preferentially induced within macrophages (e.g. *htrA*, *groE*, *spiC*, *katG*, *dps* and *ssaG*) have been used successfully to improve the immunogenicity of heterologous antigens expressed by *S. enterica* vectors (Chen and Schifferli, 2003; Dunstan *et al.*, 1999; Everest *et al.*, 1995; Marshall *et al.*, 2000; McKelvie et al 2004; Roberts *et al.*, 1998).

13.3.3 Chromosomal integration of foreign genes

Integration of the heterologous gene into the bacterial chromosome can result in stabilization of expression of the foreign antigen. The foreign gene is introduced into the *S. enterica* vaccine on a suicide-vector (incapable of replication in the bacterial host) with flanking regions that allow incorporation into the *S. enterica* chromosome by homologous recombination into a predetermined site. One such system for the integration of cloned genes into *Salmonella* vaccine strains involves the introduction of a deletion in a region of the *his* operon in the chromosome of an attenuated *S. enterica* strain, and the construction of a suicide-vector carrying a wild-type *his* operon and the gene coding for the heterologous antigen. Selection for his^+ in bacteria cured of the plasmid allows the isolation of clones in which a recombinational event has resulted in integration of the wild-type *his* region plus the foreign gene into the chromosome of the bacteria. This system was successfully applied to the construction of a recombinant *S. enterica* expressing *E. coli* K88 fimbriae from the chromosome, which was immunogenic in mice (Hone *et al.*, 1988). Another possible approach involves the insertion of the foreign gene into a cloned *S. enterica* gene (e.g. *aroC*) whose disruption would result in attenuation of the bacterial vector. This technique has the added advantage that replacement of the wild type *aroC* gene by the one with the foreign gene inserted will introduce an attenuating mutation into the vaccine strain (Chatfield *et al.*, 1992b; Strugnell *et al.*, 1990). Using this system, the p69 antigen of *Bordetella pertussis* was expressed under the *trc* promoter from the chromosome of the *S. enterica* serovar Typhimurium vaccine strain BRD509. Mice immunized orally or intravenously with this construct showed a significant level of protection against an aerosol challenge with virulent *B. pertussis* (Strugnell *et al.*, 1992). More recently, it has been shown that a single oral dose of *S. enterica* serovar Typhimurium vaccine strain WT05 carrying a single chromosomal copy of the *Escherichia coli* heat labile toxin B subunit (LT-B), under the in vivo inducible SPI-2 promoter *ssaG*, can induce strong anti-LT-B antibody responses in mice (McKelvie *et al.*, 2004).

In another approach, the foreign gene can be inserted into the *S. enterica* chromosome and expressed as a fusion protein with the first 84 amino acids of PagC, therefore being under the control of the *pagC* promoter. Recombinant salmonellae expressing the alkaline phosphatase from *E. coli* (PhoA) using this system elicited strong humoral responses against PhoA in mice (Hohmann *et al.*, 1995).

Chromosomal integration of a heterologous gene can result in complete stabilization of expression. However, the foreign gene will be present as a single copy per bacterium, and therefore, the level of expression may be low. For example, integration of the gene encoding the fragment C of tetanus toxin (TetC) into the *aroC* gene of *S. enterica* serovar Typhimurium SL3261 increased the stability of the gene, but the amount of TetC produced was subimmunogenic (Strugnell *et al.*, 1990). Possible ways to avoid this problem may involve the integration of several copies of the foreign gene into the chromosome, or the use of stronger promoters.

13.3.4 Expression of heterologous antigens as fusion proteins

The construction of translational fusions with full length proteins or domains containing signal sequences has been used in the *S. enterica* delivery system as a strategy to direct the expression of heterologous antigens to a particular location within the bacterium or the host cell, or as an alternative to increasing the immunogenicity of "weak" antigens by fusing them to proteins that could exert a carrier effect.

Some recombinant antigens confer superior protective immune responses when expressed as secreted proteins (Kang and Curtiss, 2003). However, other studies indicate that the location in which the recombinant antigen is expressed may not be crucial for the induction of immune responses (Haddad *et al.*, 1995; Schodel and Will, 1990). Peptides from viruses, bacteria and eukaryotic parasites were immunogenic in mice when expressed in *S. enterica* as fusions with signal sequences derived from proteins that are normally translocated to the bacterial surface or that are secreted extracellularly (MalE, LamB, LT-B, OmpA or flagellin), or as fusions to secreted proteins (Cattozzo *et al.*, 1997; Charbit *et al.*, 1993; Hayes *et al.*, 1991; Jagusztyn-Krynicka *et al.*, 1993; Newton *et al.*, 1989; Newton *et al.*, 1995; Newton *et al.*, 1991; Pistor and Hobom, 1990; Schorr *et al.*, 1991; Smerdou *et al.*, 1996; Stocker, 1990; Su *et al.*, 1992; Verma *et al.*, 1995a; Wang *et al.*, 1999; Whittle *et al.*, 1997a; Wu *et al.*, 1989).

The *E. coli* haemolysin (HlyA) secretion system is fully functional in *S. enterica* and expression of heterologous antigens as fusions with the HlyA secretion signal often results in their secretion by salmonellae. This system has been successsfully tested with several different antigens (Gentschev *et al.*, 1998; Gentschev *et al.*, 1996; Hahn *et al.*, 1998; Orr *et al.*, 1999). *S. enterica* strains synthesizing either listeriolysin or p60 of *Listeria monocytogenes* fused to HlyA elicited protective immunity against *L. monocytogenes* in mice (Gentschev *et al.*, 1996). A live vaccine consisting of the p67 sporozoite antigen of *Theileria parva* expressed in secreted form by an *aroA*$^-$ mutant of *S. enterica* serovar Dublin strain, protected calves against a challenge with *T. parva* sporozoites (Gentschev *et al.*, 1998). A *S. enterica* serovar Typhimurium strain secreting bioactive human IL6 could be prepared using this system (Hahn *et al.*, 1998).

A particularly interesting system exploits the *S. enterica* type III secretion system (TTSS) to deliver heterologous antigens directly into the cytosol of the host cell (Russmann *et al.*, 1998). Peptides of interest fused to proteins bearing the type III secretion signal are efficiently processed via the MHC class I-restricted pathway. In orally immunized mice, this novel vaccination strategy has resulted in the induction of strong peptide-specific cytotoxic CD8$^+$ T-cell responses to recombinant antigens. The type III effector protein YopE from *Yersinia enterocolitica* has been used as a carrier molecule for cytosolic translocation of listeriolysin O and p60 of *Listeria monocytogenes* (Russmann, 2004). Recombinant antigens expressed in *S. enterica* as fusions with defined secretion and translocation domains of YopE were delivered to the cytosol of infected eukaryotic cells via the *Salmonella* TTSS (Igwe *et al.*, 2002; Russmann *et al.*, 2001). Fragments of the Simian Immunodeficiency Virus

(SIV) Gag protein fused to the type III-secreted SopE protein were expressed in attenuated strains of S. *enterica* serovar Typhimurium and serovar Typhi. Oral immunization with these constructs successfully primed virus-specific cytotoxic T lymphocyte responses in rhesus macaques (Evans *et al.*, 2003).

Fusion to proven carrier antigens such as tetanus toxoid can increase the immunogenicity of recombinant proteins or peptides (Khan *et al.*, 1994a; Khan *et al.*, 1994b). For example, expression of *Schistosoma mansoni* GST p28 or tandem peptide repeats from this protein as fusions to TetC in a *S. enterica* vaccine strain led to an experimental vaccine capable of protecting mice against salmonellosis, tetanus and schistosomiasis after single dose administration (Khan *et al.*, 1994b). A single dose of a *S. enterica* vaccine strain expressing tandem copies of a short peptide from glycoprotein D (gD) of Herpes Simplex Virus (HSV) as a fusion with TetC elicited high titres of specific antibodies in serum and protected mice against virus challenge (Chabalgoity *et al.*, 1996). Mucosal vaccination with a *S. enterica* vaccine strain expressing mouse Prp as fusion to TetC prevented prion infection via an oral route (Goñi *et al.*, 2005). Expression of a fatty acid binding protein from *Echinococcus granulosus* as a fusion with TetC in a dog-adapted *Salmonella* vaccine strain, proved immunogenic in dogs when delivered orally (Chabalgoity *et al.*, 2000). TetC fusions have also been used for stable expression of foreign antigens that were otherwise found to be susceptible to proteolytic degradation when expressed in the cytoplasm of *S. enterica* (Barry *et al.*, 1996; Gomez-Duarte *et al.*, 1995).

13.4 IMMUNE RESPONSES AGAINST HETEROLOGOUS ANTIGENS EXPRESSED IN *S. ENTERICA*

13.4.1 Antibody responses

Humoral immunity to a large number of recombinant antigens expressed in *S. enterica* has been induced using several expression vectors and strategies. In some cases the presence of antigen-specific antibodies in serum has been used just as an indicator of the ability of the vaccine to trigger an immune response to the heterologous antigen. However, recombinant *S. enterica* vaccines can stimulate protective humoral immunity. For example, protective antibody responses to M proteins of *S. pyogenes*, a major outer surface protein (OspA) of *Borrelia burgdorferi*, the F1 capsular antigen of *Yersinia pestis*, fragment C of tetanus toxin, *E. coli* K99 fimbriae, PspA of

S. pneumoniae, and *Campylobacter jejuni* CjaC protein have been generated by vaccination with recombinant salmonellae (Ascon *et al.*, 1998; Chatfield *et al.*, 1992a; Dunne et al., 1995; Leary *et al.*, 1997; Nayak *et al.*, 1998; Oyston *et al.*, 1995; Poirier *et al.*, 1988; Titball *et al.*, 1997; Wyszynska *et al.*, 2004). Oral immunization with recombinant *S. enterica* vaccine strains can induce mucosal antibody responses against heterologous antigens. Specific secretory antibodies have been found in the gut, in saliva and in lung washes indicating that oral immunization can potentially generate protection at distant mucosal sites (Black *et al.*, 1987; Chen and Schifferli, 2003; Dusek *et al.*, 1994; Fagan *et al.*, 1997; Fouts *et al.*, 2003; Gomez-Duarte *et al.*, 1998; Guzman *et al.*, 1991; Maskell *et al.*, 1987; Poirier *et al.*, 1988; Redman *et al.*, 1995; Simonet *et al.*, 1994; Smerdou *et al.*, 1996; Stabel *et al.*, 1990; Stabel *et al.*, 1991; Su *et al.*, 1992; Walker *et al.*, 1992; Wu *et al.*, 1997). Alternative routes for mucosal immunization like nasal, vaginal or rectal immunization also induce antibody responses to recombinant antigens (Hopkins *et al.*, 1995; Nardelli-Haefliger *et al.*, 1996; Nardelli-Haefliger *et al.*, 1997; Pasetti *et al.*, 1999; Srinivasan *et al.*, 1995).

13.4.2 The activation of T-helper lymphocytes by recombinant *S. enterica* vaccines

Specific T-cell responses against heterologous antigens expressed in *S. enterica* have been detected using in vitro lymphocyte proliferation assays, T-cell cytokine production assays and delayed type hypersensitivity (DTH) tests (Brown *et al.*, 1987; Catmull *et al.*, 1999; Chabalgoity *et al.*, 1997; Karem *et al.*, 1997; Oyston *et al.*, 1995; Pasetti *et al.*, 1999; Sadoff *et al.*, 1988; Schodel and Will, 1990; Sjostedt *et al.*, 1992; Stabel *et al.*, 1993; Steger *et al.*, 1999; Verma *et al.*, 1995a; Verma *et al.*, 1995b; Yang *et al.*, 1990). In most cases, heterologous antigens expressed in *S. enterica* elicit T-cell helper type-1 (Th1) responses as indicated by IFNγ production by T-cells and by the predominance of antigen-specific IgG2a antibodies (Schodel *et al.*, 1990; Xu *et al.*, 1995; Yang *et al.*, 1990). However, T-helper type 2 responses (Th2) characterized by high levels of antigen-specific IgG1 and IL5 production have also been reported (Brett *et al.*, 1993; Chabalgoity *et al.*, 1997; Stager *et al.*, 1997; Wu *et al.*, 1997). The mechanisms that lead to these Th2 responses are not clear given the known ability of *S. enterica* to induce Th1 immunity. A biphasic T-helper cell response elicited upon immunization with recombinant *S. enterica* has been described, with an early Th2 response that later switched to a Th1 response (Pascual *et al.*, 1999). It is also possible that the nature of the recombinant antigen may exert some influence on the type of response elicited (Stager *et al.*, 1997). Clearly more work is needed to investigate the immunological basis of the T-cell immunity to *S. enterica* and recombinant heterologous antigens. It would be particularly interesting to further explore how other variables like bacteria and host genetic background may influence on the type of T-cell response elicited.

13.4.3 Cytotoxic T-lymphocyte (CTL) responses induced by recombinant *S. enterica* vaccines

A number of reports demonstrated that specific CTL responses can be elicited against recombinant antigens expressed by *S. enterica* vaccine strains. For example, CTLs against the circumsporozoite antigens of *Plasmodium yoelii, P. berghei* and *P. falciparum,* as well as against the p27gag gene of SIV and listeriolysin of *L. monocytogenes* have been described following vaccination with recombinant salmonellae (Aggarwal *et al.*, 1990; Flynn *et al.*, 1990; Valentine *et al.*, 1996; Verma *et al.*, 1995b).

The pathways for MHC-I restricted CD8 T-cell responses to *S. enterica*-derived antigens are still a matter of debate (Houde *et al.*, 2003; Lo *et al.*, 2004; Winau *et al.*, 2004). *Salmonella*-induced apoptosis, and cross-priming through apoptotic blebs taken up by DC has been proposed as a critical prerequisite for $CD8^+$ T-cell stimulation (Winau *et al.*, 2004; Yrlid and Wick, 2000).

The mechanism of display of recombinant antigens within the host cell appears also critical for the induction of CTL responses. In some studies, antigens that were exported outside the bacterial cell (secreted antigens) proved superior at inducing immune responses to those expressed as intracellular proteins (Hess *et al.*, 1996). The subsequent route for processing and presentation of MHC Class I-associated *Salmonella* antigens may differ between different types of APC, with macrophages loading peptides on to preformed post-Golgi MHC molecules independently of the TAP/proteosome pathway used by DC (Yrlid *et al.*, 2000).

Several different strategies have been attempted to improve the ability of *S. enterica* to elicit MHC class I-restricted immune responses. Particularly, the use of the *S. enterica* TTSS to deliver heterologous antigens directly to the cytosol of the host cell has been proposed as a highly efficient strategy to increase antigen processing via the MHC class I-restricted pathway (Russmann *et al.*, 1998). This approach has been successfully used to elicit protective CD8 T-cell responses against viral and bacterial infections in mice

(Igwe *et al.*, 2002; Russmann *et al.*, 2001; Russmann *et al.*, 1998; Shams *et al.*, 2001) and in rhesus macaques (Evans *et al.*, 2003).

13.4.4 Effect of prior immunity on the protection conferred by *S. enterica* recombinant vaccines

Background immunity against *S. enterica* antigens may be high in parts of the world where infections with *S. enterica* and closely related bacteria are prevalent. This may hamper the ability of recombinant *S. enterica* vaccines to colonize the tissues of and to deliver the recombinant antigens to the immune system.

Data from animal models suggested that prior exposure to *S. enterica* strains actually enhanced subsequent responses when homologous strains of *S. enterica* were used to deliver foreign antigens (Xiong Bao and Clements, 1991). Similarly, when *S. enterica* serovar Dublin was used to deliver a viral envelope protein B-cell epitope, the antibody responses to the epitope were enhanced in mice infected 3–6 months earlier with the *Salmonella* carrier (Whittle and Verma, 1997). However, in some cases exposure to the carrier bacteria impairs the immune responses to a recombinant antigen expressed and delivered using a homologous bacterial strain. For example, prior immunity to *S. enterica* reduced the subsequent immune responses to recombinant TetC expressed in a live attenuated *S. enterica* vaccine strain, with reduced serum and fecal anti-TetC antibodies and diminished protection against challenge with tetanus toxin (Roberts *et al.*, 1999). Similarly, in experiments conducted with *S. enterica* serovar Stanley as vector for *E. coli* K88 fimbrial proteins, it was shown that priming with the vector alone 1–2 months earlier significantly impaired the antibody response to K88 when the same *S. enterica* was used to express the antigen (Attridge *et al.*, 1997). Interestingly, in those experiments it was also demonstrated that such impairment could be overcome by using a *Salmonella* carrier whose serotype differs from that used in vector priming.

Concerns have also been raised about the possibility that existing immunity to proteins such as tetanus toxoid could hamper the ability of these proteins to enhance immune responses to recombinant antigens. However, available data indicate that pre-existing immunity to tetanus toxoid can lead to increased immunogenicity of TetC fusion protein vaccines expressed by *S. enterica* (Chabalgoity *et al.*, 1995).

In some cases, pre-exposure to the carrier bacterial species does not alter the immune response to antigens expressed in heterologous bacteria. For example, when flagellin from a *S. enterica* serovar Typhi vaccine strain was used as a carrier molecule for synthetic peptides derived from the influenza virus, it was shown that the antibody and cellular responses to the peptides were not impaired by pre-immunization with either the carrier flagellin molecule alone or with *Salmonella* (Ben-Yedidia, Arnon, 1998).

In humans, the situation seems less clear. Much may depend on the population being studied and their community exposure to bacterial pathogens. For example, the response to oral immunisation with *S. enterica* serovar Typhi Ty21a is significantly impaired in individuals with pre-existing cross-reactive intestinal antibodies, although the effect could be overcome by using a higher dose of the vaccine (Forrest, 1992). However, others have reported that maximal responses occur where there has been previous mucosal priming (Ferguson and Sallam, 1992).

13.4.5 Expression of cytokines in *S. enterica* to modulate immune responses

The genes for mammalian cytokines can be expressed in *Salmonella* as functional proteins. The production and delivery of bioactive cytokines by *S. enterica* can have therapeutic applications and can modulate immune functions in the host (Carrier *et al.*, 1992; Denich *et al.*, 1993; Hahn *et al.*, 1998; Ianaro *et al.*, 1995; Whittle *et al.*, 1997b)

For example, *S. enterica* expressing human IL1β can protect mice against lethal gamma irradiation (Carrier *et al.*, 1992). Administration of *S. enterica* expressing TGFβ reduces carrageenin-induced inflammation in mice, reduces IL2 and IFNγ production in the lymph nodes that drain the sites of inflammation and enhances IL10 secretion (Ianaro *et al.*, 1995).

Expression of cytokines in *S. enterica* can also have an effect on the ability of the host to control and clear the infection. For example, *S. enterica* expressing IL4 show increased growth and extended survival in mice and the bacteria are killed less efficiently by macrophages (Denich *et al.*, 1993). Conversely expression of IL2 in *Salmonella* results in enhanced bacterial clearance from the host (al-Ramadi *et al.*, 2001).

Expression of cytokines in *S. enterica* can be used as a means to modulate antigen specific immune responses. Administration to mice of *S. enterica* expressing IL5 results in an increase in the levels of *S. enterica*-specific IgA responses induced in the animals (Whittle *et al.*, 1997b).

Delivery of cytokines by *S. enterica* vaccines may be a feasible therapeutic treatment in infectious diseases or cancer. For example, recombinant *S. enterica* expressing IL2, IFNγ and macrophage migration inhibitory factor can enhance host resistance in mice challenged with *Leishmania major* (Xu *et al.*, 1998). *S. enterica* carrying the IL4- or IL18-expressing plasmids prolonged the survival time of mice inoculated with the B16F1 melanoma (Rosenkranz *et al.*, 2003).

Thus, expression of cytokines in *Salmonella* might be an attractive tool to study the function and relative importance of cytokine networks at the sites of bacterial infection and may have potential for the treatment of infection and cancer.

13.5 *S. ENTERICA* AS A DELIVERY SYSTEM FOR DNA VACCINES

S. enterica can act as a delivery system for DNA vaccines. Plasmids encoding antigens under the control of a viral or eukaryotic promoter can be incorporated into *S. enterica*. In this system, the bacteria do not express the plasmid-encoded genes, but after phagocytosis the plasmid may be transferred to the nucleus of the host cell leading to expression of the encoded gene (Dietrich *et al.*, 2003; Weiss, 2003). The transfer of plasmid DNA to the host cell has been shown in vitro and in vivo using plasmids encoding the green fluorescent protein (GFP) driven by eukaryotic promoters. Interestingly, in vitro plasmid transfer has been shown only when using primary cell cultures but not upon infection of established cell lines (Darji *et al.*, 1997; Weiss, 2003).

An interesting feature of this system is that the host cell can process the recombinant antigen as an endogenously produced protein via the MHC class I pathway. *Salmonella*-DNA vaccination appears therefore to be particularly well suited for the generation of CTL responses against recombinant antigens. In a comparative analysis, it was shown that oral immunization with β-galactosidase delivered by *S. enterica* as a DNA vaccine elicited superior T-helper and CTL responses as compared with immunization with the same dose of *S. enterica* that expressed high amounts of the same antigen driven by a prokaryotic promoter (Darji *et al.*, 2000). These T-cell responses were long-lasting and were detected up to eight months after *S. enterica*-based DNA vaccination. However, antibody responses were only detected shortly after vaccination, suggesting that further improvements are required to generate strong and persistent antibody responses with this system (Darji *et al.*, 2000). Similarly, *S. enterica*-based oral DNA vaccination with the surface protein of hepatitis B virus as antigen, elicited CTL responses similar to those obtained with injection of the naked DNA plasmid, but antibody responses were lower after oral administration of the recombinant bacteria (Woo *et al.*, 2001).

Different results were obtained in a comparative study using an attenuated strain of *S. enterica* serovar Typhi carrying TetC encoded by either a eukaryotic or a prokaryotic expression plasmid. In this case, intranasal administration of the recombinant strains, elicited higher anti-tetanus antibody responses in animals that received the bacteria carrying the eukaryotic expression plasmid (Pasetti *et al.*, 1999).

S. enterica vectors have been tested mainly as delivery systems for DNA vaccines against viral infections and cancer, where CTL responses are critical. *S. enterica*-mediated mucosal DNA immunization was first effectively demonstrated in mice using the *aroA*$^-$ mutant, *S. enterica* serovar Typhimurium SL7207, to elicit eukaryote-specific expression of ActA and listeriolysin from *L. monocytogenes*. The development of protective CTL and antibody responses against the encoded *Listeria* antigens was observed (Darji *et al.*, 1997). A DNA vaccine encoding β-galactosidase was incorporated in *S. enterica* serovar Typhimurium SL7207 and used to elicit successful protective immunization against a murine fibrosarcoma expressing β-galactosidase as a model tumor antigen (Paglia *et al.*, 1998). Similarly, transgenic mice expressing the tumor antigen CEA had their peripheral tolerance broken by oral vaccination with *S. enterica* serovar Typhimurium SL7207 carrying a DNA plasmid that expressed CEA, resulting in protection against CEA-positive lung carcinoma cells (Niethammer *et al.*, 2001).

S. enterica has been widely assessed as a delivery system for anti-melanoma DNA vaccines. In a mouse model of melanoma, the use of an attenuated *S. enterica* serovar Typhimurium to deliver a DNA vaccine construct containing the murine ubiquitin gene fused to minigenes encoding melanoma peptide epitopes gp100 (25–33) and TRP-2(181–188) resulted in protective antitumor immunity (Xiang *et al.*, 2000). Similarly, a protective CTL response was observed using *S. enterica* carrying a plasmid encoding human gp100 under the control of the CMV promoter, in a model of mouse B16 melanoma overexpressing human gp100 (Cochlovius *et al.*, 2002). A comparative assay revealed that oral vaccination with attenuated *S. enterica* serovar Typhimurium carrying a gp100 plasmid could induce protection against B16 melanoma more effectively than subcutaneous injection with DC loaded with gp100 peptides predicted to bind to *H2-K*^{b}/*H2-D*^{b} molecules (Weth *et al.*, 2001).

Combination therapy using *S. enterica* carrying a DNA vaccine encoding LacZ, followed by a boost with protein-loaded DC, was a very potent inducer of specific CTL responses. This approach proved to be more effective than naked DNA or peptide-loaded DC immunization in protecting mice against challenge with a murine renal carcinoma cell line transfected with lacZ (Zoller and Christ, 2001).

Oral immunization with a tyrosine hydroxylase (mTH)-based DNA vaccine carried by attenuated *S. enterica* serovar Typhimurium, followed by a boost with an antibody–cytokine fusion protein that targets IL2 to the tumor microenvironment, afforded complete protection against hepatic metastases in a murine model of neuroblastoma (Pertl *et al.*, 2003). A novel oral DNA vaccine that targets proliferating endothelial cells in the tumor vasculature rather than tumor cells, was constructed using *S. enterica* as a vector for a DNA vaccine encoding the vascular-endothelial growth factor (VEGF). Proliferating endothelial cells in the tumor vasculature upregulate vascular-endothelial growth factor receptor 2 (FLK-1), and the vaccine acted by suppressing angiogenesis in the tumor vasculature by CTL-mediated killing of endothelial cells (Niethammer *et al.*, 2002; Reisfeld *et al.*, 2004). Recently it has been reported that a DNA vaccine encoding the chemokine CCL21 and the inhibitor of apoptosis protein surviving delivered by the oral route by a *S. enterica* vaccine strain induced eradication or suppression of pulmonary metastases of non-small cell lung carcinoma (Xiang *et al.*, 2005).

The use of *S. enterica* as vector for DNA vaccines has also been assessed for the construction of vaccines against viral infections. A number of reports have demonstrated the feasibility of oral vaccination against HSV, hepatitis C, hepatitis B, HIV, pseudorabies and measles (Flo *et al.*, 2001; Pasetti *et al.*, 2003; Shata *et al.*, 2001; Shiau *et al.*, 2001; Wedemeyer *et al.*, 2001; Woo *et al.*, 2001; Zheng *et al.*, 2001). In most cases, the use of this system resulted in increased anti-viral CTL responses and activation of $CD8^+$ T-cells at mucosal sites (Shata *et al.*, 2001). Immune protection of mucosal sites was also demonstrated after oral immunization with a *S. enterica* strain carrying a DNA vaccine against HSV that afforded protection against intravaginal challenge with HSV (Flo *et al.*, 2001).

13.6 NEW EMERGING APPLICATIONS OF *S. ENTERICA* AS A VACCINE VECTOR

13.6.1 *S. enterica* as a vector for gene therapy

The finding that *S. enterica* can transfer eukaryotic expression plasmids to infected cells opened up the exciting possibility of oral gene therapy using *S. enterica* as a vector. *S. enterica* naturally infect DC and macrophages, and thus this system could be applied to correct in vivo a genetic defect of immune cells. For example, *S. enterica* carrying a eukaryotic expression vector encoding the murine IFNγ gene, was able to restore the production of this cytokine in macrophages of IFNγ-deficient mice. Administration of the bacteria containing the IFNγ expressing DNA vector restored natural resistance to bacterial infections in IFNγ deficient animals (Paglia *et al.*, 2000).

Given the central role that DC play in triggering effective anti-tumor immunity, this approach has also been attempted for the development of new immunotherapies against cancer. The effectiveness of *S. enterica*-based oral immune-gene therapy for the treatment of cancer was demonstrated in a mouse model of B-cell lymphoma (BCL) using an *aroA*$^-$ strain of *S. enterica* serovar Typhimurium as a delivery vector for the CD40 ligand (CD40L) gene expressed in a DNA vaccine. In this model the mice receiving the bacteria with the DNA vaccine encoding CD40L were protected against BCL (Urashima *et al.*, 2000). In a different experimental setting, a recombinant *S. enterica* harbouring a chimeric DNA vaccine containing the entire CEA extracellular domain fused to the C-terminus of murine CD40L conferred 100% protection against murine colon adenocarcinoma in CEA-transgenic mice (Xiang *et al.*, 2001).

Cytokine therapy has potential for the treatment of cancer due to the ability of cytokines to modulate functions of the immune system. Cytokine gene therapy could be superior to direct protein administration since it may retain its modulating effect on the immune system, without the limitations of short half-life and severe side effects commonly observed with direct administration of cytokines. Oral cytokine gene therapy using an attenuated *S. enterica* that carried a DNA vaccine encoding IL12 and GM-CSF resulted in a high degree of protection against transplantable murine tumors. *S. enterica* carrying the GM-CSF-encoding DNA vaccine was the most effective in prolonging survival time of tumor bearing-mice. Surprisingly, the combination of *S. enterica* carrying the GM-CSF-encoding DNA vaccine with *S. enterica* carrying the IL12-encoding DNA vaccine did not result in a synergistic effect (Yuhua *et al.*, 2001).

Oral cytokine gene therapy has also been attempted in the mouse model of melanoma using two different cytokines with previously proven antitumoral effect. Administration of an attenuated *S. enterica* carrying DNA vaccines encoding either IL4 or IL18 significantly prolonged the mean survival time of B16-bearing mice (Rosenkranz *et al.*, 2003).

S. enterica delivery systems have also been used for DC-based immunotherapies. Therapeutic vaccination using DC loaded ex vivo with bacteria expressing tyrosinase-related protein-2 (Trp-2), elicited protection against B16 melanoma in mice (Rescigno *et al.*, 2001).

13.6.2 Use of *S. enterica* strains that replicate preferentially in distinct anatomical sites

One of the most important characteristics of bacteria used as tools for immunotherapy of cancer is their ability to preferentially localize and accumulate within tumors in vivo. *S. enterica* has been used for this purpose since it has been shown that when administered to tumor-bearing mice, the bacteria will preferentially accumulate at the tumor site and in some cases within metastases (Chabalgoity *et al.*, 2002; Pawelek *et al.*, 2003; Yu *et al.*, 2004). An initial report showed that wild type strains of *S. enterica* serovar Typhimurium had the property of preferential replication at tumor sites in mice bearing melanoma (Pawelek *et al.*, 1997). However, the *S. enterica* infection itself was lethal for the mice. Attenuated *S. enterica* strains with mutations in the *purI* gene that render the strain auxotrophic for adenine, also showed preferential replication at tumor sites, with bacterial loads in the tumor being 9000-fold higher than those found in the liver. Interestingly, inoculation of the *S. enterica purI* strains in melanoma-bearing mice resulted in suppression of tumor growth and prolonged survival. Furthermore, the bacteria could efficiently deliver a DNA vaccine containing the HSV-thymidine kinase gene to the tumor resulting in the *in situ* activation of the antiviral drug ganciclovir (Pawelek *et al.*, 1997). Since that first publication, several reports have dealt with the use of *S. enterica* as an anticancer agent. A derivative of the original *purI* strain was constructed by introducing a deletion in the *msbB* gene involved in secondary acylation reactions that complete the biosynthesis of lipid A. Exposure of experimental animals or cultured cells to *S. enterica msbB* mutants results in reduced release of TNFα and IL1β, and therefore in reduced tissue inflammation, and decreased potential for the development of cytokine mediated septic shock (Khan *et al.*, 1998). The new *purI msbB* double mutant strain of *S. enterica* (named VNP20009) maintained the tumor targeting capacity and the antitumor effect observed with the parental strain in murine models of melanoma, as well as in other transplantable tumor models in mice (Low *et al.*, 1999; Rosenberg *et al.*, 2002; Zheng *et al.*, 2000). The antitumor efficacy of VNP20009 has also been evaluated in combination with X-irradiation in melanoma-bearing mice. Although each of these treatments slowed tumor growth and prolonged survival, the combined treatment produced synergistic antitumor effects, suggesting that combination of these genetically engineered *S. enterica* with radiotherapy could provide a new beneficial treatment for solid tumors (Platt *et al.*, 2000). *S. enterica* VNP20009 is also being assessed as a vector to target anticancer proteins or genes, such as cytosine deaminase (CD), to the tumor site. This use of VNP20009 has been designated Tumor Amplified Protein Expression (TAPET). It has been demonstrated that administration of TAPET-cytosine deaminase into tumor-bearing mice resulted in bacterial accumulation and CD expression in the tumors for up to 14 days after a single bolus intravenous administration of the bacteria (Zheng *et al.*, 2000). Biodistribution of VNP20009 in non-human primates and monkeys has also been studied to assess its safety (Clairmont *et al.*, 2000) and it has been proposed for parenteral administration in humans (Low *et al.*, 2004).

Results from a small phase I clinical trial in patients with metastatic melanoma have recently been published. Patients were treated with different doses of VNP20009 and, although the trial showed that the strain is safe for administration to humans, no anti-tumoral effect could be observed and tumor targeting was only seen in a small percentage of patients treated with the larger dose of bacteria (Toso *et al.*, 2002).

The factors that contribute to the preferential tumor accumulation of *S. enterica* are not fully defined. There is not a satisfactory explanation for the tumor targeting capacity observed in mice and this phenomenon is not observed with all *S. enterica* strains. It has been hypothesized that bacteria are likely to multiply better in tumors than in normal tissues because of the nutrients provided by rapidly growing cells and high percentage of necrotic cells (Forbes *et al.*, 2003). Tumors are vascularized tissues favoring entry and entrapment of bacteria. Moreover, granulocytes are rarely observed within tumors, suggesting a highly permissive environment for bacterial survival and replication. A different factor that might also contribute to accumulation of bacteria in tumors could be the capacity of *Salmonella* to invade and replicate within tumor cells. However, we have compared the biodistribution of a *Salmonella enterica* serovar Typhimurium with a derivative that showed increased capacity for in vitro infection of cultured melanoma cells, and we did not find higher numbers of bacteria within the tumor in melanoma-bearing mice that received the hyper-invasive strain (our unpublished results). Further, reports using electron or intravital microscopy to evaluate in vivo localization of *Salmonella* sistemically injected into tumor-bearing mice, showed that most bacteria were located in necrotic areas of the tumor (Forbes *et al.*, 2003). The bacterial SPI-2 pathogenicity island has been identified as essential for the antitumor effects of *S. enterica*, but further details are required (Pawelek *et al.*, 2002).

An alternative approach for the treatment of tumors used direct intratumoral injection of *S. enterica*. A pilot trial with TAPET- cytosine deaminase in cancer patients, using *S. enterica* expressing the *E. coli* cytosine deaminase gene, was recently completed. A total of three patients received three dose levels of TAPET- cytosine deaminase via intratumoral injection once every 28 days, provided the disease did not progress and no intolerable toxicity was observed. Between days 4 to 14 of each 28 day cycle, patients also received 5-fluorocytosine (5-FC). Six cycles of treatment were administered. No significant adverse effects clearly attributable to TAPET- cytosine deaminase were demonstrated and two patients had evidence of intratumor bacterial colonization with TAPET- cytosine deaminase, which persisted for at least 15 days after initial injection. Conversion of 5-fluorocytosine to 5-fluorouracil (the active antitumoral drug) as a result of cytosine deaminase expression was demonstrated in these two patients (Nemunaitis *et al.*, 2003). A different approach for cancer immunotherapy has recently been reported where tumor cells are infected in vivo with a live attenuated *Salmonella* to become targets for anti-*Salmonella* specific T-cells (Avogadri *et al.*, 2005).

13.7 CONCLUSIONS

More than two decades of research have firmly established that live attenuated salmonellae have great potential as delivery systems for heterologous molecules. Multivalent vaccines that could have broad applications in human and veterinary medicine have been developed using *S. enterica* as vector. Some new developments in the field, like the possibility of oral delivery of DNA vaccines, gene therapy, modulation of the immune responses to infection with *S. enterica* encoding cytokines, or tumor-targeted protein expression, keep expanding the potential of delivery systems based on *S. enterica*.

Less information is available regarding the potential use of *S. enterica*-based recombinant vaccines in humans and domestic animals. Well defined clinical trials will ultimately prove if approaches that have so far been successful in experimental models, have enough merit to become standard clinical practices in humans and domestic or companion animals.

REFERENCES

Aggarwal, A., Kumar, S., Jaffe, R. *et al.* (1990). Oral *Salmonella*: malaria circumsporozoite recombinants induce specific $CD8^+$ cytotoxic T-cells. *J Exp Med*, **172**, 1083–90.

al-Ramadi, B. K., Al-Dhaheri, M. H., Mustafa, N. *et al.* (2001). Influence of vector-encoded cytokines on anti-*Salmonella* immunity: divergent effects of interleukin-2 and tumor necrosis factor alpha. *Infect Immun*, **69**, 3980–8.

Ascon, M. A., Hone, D. M., Walters, N. and Pascual, D. W. (1998). Oral immunization with a *Salmonella typhimurium* vaccine vector expressing recombinant enterotoxigenic *Escherichia coli* K99 fimbriae elicits elevated antibody titers for protective immunity. *Infect Immun*, **66**, 5470–6.

Attridge, S. R., Davies, R. and LaBrooy, J. T. (1997). Oral delivery of foreign antigens by attenuated *Salmonella*: consequences of prior exposure to the vector strain. *Vaccine*, **15**, 155–62.

Avogadri, F., Martinoli, Ch., Petrovska, L. *et al.* (2005). Cancer immunotherapy based on killing of *Salmonella*-infected tumor cells. *Cancer Res.*, **65**, 3920–7.

Barry, E. M., Gomez-Duarte, O., Chatfield, S. *et al.* (1996). Expression and immunogenicity of pertussis toxin S1 subunit – tetanus toxin fragment C fusions in *Salmonella typhi* vaccine strain CVD 908. *Infect Immun*, **64**, 4172–81.

Ben-Yedidia, T. and Arnon, R. (1998). Effect of pre-existing carrier immunity on the efficacy of synthetic influenza vaccine. *Immunol Lett*, **64**, 9–15.

Black, R. E., Levine, M. M., Clements, M. L., Losonsky *et al.* (1987). Prevention of shigellosis by a *Salmonella typhi – Shigella sonnei* bivalent vaccine. *J Infect Dis*, **155**, 1260–5.

Brayton, K. A., van der Walt, M., Vogel, S. W. and Allsopp, B. A. (1998). A partially protective clone from *Cowdria ruminantium* identified by using a *Salmonella* vaccine delivery system. *Ann N Y Acad Sci*, **849**, 247–52.

Brett, S. J., Dunlop, L., Liew, F. Y. and Tite, J. P. (1993). Influence of the antigen delivery system on immunoglobulin isotype selection and cytokine production in response to influenza A nucleoprotein. *Immunology*, **80**, 306–12.

Brown, A., Hormaeche, C. E., Demarco-de-Hormaeche, R. *et al.* (1987). An attenuated *aroA Salmonella* typhimurium vaccine elicits humoral and cellular immunity to cloned beta-galactosidase in mice. *J Infect Dis*, **155**, 86–92.

Capozzo, A.V., Cuberos, L., Levine, M. M. and Pasetti, M. F. (2004). Mucosally delivered *Salmonella* live vector vaccines elicit potent immune responses against a foreign antigen in neonatal mice born to naive and immune mothers. *Infect Immun*, **72**, 4637–46.

Carrier, M. J., Chatfield, S. N., Dougan, G. *et al.* (1992). Expression of human IL1 beta in *Salmonella typhimurium*. A model system for the delivery of recombinant therapeutic proteins *in vivo*. *J Immunol*, **148**, 1176–81.

Catmull, J., Wilson, M. E., Kirchhoff, L. V., Metwali, A. and Donelson, J. E. (1999). Induction of specific cell-mediated immunity in mice by oral immunization with *Salmonella* expressing *Onchocerca volvulus* glutathione S-transferase. *Vaccine*, **17**, 31–9.

Cattozzo, E. M., Stocker, B. A., Radaelli, A., De Giuli Morghen, C. and Tognon, M. (1997). Expression and immunogenicity of V3 loop epitopes of HIV-1, isolates SC and WMJ2, inserted in *Salmonella* flagellin. *J Biotechnol*, **56**, 191–203.

Chabalgoity, J. A., Dougan, G., Mastroeni, P. and Aspinall, R. J. (2002). Live bacteria as the basis for immunotherapies against cancer. *Expert Rev Vaccines*, **1**, 495–505.

Chabalgoity, J. A., Harrison, J. A., Esteves, A. *et al.* (1997). Expression and immunogenicity of an *Echinococcus granulosus* fatty acid-binding protein in live attenuated *Salmonella* vaccine strains. *Infect Immun*, **65**, 2402–12.

Chabalgoity, J. A., Khan, C. M., Nash, A. A. and Hormaeche, C. E. (1996). A *Salmonella typhimurium htrA* live vaccine expressing multiple copies of a peptide comprising amino acids 8–23 of herpes simplex virus glycoprotein D as a genetic fusion to tetanus toxin fragment C protects mice from herpes simplex virus infection. *Mol Microbiol*, **19**, 791–801.

Chabalgoity, J. A., Moreno, M., Carol, H., Dougan, G. and Hormaeche, C. E. (2000). A dog-adapted *Salmonella typhimurium* strain as a basis for a Live oral *Echinococcus granulosus* vaccine. *Vaccine*, **19**, 460–9.

Chabalgoity, J. A., Villareal-Ramos, B., Khan, C. M. *et al.* (1995). Influence of preimmunization with tetanus toxoid on immune responses to tetanus toxin fragment C-guest antigen fusions in a *Salmonella* vaccine carrier. *Infect Immun*, **63**, 2564–9.

Charbit, A., Martineau, P., Ronco, J. *et al.* (1993). Expression and immunogenicity of the V3 loop from the envelope of human immunodeficiency virus type 1 in an attenuated *aroA* strain of *Salmonella typhimurium* upon genetic coupling to two *Escherichia coli* carrier proteins. *Vaccine*, **11**, 1221–8.

Chatfield, S. N., Charles, I. G., Makoff, A. J. *et al.* (1992a). Use of the *nirB* promoter to direct the stable expression of heterologous antigens in *Salmonella* oral vaccine strains: development of a single-dose oral tetanus vaccine. *Biotechnology N Y*, **10**, 888–92.

Chatfield, S. N., Fairweather, N., Charles, I. *et al.* (1992b). Construction of a genetically defined *Salmonella typhi* Ty2 *aroA, aroC* mutant for the engineering of a candidate oral typhoid–tetanus vaccine. *Vaccine*, **10**, 53–60.

Chen, H. and Schifferli, D. M. (2001). Enhanced immune responses to viral epitopes by combining macrophage-inducible expression with multimeric display on a *Salmonella* vector. *Vaccine*, **19**, 3009–18.

(2003). Construction, characterization, and immunogenicity of an attenuated *Salmonella enterica* serovar Typhimurium *pgtE* vaccine expressing fimbriae with integrated viral epitopes from the spiC promoter. *Infect Immun*, **71**, 4664–73.

Cicin-Sain, L., Brune, W., Bubic, I., Jonjic, S. and Koszinowski, U. H. (2003). Vaccination of mice with bacteria carrying a cloned herpes virus genome reconstituted *in vivo*. *J Virol*, **77**, 8249–55.

Clairmont, C., Lee, K. C., Pike, J. *et al.* (2000). Biodistribution and genetic stability of the novel antitumor agent VNP20009, a genetically modified strain of *Salmonella typhimurium*. *J Infect Dis*, **181**, 1996–2002.

Cochlovius, B., Stassar, M. J., Schreurs, M. W., Benner, A. and Adema, G. J. (2002). Oral DNA vaccination: antigen uptake and presentation by dendritic cells elicits protective immunity. *Immunol Lett*, **80**, 89–96.

Coulson, N. M., Fulop, M. and Titball, R. W. (1994a). Bacillus anthracis protective antigen, expressed in *Salmonella typhimurium* SL 3261, affords protection against anthrax spore challenge. *Vaccine*, **12**, 1395–401.

(1994b). Effect of different plasmids on colonization of mouse tissues by the aromatic amino acid dependent *Salmonella typhimurium* SL 3261. *Microb Pathog*, **16**, 305–11.

Covone, M. G., Brocchi, M., Palla, E. *et al.* (1998). Levels of expression and immunogenicity of attenuated *Salmonella enterica* serovar Typhimurium strains expressing *Escherichia coli* mutant heat-labile enterotoxin. *Infect Immun*, **66**, 224–31.

Darji, A., Guzman, C. A., Gerstel, B. *et al.* (1997). Oral somatic transgene vaccination using attenuated *S. typhimurium*. *Cell*, **91**, 765–75.

Darji, A., zur Lage, S., Garbe, A. I., Chakraborty, T. and Weiss, S. (2000). Oral delivery of DNA vaccines using attenuated *Salmonella typhimurium* as carrier. *FEMS Immunol Med Microbiol*, **27**, 341–9.

Denich, K., Borlin, P., O'Hanley, P. D., Howard, M. and Heath, A. W. (1993). Expression of the murine interleukin-4 gene in an attenuated *aroA* strain of *Salmonella typhimurium*: persistence and immune response in BALB/c mice and susceptibility to macrophage killing. *Infect Immun*, **61**, 4818–27.

Dietrich, G., Spreng, S., Favre, D., Viret, J. F. and Guzman, C. A. (2003). Live attenuated bacteria as vectors to deliver plasmid DNA vaccines. *Curr Opin Mol Ther*, **5**, 10–19.

Dunne, M., al-Ramadi, B. K., Barthold, S. W., Flavell, R. A. and Fikrig, E. (1995). Oral vaccination with an attenuated *Salmonella typhimurium* strain expressing *Borrelia burgdorferi* OspA prevents murine Lyme borreliosis. *Infect Immun*, **63**, 1611–14.

Dunstan, S. J., Simmons, C. P. and Strugnell, R. A. (1999). Use of *in vivo*-regulated promoters to deliver antigens from attenuated *Salmonella* enterica var. Typhimurium. *Infect Immun*, **67**, 5133–41.

Dusek, D. M., Progulske-Fox, A. and Brown, T. A. (1994). Systemic and mucosal immune responses in mice orally immunized with avirulent *Salmonella typhimurium* expressing a cloned *Porphyromonas gingivalis* hemagglutinin. *Infect Immun*, **62**, 1652–7.

Ervin, S. E., Small, P., Jr. and Gulig, P. A. (1993). Use of incompatible plasmids to control expression of antigen by *Salmonella typhimurium* and analysis of immunogenicity in mice. *Microb Pathog*, **15**, 93–101.

Evans, D. T., Chen, L. M., Gillis, J. *et al.* (2003). Mucosal priming of simian immunodeficiency virus-specific cytotoxic T-lymphocyte responses in rhesus macaques by the *Salmonella* type III secretion antigen delivery system. *J Virol*, **77**, 2400–9.

Everest, P., Frankel, G., Li, J. *et al.* (1995). Expression of LacZ from the *htrA*, *nirB* and *groE* promoters in a *Salmonella* vaccine strain: influence of growth in mammalian cells. *FEMS Microbiol Lett*, **126**, 97–101.

Fagan, P. K., Djordjevic, S. P., Chin, J., Eamens, G. J. and Walker, M. J. (1997). Oral immunization of mice with attenuated *Salmonella typhimurium aroA* expressing a recombinant *Mycoplasma hyopneumoniae* antigen (NrdF). *Infect Immun*, **65**, 2502–7.

Ferguson, A. and Sallam, J. (1992). Mucosal immunity to oral vaccines. *Lancet*, **339**, 179.

Flo, J., Tisminetzky, S. and Baralle, F. (2001). Oral transgene vaccination mediated by attenuated salmonellae is an effective method to prevent *Herpes simplex* virus-2 induced disease in mice. *Vaccine*, **19**, 1772–82.

Flynn, J. L., Weiss, W. R., Norris, K. A. *et al.* (1990). Generation of a cytotoxic T-lymphocyte response using a *Salmonella* antigen-delivery system. *Mol Microbiol*, **4**, 2111–18.

Forbes, N. S., Munn, L. L., Fukumura, D. and Jain, R. K. (2003). Sparse initial entrapment of systemically injected *Salmonella typhimurium* leads to heterogeneous accumulation within tumors. *Cancer Res*, **63**, 5188–93.

Forrest, B. D. (1992). Impairment of immunogenicity of *Salmonella typhi* Ty21a due to preexisting cross-reacting intestinal antibodies. *J Infect Dis*, **166**, 210–12.

Fouts, T. R., DeVico, A. L., Onyabe, D. Y. *et al.* (2003). Progress toward the development of a bacterial vaccine vector that induces high-titer long-lived broadly neutralizing antibodies against HIV-1. *FEMS Immunol Med Microbiol*, **37**, 129–34.

Galan, J. E., Nakayama, K. and Curtiss, R., III (1990). Cloning and characterization of the asd gene of *Salmonella typhimurium*: use in stable maintenance of recombinant plasmids in *Salmonella* vaccine strains. *Gene*, **94**, 29–35.

Galen, J. E., Nair, J., Wang, J. Y. *et al.* (1999). Optimization of plasmid maintenance in the attenuated live vector vaccine strain *Salmonella typhi* CVD 908-*htrA*. *Infect Immun*, **67**, 6424–33.

Gentschev, I., Glaser, I., Goebel, W. *et al.* (1998). Delivery of the p67 sporozoite antigen of *Theileria parva* by using recombinant *Salmonella dublin*: secretion of the product enhances specific antibody responses in cattle. *Infect Immun*, **66**, 2060–4.

Gentschev, I., Mollenkopf, H., Sokolovic, Z. *et al.* (1996). Development of antigen-delivery systems, based on the *Escherichia coli* hemolysin secretion pathway. *Gene*, **179**, 133–40.

Gomez-Duarte, O. G., Galen, J., Chatfield, S. N. *et al.* (1995). Expression of fragment C of tetanus toxin fused to a carboxyl-terminal fragment of diphtheria toxin in *Salmonella typhi* CVD 908 vaccine strain. *Vaccine*, **13**, 1596–602.

Gomez-Duarte, O. G., Lucas, B., Yan, Z. X. *et al.* (1998). Protection of mice against gastric colonization by *Helicobacter pylori* by single oral dose immunization with attenuated *Salmonella typhimurium* producing urease subunits A and B. *Vaccine*, **16**, 460–71.

Goñi, F., Knudsen, E., Schreiber, F. *et al.* (2005). Mucosal vaccination delays or prevents prion infection via an oral route. *Neuroscience*, **133** (2), 413–21.

Guzman, C. A., Brownlie, R. M., Kadurugamuwa, J., Walker, M. J. and Timmis, K. N. (1991). Antibody responses in the lungs of mice following oral immunization with *Salmonella typhimurium aroA* and invasive *Escherichia coli* strains expressing the filamentous hemagglutinin of *Bordetella pertussis*. *Infect Immun*, **59**, 4391–7.

Haddad, D., Liljeqvist, S., Kumar, S. *et al.* (1995). Surface display compared to periplasmic expression of a malarial antigen in *Salmonella typhimurium* and its implications for immunogenicity. *FEMS Immunol Med Microbiol*, **12**, 175–86.

Hahn, H. P., Hess, C., Gabelsberger, J., Domdey, H. and von Specht, B. U. (1998). A *Salmonella typhimurium* strain genetically engineered to secrete effectively a bioactive human interleukin (hIL)-6 via *the Escherichia coli* hemolysin secretion apparatus. *FEMS Immunol Med Microbiol*, **20**, 111–19.

Hayes, L. J., Conlan, J. W., Everson, J. S., Ward, M. E. and Clarke, I. N. (1991). *Chlamydia trachomatis* major outer membrane protein epitopes expressed as fusions with LamB in an attenuated *aroA* strain of *Salmonella typhimurium*; their application as potential immunogens. *J Gen Microbiol*, **137**, 1557–64.

Hess, J., Gentschev, I., Miko, D. *et al.* (1996). Superior efficacy of secreted over somatic antigen display in recombinant *Salmonella* vaccine induced protection against listeriosis. *Proc Natl Acad Sci USA*, **93**, 1458–63.

Hohmann, E. L., Oletta, C. A., Loomis, W. P. and Miller, S. I. (1995). Macrophage-inducible expression of a model antigen in *Salmonella typhimurium* enhances immunogenicity. *Proc Natl Acad Sci USA*, **92**, 2904–8.

Hone, D., Attridge, S., van-den-Bosch, L. and Hackett, J. (1988). A chromosomal integration system for stabilization of heterologous genes in *Salmonella* based vaccine strains. *Microb Pathog*, **5**, 407–18.

Hopkins, S., Kraehenbuhl, J. P., Schodel, F. *et al.* (1995). A recombinant *Salmonella typhimurium* vaccine induces local immunity by four different routes of immunization. *Infect Immun*, **63**, 3279–86.

Houde, M., Bertholet, S., Gagnon, E. *et al.* (2003). Phagosomes are competent organelles for antigen cross-presentation. *Nature*, **425**, 402–6.

Ianaro, A., Xu, D., O'Donnell, C. A., Di-Rosa, M. and Liew, F. Y. (1995). Expression of TGF-beta in attenuated *Salmonella typhimurium*: oral administration leads to the reduction of inflammation, IL2 and IFNγ, but enhancement of IL10, in carrageenin-induced oedema in mice. *Immunology*, **84**, 8–15.

Igwe, E. I., Geginat, G. and Russmann, H. (2002). Concomitant cytosolic delivery of two immunodominant listerial antigens by *Salmonella enterica* serovar Typhimurium confers superior protection against murine listeriosis. *Infect Immun*, **70**, 7114–19.

Jagusztyn-Krynicka, E. K., Clark-Curtiss, J. E. and Curtiss, R., III (1993). *Escherichia coli* heat-labile toxin subunit B fusions with *Streptococcus sobrinus* antigens expressed by *Salmonella typhimurium* oral vaccine strains: importance of the linker for antigenicity and biological activities of the hybrid proteins. *Infect Immun*, **61**, 1004–15.

Kang, H. Y. and Curtiss, R., III (2003). Immune responses dependent on antigen location in recombinant attenuated *Salmonella typhimurium* vaccines following oral immunization. *FEMS Immunol Med Microbiol*, **37**, 99–104.

Karem, K. L., Bowen, J., Kuklin, N. and Rouse, B. T. (1997). Protective immunity against herpes simplex virus (HSV) type 1 following oral administration of recombinant *Salmonella typhimurium* vaccine strains expressing HSV antigens. *J Gen Virol*, **78**, 427–34.

Karem, K. L., Chatfield, S., Kuklin, N. and Rouse, B. T. (1995). Differential induction of carrier antigen-specific immunity by *Salmonella typhimurium* live-vaccine strains after single mucosal or intravenous immunization of BALB/c mice. *Infect Immun*, **63**, 4557–63.

Khan, C. M., Villarreal-Ramos, B., Pierce, R. J. *et al.* (1994a). Construction, expression, and immunogenicity of multiple tandem copies of the *Schistosoma mansoni* peptide 115–131 of the P28 glutathione S-transferase expressed as C-terminal fusions to tetanus toxin fragment C in a live aro-attenuated vaccine strain of *Salmonella*. *J Immunol*, **153**, 5634–42.

(1994b). Construction, expression, and immunogenicity of the *Schistosoma mansoni* P28 glutathione S-transferase as a genetic fusion to tetanus toxin fragment C in a live *aro* attenuated vaccine strain of *Salmonella*. *Proc Natl Acad Sci USA*, **91**, 11261–5.

Khan, S. A., Everest, P., Servos, S. *et al.* (1998). A lethal role for lipid A in *Salmonella* infections. *Mol Microbiol*, **29**, 571–9.

Kochi, S. K., Killeen, K. P. and Ryan, U. S. (2003). Advances in the development of bacterial vector technology. *Expert Rev Vaccines*, **2**, 31–43.

Leary, S. E. C., Griffin, K. F., Garmory, H. S., Williamson, E. D. and Titball, R. W. (1997). Expression of an F1/V fusion protein in attenuated *Salmonella typhimurium* and protection of mice against plague. *Microb Pathog*, **23**, 167–79.

Lo, W. F., Dunn, C. D., Ong, H., Metcalf, E. S. and Soloski, M. J. (2004). Bacterial and host factors involved in the major histocompatibility complex class Ib-restricted presentation of *Salmonella* Hsp 60: novel pathway. *Infect Immun*, **72**, 2843–9.

Londono, L. P., Chatfield, S., Tindle, R. W. *et al.* (1996). Immunisation of mice using *Salmonella typhimurium* expressing human papillomavirus type 16 E7 epitopes inserted into hepatitis B virus core antigen. *Vaccine*, **14**, 545–52.

Low, K. B., Ittensohn, M., Le, T. *et al.* (1999). Lipid A mutant *Salmonella* with suppressed virulence and TNFα induction retain tumor-targeting *in vivo*. *Nat Biotechnol*, **17**, 37–41.

Low, K. B., Ittensohn, M., Luo, X. *et al.* (2004). Construction of VNP20009: a novel, genetically stable antibiotic-sensitive strain of tumor-targeting *Salmonella* for parenteral administration in humans. *Methods Mol Med*, **90**, 47–60.

Marshall, D. G., Haque, A., Fowler, R. *et al.* (2000). Use of the stationary phase inducible promoters, *spv* and *dps*, to drive heterologous antigen expression in *Salmonella* vaccine strains. *Vaccine*, **18**, 1298–306.

Maskell, D. J., Sweeney, K. J., O'Callaghan, D. *et al.* (1987). *Salmonella typhimurium aroA* mutants as carriers of the *Escherichia coli* heat-labile enterotoxin B subunit to the murine secretory and systemic immune systems. *Microb Pathog*, **2**, 211–21.

Mastroeni, P., Chabalgoity, J. A., Dunstan, S. J., Maskell, D. J. and Dougan, G. (2001). *Salmonella*: immune responses and vaccines. *Vet J*, **161**, 132–64.

McKelvie, N. D., Stratford, R., Wu, T. *et al.* (2004). Expression of heterologous antigens in *Salmonella typhimurium* vaccine vectors using the in vivo-inducible, SPI-2 promoter, *ssaG*. *Vaccine*, **22**, 3243–55.

McSorley, S. J., Xu, D. and Liew, F. Y. (1997). Vaccine efficacy of *Salmonella* strains expressing glycoprotein 63 with different promoters. *Infect Immun*, **65**, 171–8.

Molina, N. C. and Parker, C. D. (1990). Murine antibody response to oral infection with live *aroA* recombinant *Salmonella dublin* vaccine strains expressing filamentous hemagglutinin antigen from *Bordetella pertussis*. *Infect Immun*, **58**, 2523–8.

Nakayama, K., Kelly, S. M. and Curtiss, R., III (1988). Construction of an Asd^+ expression cloning vector: stable maintenance and high level expression of cloned genes in a *Salmonella* vaccine strain. *Bio/Technology*, **6**, 693–7.

Nardelli-Haefliger, D., Kraehenbuhl, J. P., Curtiss, R., III *et al.* (1996). Oral and rectal immunization of adult female volunteers with a recombinant attenuated *Salmonella typhi* vaccine strain. *Infect Immun*, **64**, 5219–24.

Nardelli-Haefliger, D., Roden, R. B., Benyacoub, J. *et al.* (1997). Human papillomavirus type 16 virus-like particles expressed in attenuated *Salmonella typhimurium* elicit mucosal and systemic neutralizing antibodies in mice. *Infect Immun*, **65**, 3328–36.

Nayak, A. R., Tinge, S. A., Tart, R. C. *et al.* (1998). A live recombinant avirulent oral *Salmonella* vaccine expressing pneumococcal surface protein A induces protective responses against *Streptococcus pneumoniae*. *Infect Immun*, **66**, 3744–51.

Nemunaitis, J., Cunningham, C., Senzer, N. *et al.* (2003). Pilot trial of genetically modified, attenuated *Salmonella* expressing the *E. coli* cytosine deaminase gene in refractory cancer patients. *Cancer Gene Ther*, **10**, 737–44.

Newton, S. M., Jacob, C. O. and Stocker, B. A. (1989). Immune response to cholera toxin epitope inserted in *Salmonella* flagellin. *Science*, **244**, 70–2.

Newton, S. M., Joys, T. M., Anderson, S. A. *et al.* (1995). Expression and immunogenicity of an 18-residue epitope of HIV1 gp41 inserted in the flagellar protein of a *Salmonella* live vaccine. *Res Microbiol*, **146**, 203–16.

Newton, S. M., Kotb, M., Poirier, T. P., Stocker, B. A. and Beachey, E. H. (1991). Expression and immunogenicity of a streptococcal M protein epitope inserted in *Salmonella* flagellin. *Infect Immun*, **59**, 2158–65.

Niethammer, A. G., Xiang, R., Becker, J. C. *et al.* (2002). A DNA vaccine against VEGF receptor 2 prevents effective angiogenesis and inhibits tumor growth. *Nat Med*, **8**, 1369–75.

Niethammer, A. G., Xiang, R., Ruehlmann, J. M. *et al.* (2001). Targeted interleukin 2 therapy enhances protective immunity induced by an autologous oral DNA vaccine against murine melanoma. *Cancer Res*, **61**, 6178–84.

Orr, N., Galen, J. E. and Levine, M. M. (1999). Expression and immunogenicity of a mutant diphtheria toxin molecule, CRM(197), and its fragments in *Salmonella typhi* vaccine strain CVD 908-*htrA*. *Infect Immun*, **67**, 4290–4.

(2001). Novel use of anaerobically induced promoter, *dmsA*, for controlled expression of fragment C of tetanus toxin in live attenuated *Salmonella enterica* serovar Typhi strain CVD 908-*htrA*. *Vaccine*, **19**, 1694–700.

Oyston, P. C., Williamson, E. D., Leary, S. E. *et al.* (1995). Immunization with live recombinant *Salmonella typhimurium aroA* producing F1 antigen protects against plague. *Infect Immun*, **63**, 563–8.

Paglia, P., Medina, E., Arioli, I., Guzman, C. A. and Colombo, M. P. (1998). Gene transfer in dendritic cells, induced by oral DNA vaccination with *Salmonella typhimurium*, results in protective immunity against a murine fibrosarcoma. *Blood*, **92**, 3172–6.

Paglia, P., Terrazzini, N., Schulze, K., Guzman, C. A. and Colombo, M. P. (2000). In vivo correction of genetic defects of monocyte/macrophages using attenuated *Salmonella* as oral vectors for targeted gene delivery. *Gene Ther*, **7**, 1725–30.

Pascual, D. W., Hone, D. M., Hall, S. *et al.* (1999). Expression of recombinant enterotoxigenic *Escherichia coli* colonization factor antigen I by *Salmonella typhimurium* elicits a biphasic T-helper cell response. *Infect Immun*, **67**, 6249–56.

Pasetti, M. F., Anderson, R. J., Noriega, F. R., Levine, M. M. and Sztein, M. B. (1999). Attenuated Δ*guaBA Salmonella typhi* vaccine strain CVD 915 as a live vector utilizing prokaryotic or eukaryotic expression systems to deliver foreign antigens and elicit immune responses. *Clin Immunol*, **92**, 76–89.

Pasetti, M. F., Barry, E. M., Losonsky, G. *et al.* (2003). Attenuated *Salmonella enterica* serovar Typhi and *Shigella flexneri* 2a strains mucosally deliver DNA vaccines encoding measles virus hemagglutinin, inducing specific immune responses and protection in cotton rats. *J Virol*, **77**, 5209–17.

Pawelek, J. M., Low, K. B. and Bermudes, D. (1997). Tumor-targeted *Salmonella* as a novel anticancer vector. *Cancer Res*, **57**, 4537–44.

(2003). Bacteria as tumour-targeting vectors. *Lancet Oncol*, **4**, 548–56.

Pawelek, J. M., Sodi, S., Chakraborty, A. K. *et al.* (2002). *Salmonella* pathogenicity island-2 and anticancer activity in mice. *Cancer Gene Ther*, **9**, 813–18.

Pertl, U., Wodrich, H., Ruehlmann, J. M. *et al.* (2003). Immunotherapy with a posttranscriptionally modified DNA vaccine induces complete protection against metastatic neuroblastoma. *Blood*, **101**, 649–54.

Pistor, S. and Hobom, G. (1990). OmpA-Haemagglutinin fusion proteins for oral immunization with live attenuated *Salmonella*. *Res Microbiol*, **141**, 879–81.

Platt, J., Sodi, S., Kelley, M. *et al.* (2000). Antitumour effects of genetically engineered *Salmonella* in combination with radiation. *Eur J Cancer*, **36**, 2397–402.

Poirier, T. P., Kehoe, M. A. and Beachey, E. H. (1988). Protective immunity evoked by oral administration of attenuated *aroA Salmonella typhimurium* expressing cloned streptococcal M protein. *J Exp Med*, **168**, 25–32.

Redman, T. K., Harmon, C. C., Lallone, R. L. and Michalek, S. M. (1995). Oral immunization with recombinant *Salmonella typhimurium* expressing surface protein antigen A of *Streptococcus sobrinus*: dose response and induction of protective humoral responses in rats. *Infect Immun*, **63**, 2004–11.

Redman, T. K., Harmon, C. C. and Michalek, S. M. (1996). Oral immunization with recombinant *Salmonella typhimurium* expressing surface protein antigen A (SpaA) of *Streptococcus sobrinus*: effects of the *Salmonella* virulence plasmid on the induction of protective and sustained humoral responses in rats. *Vaccine*, **14**, 868–78.

Reisfeld, R. A., Niethammer, A. G., Luo, Y. and Xiang, R. (2004). DNA vaccines suppress tumor growth and metastases by the induction of anti-angiogenesis. *Immunol Rev*, **199**, 181–90.

Rescigno, M., Valzasina, B., Bonasio, R., Urbano, M. and Ricciardi-Castagnoli, P. (2001). Dendritic cells, loaded with recombinant bacteria expressing tumor antigens, induce a protective tumor-specific response. *Clin Cancer Res*, **7**, 865s–870s.

Roberts, M., Bacon, A., Li, J. and Chatfield, S. (1999). Prior immunity to homologous and heterologous *Salmonella* serotypes suppresses local and systemic anti-fragment C antibody responses and protection from tetanus toxin in mice immunized with *Salmonella* strains expressing fragment C. *Infect Immun*, **67**, 3810–15.

Roberts, M., Li, J., Bacon, A. and Chatfield, S. (1998). Oral vaccination against tetanus: comparison of the immunogenicities of *Salmonella* strains expressing fragment C from the *nirB* and *htrA* promoters [published erratum appears in *Infect Immun* (1999), **67**, 468]. *Infect Immun*, **66**, 3080–7.

Rosenberg, S. A., Spiess, P. J. and Kleiner, D. E. (2002). Antitumor effects in mice of the intravenous injection of attenuated *Salmonella typhimurium*. *J Immunother*, **25**, 218–25.

Rosenkranz, C. D., Chiara, D., Agorio, C. *et al.* (2003). Towards new immunotherapies: targeting recombinant cytokines to the immune system using live attenuated *Salmonella*. *Vaccine*, **21**, 798–801.

Russmann, H. (2004). Inverted pathogenicity: the use of pathogen-specific molecular mechanisms for prevention or therapy of disease. *Int J Med Microbiol*, **293**, 565–9.

Russmann, H., Igwe, E. I., Sauer, J. *et al.* (2001). Protection against murine listeriosis by oral vaccination with recombinant *Salmonella* expressing hybrid *Yersinia* type III proteins. *J Immunol*, **167**, 357–65.

Russmann, H., Shams, H., Poblete, F. *et al.* (1998). Delivery of epitopes by the *Salmonella* type III secretion system for vaccine development. *Science*, **281**, 565–8.

Sadoff, J. C., Ballou, W. R., Baron, L. S. *et al.* (1988). Oral *Salmonella typhimurium* vaccine expressing circumsporozoite protein protects against malaria. *Science,* **240**, 336–8.

Schodel, F. and Will, H. (1990). Expression of hepatitis B virus antigens in attenuated salmonellae for oral immunization. *Res Microbiol,* **141**, 831–7.

Schodel, F., Milich, D. R. and Will, H. (1990). Hepatitis B virus nucleocapsid/pre-S2 fusion proteins expressed in attenuated *Salmonella* for oral vaccination. *J Immunol,* **145**, 4317–21.

Schorr, J., Knapp, B., Hundt, E., Kupper, H. A. and Amann, E. (1991). Surface expression of malarial antigens in *Salmonella typhimurium*: induction of serum antibody response upon oral vaccination of mice. *Vaccine,* **9**, 675–81.

Shams, H., Poblete, F., Russmann, H., Galan, J. E. and Donis, R. O. (2001). Induction of specific $CD8^+$ memory T-cells and long lasting protection following immunization with *Salmonella typhimurium* expressing a lymphocytic choriomeningitis MHC class I-restricted epitope. *Vaccine,* **20**, 577–85.

Shata, M. T., Reitz, M. S., Jr., DeVico, A. L., Lewis, G. K. and Hone, D. M. (2001). Mucosal and systemic HIV-1 Env-specific $CD8^+$ T-cells develop after intragastric vaccination with a *Salmonella* Env DNA vaccine vector. *Vaccine,* **20**, 623–9.

Shiau, A. L., Chu, C. Y., Su, W. C. and Wu, C. L. (2001). Vaccination with the glycoprotein D gene of pseudorabies virus delivered by nonpathogenic *Escherichia coli* elicits protective immune responses. *Vaccine,* **19**, 3277–84.

Simonet, M., Fortineau, N., Beretti, J. L. and Berche, P. (1994). Immunization with live *aroA* recombinant *Salmonella typhimurium* producing invasin inhibits intestinal translocation of *Yersinia pseudotuberculosis. Infect Immun,* **62**, 863–7.

Sjostedt, A., Sandstrom, G. and Tarnvik, A. (1992). Humoral and cell-mediated immunity in mice to a 17-kilodalton lipoprotein of *Francisella tularensis* expressed by *Salmonella typhimurium. Infect Immun,* **60**, 2855–62.

Smerdou, C., Anton, I. M., Plana, J., Curtiss, R., III and Enjuanes, L. (1996). A continuous epitope from transmissible gastroenteritis virus S protein fused to *E. coli* heat-labile toxin B subunit expressed by attenuated *Salmonella* induces serum and secretory immunity. *Virus Res,* **41**, 1–9.

Srinivasan, J., Tinge, S., Wright, R., Herr, J. C. and Curtiss, R., III (1995). Oral immunization with attenuated *Salmonella* expressing human sperm antigen induces antibodies in serum and the reproductive tract. *Biol Reprod,* **53**, 462–71.

Stabel, T. J., Mayfield, J. E., Morfitt, D. C. and Wannemuehler, M. J. (1993). Oral immunization of mice and swine with an attenuated *Salmonella choleraesuis* [Δ*cya*-12 Δ*crp-cdt*19] mutant containing a recombinant plasmid. *Infect Immun*, **61**, 610–18.

Stabel, T. J., Mayfield, J. E., Tabatabai, L. B. and Wannemuehler, M. J. (1990). Oral immunization of mice with attenuated *Salmonella typhimurium* containing a recombinant plasmid which codes for production of a 31-kilodalton protein of *Brucella abortus*. *Infect Immun*, **58**, 2048–55.

(1991). Swine immunity to an attenuated *Salmonella typhimurium* mutant containing a recombinant plasmid which codes for production of a 31-kilodalton protein of Brucella abortus. *Infect Immun*, **59**, 2941–7.

Stager, S., Gottstein, B. and Muller, N. (1997). Systemic and local antibody response in mice induced by a recombinant peptide fragment from *Giardia lamblia* variant surface protein (VSP) H7 produced by a *Salmonella typhimurium* vaccine strain. *Int J Parasitol*, **27**, 965–71.

Steger, K. K., Valentine, P. J., Heffron, F., So, M. and Pauza, C. D. (1999). Recombinant, attenuated *Salmonella typhimurium* stimulate lymphoproliferative responses to SIV capsid antigen in rhesus macaques. *Vaccine*, **17**, 923–32.

Stocker, B. A. (1990). Aromatic-dependent *Salmonella* as live vaccine presenters of foreign epitopes as inserts in flagellin. *Res Microbiol*, **141**, 787–96.

Strugnell, R., Dougan, G., Chatfield, S. *et al.* (1992). Characterization of a *Salmonella typhimurium aro* vaccine strain expressing the P.69 antigen of *Bordetella pertussis*. *Infect Immun*, **60**, 3994–4002.

Strugnell, R. A., Maskell, D., Fairweather, N. *et al.* (1990). Stable expression of foreign antigens from the chromosome of *Salmonella typhimurium* vaccine strains. *Gene*, **88**, 57–63.

Su, G. F., Brahmbhatt, H. N., Wehland, J., Rohde, M. and Timmis, K. N. (1992). Construction of stable LamB-Shiga toxin B subunit hybrids: analysis of expression in *Salmonella typhimurium aroA* strains and stimulation of B subunit-specific mucosal and serum antibody responses. *Infect Immun*, **60**, 3345–59.

Titball, R. W., Howells, A. M., Oyston, P. C. and Williamson, E. D. (1997). Expression of the *Yersinia pestis* capsular antigen (F1 antigen) on the surface of an *aroA* mutant of *Salmonella typhimurium* induces high levels of protection against plague. *Infect Immun*, **65**, 1926–30.

Toso, J. F., Gill, V. J., Hwu, P. *et al.* (2002). Phase I study of the intravenous administration of attenuated *Salmonella typhimurium* to patients with metastatic melanoma. *J Clin Oncol*, **20**, 142–52.

Turner, S. J., Carbone, F. R. and Strugnell, R. A. (1993). *Salmonella typhimurium* Δ*aroA* Δ*aroD* mutants expressing a foreign recombinant protein induce specific major histocompatibility complex class I-restricted cytotoxic T-lymphocytes in mice. *Infection And Immunity*, **61**, 5374–80.

Urashima, M., Suzuki, H., Yuza, Y. *et al.* (2000). An oral CD40 ligand gene therapy against lymphoma using attenuated *Salmonella typhimurium*. *Blood*, **95**, 1258–63.

Valentine, P. J., Meyer, K., Rivera, M. M. *et al.* (1996). Induction of SIV capsid-specific CTL and mucosal sIgA in mice immunized with a recombinant *S. typhimurium aroA* mutant. *Vaccine*, **14**, 138–46.

Verma, N. K., Ziegler, H. K., Stocker, B. A. and Schoolnik, G. K. (1995a). Induction of a cellular immune response to a defined T-cell epitope as an insert in the flagellin of a live vaccine strain of *Salmonella*. *Vaccine*, **13**, 235–44.

Verma, N. K., Ziegler, H. K., Wilson, M., Khan, M. *et al.* (1995b). Delivery of class I and class II MHC-restricted T-cell epitopes of listeriolysin of *Listeria monocytogenes* by attenuated *Salmonella*. *Vaccine*, **13**, 142–50.

Vrtala, S., Grote, M., Ferreira, F. *et al.* (1995). Humoral immune responses to recombinant tree pollen allergens (Bet v I and Bet v II) in mice: construction of a live oral allergy vaccine. *Int Arch Allergy Immunol*, **107**, 290–4.

Walker, M. J., Rohde, M., Timmis, K. N. and Guzman, C. A. (1992). Specific lung mucosal and systemic immune responses after oral immunization of mice with *Salmonella typhimurium aroA*, *Salmonella typhi* Ty21a, and invasive *Escherichia coli* expressing recombinant pertussis toxin S1 subunit. *Infect Immun*, **60**, 4260–8.

Wang, J., Michel, V., Leclerc, C., Hofnung, M. and Charbit, A. (1999). Immunogenicity of viral B-cell epitopes inserted into two surface loops of the *Escherichia coli* K12 LamB protein and expressed in an attenuated *aroA* strain of *Salmonella typhimurium*. *Vaccine*, **17**, 1–12.

Wedemeyer, H., Gagneten, S., Davis, A. *et al.* (2001). Oral immunization with HCV-NS3-transformed *Salmonella*: induction of HCV-specific CTL in a transgenic mouse model. *Gastroenterology*, **121**, 1158–66.

Weiss, S. (2003). Transfer of eukaryotic expression plasmids to mammalian hosts by attenuated *Salmonella* spp. *Int J Med Microbiol*, **293**, 95–106.

Weth, R., Christ, O., Stevanovic, S. and Zoller, M. (2001). Gene delivery by attenuated *Salmonella typhimurium*: comparing the efficacy of helper versus cytotoxic T-cell priming in tumor vaccination. *Cancer Gene Ther*, **8**, 599–611.

Whittle, B. L. and Verma, N. K. (1997). The immune response to a B-cell epitope delivered by *Salmonella* is enhanced by prior immunological experience. *Vaccine*, **15**, 1737–40.

Whittle, B. L., Lee, E., Weir, R. C. and Verma, N. K. (1997a). Immune response to a Murray Valley encephalitis virus epitope expressed in the flagellin of an attenuated strain of *Salmonella*. *J Med Microbiol*, **46**, 129–38.

Whittle, B. L., Smith, R. M., Matthaei, K. I., Young, I. G. and Verma, N. K. (1997b). Enhancement of the specific mucosal IgA response *in vivo* by interleukin-5 expressed by an attenuated strain of *Salmonella* serotype Dublin. *J Med Microbiol*, **46**, 1029–38.

Winau, F., Kaufmann, S. H. and Schaible, U. E. (2004). Apoptosis paves the detaur path for CD8 T-cell activation against intracellular bacteria. *Cell Microbiol*, **6**, 599–607.

Woo, P. C., Wong, L. P., Zheng, B. J. and Yuen, K. Y. (2001). Unique immunogenicity of hepatitis B virus DNA vaccine presented by live-attenuated *Salmonella typhimurium*. *Vaccine*, **19**, 2945–54.

Wu, J. Y., Newton, S., Judd, A., Stocker, B. and Robinson, W. S. (1989). Expression of immunogenic epitopes of hepatitis B surface antigen with hybrid flagellin proteins by a vaccine strain of *Salmonella*. *Proc Natl Acad Sci USA*, **86**, 4726–30.

Wu, S., Pascual, D. W., Lewis, G. K. and Hone, D. M. (1997). Induction of mucosal and systemic responses against human immunodeficiency virus type 1 glycoprotein 120 in mice after oral immunization with a single dose of a *Salmonella*-HIV vector. *AIDS Res Hum Retroviruses*, **13**, 1187–94.

Wyszynska, A., Raczko, A., Lis, M. and Jagusztyn-Krynicka, E. K. (2004). Oral immunization of chickens with avirulent *Salmonella* vaccine strain carrying *C. jejuni* 72D_3/92 cjaA gene elicits specific humoral immune response associated with protection against challenge with wild-type *Campylobacter*. *Vaccine*, **22**, 1379–89.

Xiang, R., Lode, H. N., Chao, T. H. *et al.* (2000). An autologous oral DNA vaccine protects against murine melanoma. *Proc Natl Acad Sci USA*, **97**, 5492–7.

Xiang, R., Mizutani, N., Luo, Y. *et al.* (2005). A DNA vaccine targeting survivin combines apoptosis with suppression of angiogenesis in lung tumor eradication. *Cancer Res.*, **65**, 553–61.

Xiang, R., Primus, F. J., Ruehlmann, J. M. *et al.* (2001). A dual-function DNA vaccine encoding carcinoembryonic antigen and CD40 ligand trimer induces T-cell-mediated protective immunity against colon cancer in carcinoembryonic antigen-transgenic mice. *J Immunol*, **167**, 4560–5.

Xiong Bao, J. and Clements, J. D. (1991). Prior immunologic experience potentiates the subsequent antibody response when *Salmonella* strains are used as vaccine carriers. *Infect Immun*, **59**, 3841–5.

Xu, D., McSorley, S. J., Chatfield, S. N., Dougan, G. and Liew, F. Y. (1995). Protection against i infection in genetically susceptible BALB/c mice by gp63 delivered orally in attenuated *Salmonella typhimurium* (*aroA*$^-$ *aroD*$^-$). *Immunology,* **85**, 1–7.

Xu, D., McSorley, S. J., Tetley, L. *et al.* (1998). Protective effect on *Leishmania major* infection of migration inhibitory factor, TNFα, and IFNγ administered orally via attenuated *Salmonella typhimurium. J Immunol,* **160**, 1285–9.

Yang, D. M., Fairweather, N., Button, L. L. *et al.* (1990). Oral *Salmonella typhimurium* (*aroA*$^-$) vaccine expressing a major leishmanial surface protein (gp63) preferentially induces T-helper 1 cells and protective immunity against leishmaniasis. *J Immunol,* **145**, 2281–5.

Yrlid, U. and Wick, M. J. (2000). *Salmonella*-induced apoptosis of infected macrophages results in presentation of a bacteria-encoded antigen after uptake by bystander dendritic cells. *J Exp Med,* **191**, 613–24.

Yrlid, U., Svensson, M., Johansson, C. and Wick, M. J. (2000). *Salmonella* infection of bone marrow-derived macrophages and dendritic cells: influence on antigen presentation and initiating an immune response. *FEMS Immunol Med Microbiol,* **27**, 313–20.

Yu, Y. A., Shabahang, S., Timiryasova, T. M. *et al.* (2004). Visualization of tumors and metastases in live animals with bacteria and vaccinia virus encoding light-emitting proteins. *Nat Biotechnol,* **22**, 313–20.

Yuhua, L., Kunyuan, G., Hui, C. *et al.* (2001). Oral cytokine gene therapy against murine tumor using attenuated *Salmonella typhimurium. Int J Cancer,* **94**, 438–43.

Zheng, B., Woo, P. C., Ng, M. *et al.* (2001). A crucial role of macrophages in the immune responses to oral DNA vaccination against hepatitis B virus in a murine model. *Vaccine,* **20**, 140–7.

Zheng, L. M., Luo, X., Feng, M. *et al.* (2000). Tumor amplified protein expression therapy: *Salmonella* as a tumor-selective protein delivery vector. *Oncol Res,* **12**, 127–35.

Zoller, M. and Christ, O. (2001). Prophylactic tumor vaccination: comparison of effector mechanisms initiated by protein versus DNA vaccination. *J Immunol,* **166**, 3440–50.

Index